UTOPIAS IN LATIN AMERICA

Past and Present

Sussex Library of Study

New Historical and Comparative Perspectives on Latin America

SOCIETY, POLITICS, AND CULTURE

This Sussex series, organized in cooperation with the Center for Iberian and Latin American Studies (CILAS) at the University of California, San Diego, is entitled "New Historical and Comparative Perspectives on Latin America: Society, Politics, and Culture." The series will focus on the interdisciplinary study of Latin America, bringing together different viewpoints from the social sciences and the humanities.

The Editorial Board is chaired by Carlos H. Waisman, Professor of Sociology, and made up by faculty members in the departments of History, Literature, Anthropology, and other social sciences at UC San Diego. The series publishes original monographs and contributed works from scholars in the United States, Latin America, and Europe, as well as papers drawn from CILAS projects and research conferences.

UTOPIAS IN LATIN AMERICA
Past and Present

EDITED BY JUAN PRO

sussex
ACADEMIC
PRESS
Brighton • Chicago • Toronto

2 4 6 8 10 9 7 5 3 1
First published in 2018 in Great Britain by
SUSSEX ACADEMIC PRESS
P.O. Box 139
Eastbourne BN24 9BP

Distributed in the United States of America by
SUSSEX ACADEMIC PRESS
Independent Publishers Group
814 N. Franklin Street, Chicago, IL 60610

British Library Cataloguing in Publication Data
A CIP catalogue record for this book is available from the British Library.

Library of Congress Cataloging-in-Publication Data
Names: Pro Ruiz, Juan, editor.
Title: Utopias in Latin America : past and present / edited by Juan Pro.
Description: Brighton ; Portland : Sussex Academic Press, [2018] | Series: Center for Iberian and Latin American Studies (CILAS)/Sussex Academic Latin American library
Includes bibliographical references and index.
Identifiers: LCCN 2017057086 | ISBN 9781845199227 (hb : alk. paper)
Subjects: LCSH: Utopias—Latin America—History.
Classification: LCC HX806 .U775164 2018 | DDC 355/.02098—dc23
LC record available at https://lccn.loc.gov/2017057086

MIX
Paper | Supporting responsible forestry
FSC® C013604

Typeset and designed by Sussex Academic Press, Brighton & Eastbourne.
Printed and bound by CPI Group (UK) Ltd, Croydon, CR0 4YY

Contents

Series Editor's Preface

This volume brings together papers that explore Latin America as a setting, for both Europeans and Latin Americans, of utopian thought and action.

Utopian visions or practices have existed since the beginning of recorded history, but it is with modernity that the imagining and the attempts to establish alternative forms of social organization have become institutionalized. The essence of modernity was captured by Marx's famous dictum, "all that is solid melts into air, all that is holy is profaned . . ." [1] In pre-capitalist and pre-Enlightenment societies, utopian thought or attempts to establish utopian communities were the product of either unusual intellectual foresight and audacity, such as Plato's or Thomas More's, or extreme circumstances, such as religious persecution of "heretic" medieval sects. These societies were not composed of social classes in the modern sense of the term, but of rigid status groups, practically akin to castes. Social mobility, both vertical and horizontal, was rare, except as a consequence of wars or other calamities. And the prevailing belief was that God had dictated both social organization and political institutions, which were therefore inalterable.

Modernity destroyed these social structures and ideologies: The establishment of capitalism meant, in terms of social structure, the abolition of serfdom and the instauration of markets for goods and labor, with the ensuing potential for large-scale vertical and horizontal mobility; and the Enlightenment set the stage for the understanding that institutions and forms of social organization were contingent, and subject to the human will. These changes created the conditions for the massive mobilization of utopian thought and action that has characterized Western politics and culture since the "Great Transformation": from the very idea of a market society which, as Karl Polanyi has pointed out,[2] was itself utopian, to the liberal and democratic institutions understood as "government by the people"; anarchism (itself an extreme form of liberalism), and the totalitarian ideologies of the twentieth century.

As this book makes clear, Latin America was strongly involved in this process of utopian theorizing and implementation. First, as several of these essays show, the region was conceived as a social "*tabula rasa*" in the imaginary of European utopians: an ideal environment for carrying out new social designs, at both the micro and macro levels. In the second place, Latin America has been a milieu in which both elites and anti-status quo revolutionaries have engaged in envisioning and bringing, or trying to bring into being, alternative institutional formulae for the organization of social or political life, sometimes at the level of local communities, others as the result of the thorough transformation of societies. nineteenth-century constitutions, as this book shows, contained strong utopian elements, and Latin America has been a fertile ground for the appearance and development of radical movements, political and religious, reformist and revolutionary, some democratic and others totalitarian.

Why this propensity? A central trigger for utopianism, both North and South, has been an intense dissatisfaction with the status quo, coupled with the conviction that other institutional arrangements, more consistent with the utopist's values, are possible. Sources of collective dissatisfaction have been plenty in Latin America since independence: economic backwardness with respect to other societies; the widespread belief that the region is dependent or directly controlled by foreign powers, responsible for this state of affairs; glaring inequality along economic and in many societies ethnic lines; prevalence, for much of the twentieth century, of authoritarian rule, sometimes very repressive; marginality and economic and political exclusion of large segments of the population; extensive governmental corruption; etc. At the same time, political ideologies, usually of European origin but locally re-interpreted, have served as a basis for generating alternative institutional frameworks: from conventional Marxist-Leninist formulae to original local products, such as theology of liberation or Hugo Chavez's "Twentieth-Century Socialism" and other variants of contemporary populism.

Many of these movements or actual experiments have failed and others are trailing badly, but as long as the two determinants of utopianism mentioned above — dissatisfaction with the current social organization and the hunger for alternative formulae — remain, utopian movements and organizations will be recycled and re-constituted, and they and the societal responses they generate will continue making up one of the central dimensions of the social and political dynamics of Latin American societies.

---

1 Karl Marx and Friedrich Engels, *Manifesto of the Communist Party*, Ch. 1.

2 Karl Polanyi, *The Great Transformation* (New York: Farrar & Rinehart, 1944).

Editor's Acknowledgments

The texts in this volume reflect part of the work of the Transatlantic Network of Utopian Studies (*Red Trasatlántica de Estudio de las Utopías – Rede Trasatlântica de Estudo das Utopias*), which was created in 2015 for the specific purpose of putting scholars of utopianism in the Iberian and Latin American spheres in touch with each other and, in turn, promoting the establishment of a field of Utopian Studies in the Spanish- and Portuguese-speaking world. The Network was set up in 2015–2016 with financial assistance from the Santander Bank as a collaborative inter-university project between the Universidad Autónoma de Madrid and Latin American universities (Project CEAL-AL/2015-26). At present, it comprises more than a hundred researchers from all over the world interested in the topic of utopia in the Atlantic area.

The texts have all been presented and discussed in a variety of academic forums, such as the International Conference, *Utopian Imaginaries: Past, Present and Future* (Madrid, 30 September, 1 and 2 October 2015), which is where the idea of publishing them in a collective volume arose. That conference was organized by the IMAGEST group at the Universidad Autónoma de Madrid (*Imaginaries of the State: Models, Utopias and Dystopias in the Nation-State Building Process in Spain from a Comparative Perspective, 18th–20th Centuries*, HAR2012-32713 MINECO-FEDER project of the National R & D Plan of Spain, 2013–2015). Subsequent discussion of these topics, always within the framework of the Transatlantic Network mentioned above, has continued under the aegis of the HISTOPIA group (*History of the Future: Utopia and its Alternatives in the Modern Horizons of Expectation, 19th–21st centuries*, HAR2015-65957-P MINECO-FEDER project of the National R&D Plan of Spain, 2016–2018). Chapters 1, 5, 6, 7, 8 and 11 come from the research carried out by their authors in the HISTOPIA project. This book has been made possible thanks to the financial support of this research project.

The authors are grateful for the criticisms and suggestions received from colleagues who took part in the 2015 Conference, as well as at other meetings, including the International Utopian Studies Seminar (*Seminário Internacional de Estudos Utópicos*) at the Universidade Estadual de Campinas (São Paulo, Brazil, November 2015), the Seminar on Contemporary Social and Cultural History at the Universidad Autónoma de Madrid (September 2016), the Third Conference on Latin American Intellectual History: Forms of Intellectual History: Theory and Praxis (*Formas de Historia Intelectual. Teoría y Praxis*, El Colegio de México, November 2016) and the Colloquium Utopia: A Historical Approach to a Philosophical Notion (*La utopía: enfoque histórico de una noción filosófica*, Madrid,

Casa de Velázquez, February 2017). Nevertheless, any errors are obviously the sole responsibility of the authors.

We specially wish to thank Janet and Anthony Dawson, who translated in English the original Spanish texts of most of the book (Introduction and chapters 1, 2, 3, 4, 5, 6, 7, 8, 9, 10, 11 and 12) and revised chapter 13. Their help in the preparation of this edition has been invaluable.

Introduction

Juan Pro

This book considers the question of utopia in Latin America. It is plural, like the topic of utopia itself and the part of the world to which it refers. The collection of thirteen texts, which focus on thirteen cases or specific aspects of the utopian phenomenon in Latin America, shows the importance of utopia in the modern world. Because of the characteristics of our subject of study, we have decided not to include a specific chapter of conclusions. Utopia is an open space — a Not Yet, as Ernst Bloch would say[1] — the results of which can never be evaluated completely, since each generation reads them afresh, offering new interpretations and drawing new conclusions.

We have therefore refrained from coming to any conclusions as such. Perhaps arriving at a final conclusion based solely on a few specific cases would have been a step too far, given the vastness of Latin American space and its rich history. Nevertheless, the cases presented here are sufficient to suggest a positive overall interpretation of Latin American utopias, without allowing enthusiasm to cloud the critical sense that ought to guide all research. It is left then to the readers to draw their own conclusions and to nurture their own utopias. Our hope is that the historical cases that we present here will serve as an invitation to think about all that the concept of utopia contains: the promises that it has symbolized in the past for Latin America, and those that it can continue to offer in the present.

Utopia is a burning issue of the kind that cannot be disregarded. The recent commemoration of the fifth centenary of the first edition of *Utopia* by Thomas More, who coined the word, serves to remind us that this concept has been with us since Early Modern times.[2] Indeed, it is an essential part of Western modernity, which began with the discovery and colonization of America. This does not mean that utopia makes any claim to spatio-temporal exclusivity. Some scholars have plausibly argued that utopian thinking and mechanisms existed in Antiquity and the Middle Ages, well before the term was invented, and there are those who have found parallels in other cultural contexts, such as the Islamic world, India and China, even to the point of suggesting that it is an anthropological universal.

Nevertheless, what nobody has disputed is the centrality of utopianism to the development of the West in modern times, from the Renaissance to the present day. The break with age-old traditions, confidence that human rationality could find out about the world and organize it, the criticism of inherited conventions — all these opened the door to the imagining of possible worlds in search of a better future. Geographical discovery and voyages of exploration provided a powerful stimulus for thinking about how many alternative ways of organizing society and distributing power were potentially available. Even More's *Utopia* purported to be the account of a Portuguese seafarer, Raphael Hythloday, who had sailed with Amerigo Vespucci and recounted what he had seen in remote lands, lands that some have tried to identify with specific real places in the Antilles or on the American continent.

At the heart of this book — and this would be the main conclusion, if it were necessary to propose one — is the idea that there has been a constant relationship between utopia and America (Latin America in particular), which has helped to clarify the concept of both *Utopia* and *Latin America.* The one cannot be understood without the other.

From the very beginning of modern history at the end of the fifteenth century, down to the most recent political experiments currently being developed in a number of countries, the destinies of this part of the world have been bound up with various forms of utopia. The Latin American experience helps us understand that the impulse that lay behind Thomas More's new literary genre in 1516 was the same one that prompted nineteenth-century European socialists and anarchists to emigrate to the Americas to put their ideal communities into practice, and also inspired the mobilization of the revolutionary guerrillas and populist masses in the twentieth century. This impulse of a bold imagination eager to build the future is what can be called *utopia* in its many and diverse manifestations. The idea that we are putting forward is that this powerful cultural force that encourages change has been, and continues to be an essential feature of the history of Latin America. In other words, it is not only the physical Latin America that is reflected in statistics, but also the one dreamt of by its inhabitants or imagined by foreign observers, which are just as "real". What is more, we are not the first to say this, since the same idea has been expressed before by important Latin American philosophers and literary essayists such as the Mexican, Alfonso Reyes.[3] The link between America and utopia has also been the subject of academic studies, although basically from the standpoint of cultural studies or literary analysis, or by concentrating exclusively on present-day social struggles and mobilizations in the region.[4] In this book, we put forward the view that it is essentially a historical topic, the extent and depth of which can only be understood by expanding the focus to include the most diverse manifestations of human activity and dealing with them in a temporal framework that connects past, present and future.

Myths and legends of utopian content already proliferated at the time of the voyages of exploration, spurring on the conquistadors, while the knowledge gap about lands awaiting discovery was filled with stories about utopias such as El Dorado or the land of the Amazons.[5] From the sixteenth century onwards, the America that the Spanish and Portuguese had discovered became the space in which it was possible to envision the widest variety of forms of human coexistence. This was partly due to the discovery of indigenous cultures and the sheer lushness and profusion of their natural environments, which provided the raw material for thinking about diversity and variety and for comparing and expanding the boundaries of the possible, and partly to the colonial tendency to consider the vast American territories as empty space in which it was possible to start afresh. Starting from scratch and founding towns and cities was the way to set utopias in motion, to imagine the best of worlds without being subject to the inertia and constraints imposed by history and the institutions and customs of the Old World. This sort of utopianism was inscribed into the shapes of Latin American towns and cities, their toponymy, the programmatic documents of the conquest, as well as the proposals critical of the way it should be carried out, those by Vasco de Quiroga, for example (which are discussed in Chapter 3 of this book).

In the eighteenth century, the Enlightenment multiplied that world of possibilities and America became the place where full rationality, which continued to be fettered by tradition and fanaticism in Europe, was once more possible. There can be little doubt about the utopian content of the Jesuit Reductions in Paraguay or California, or the Franciscan projects. In the same century, idealization of the Inca monarchy, which was by then extinct, gave rise to a utopia based on its future restoration.

Utopias infused by enlightened rationalism flourished, but they were at the same time changing tone, and abandoning the classic genre of imagining ideal societies situated in remote, hitherto undiscovered, places. The exploration of virtually the whole planet made this literary device less plausible. The future, on the other hand, was unfolding like an endless virgin territory that could be colonized by uchronias, that is, timeless utopias.[6] The revolutions in the United States (1776) and France (1789) finally breached the last dikes that held back this utopianism of the future. Henceforth, the future would no longer be predetermined. It would not simply be the continuation of the present and the past, but a blank slate that nations and peoples could fill with their own decisions and struggles. Utopia became commonplace and ubiquitous, for in such conditions it was essential to explore the diverse possibilities of the future, compare alternative developments and their consequences, and to have a compass to set a course for that voyage into the unknown.

Revolution came early to Spanish and Portuguese America, right from the beginning of the nineteenth century. And it came to stay, for it became a fundamental ingredient in nineteenth- and twentieth-century Latin American history. Revolution and the utopian proposals associated with it not only marked the processes of emancipation in the new Latin American nations, but also the subsequent creation and consolidation of the independent states, as well as the history of the entire region (as Chapter 5 shows in the specific case of Chile). The significance of constitutions as, in some respects, utopian texts should be indicated here, since they represent designs for an ideal state that must be achieved in the future, rather than an actual state of affairs that needs to be institutionally defined. This is truer of Latin America than of Europe, although there is a certain utopian component in the history of all modern constitutionalism. In the case of the Latin American constitutions, most of which were born of revolutionary processes, the aspirations of their authors were frequently written into them. Such aspirations for the future were achieved later, either in part or not at all, in a climate of marked political instability, which quickly made the utopias of their forebears obsolete; one needs only to think of the 1917 Mexican Constitution, a lengthy compendium of the dreams of the Revolution, some of which have still not been realized a hundred years later.

After Independence, Latin America continued to be fertile territory for political and social experimentation, in which revolutions and counterrevolutions, constitutions, republics, movements and alternatives could all take root. It was an extremely fluid history, punctuated with utopias both homegrown and imported. Those that were imported are better known. For many European intellectuals and activists, Latin America was the ideal territory for testing utopian proposals that ran into insurmountable problems in Europe. The so-called "utopian socialists", for example, frequently looked to Latin America for places to set up their model communities with greater success than in similar experiments undertaken on the other side of the Atlantic. Some of the utopian socialists, such as Considerant, Chevalier, Saint-Simon and Cabet, actually went to America to try and put their ideas into practice. Nonetheless, it was mainly ideas, rather than people, that flowed from one side of the Atlantic to the other, looking to Latin America for suitable soil in which they could germinate. Chapters 6, 7 and 8 of the book give an account of the thrust of this current in various time frames, territories and ideologies.

After the failure of the 1848 revolutions in Europe, Latin America became a destination for exiles, where freedom of thought and the opportunity to experiment with democratic and socialist utopias could be recovered. America was also the utopian dream of many thousands of emigrants from old Europe, emigrants who expected to live the rich and fulfilling lives that they had not been able to enjoy in their places of origin, whether through political or religious persecution, or the distress of being part of a minority that was discriminated against, or simply because of extreme poverty, exacerbated at moments of agricultural crisis, like those experienced at the end of the nineteenth century. Faced with the same old world, fraught with suffering, America stood as the idealized image of an earthly paradise, full of freedom, opportunities, promises: a true utopia. Chapter 9 shows an extreme case of Latin America as a refuge for utopian experimentation in the specific case of Italian anarchism transplanted to Brazil.

Two major stereotypes around that time explained the special link between Latin America and utopia. The first one emphasized the nature of Latin America as a *terra nova*, a new land in which the way of organizing future society had not yet been determined. It was also free of the vices and the straitjacket of traditions, prejudices, and established powers imposed by European history, which meant that utopian proposals that faced insurmountable obstacles back in Europe were now seen as practicable. In the second stereotype, European Romantics regarded the luxuriant vegetation and abundance of nature in some parts of America, especially the tropics, as motivating the wealth of possibilities that the land offered for different forms of social organization. The colossal scale of the natural setting of Latin America, which the Europeans found so breathtaking, excited their imagination, suggesting not only that previously unimaginable forms of nature were possible there, but also new-style social institutions and practices, the fruit of an equally fertile utopian imagination.[7]

The special link in the minds of Europeans between utopia and America in general, and Latin America in particular, became more apparent over time. The idealization of everything Latin American went on to turn it into a symbol of utopia during the twentieth century, with the mythologizing of guerrilla struggle, the Cuban revolution, the figure of *Che* Guevara, Latin American protest songs, and so on. The existence of an endogenous, homegrown Latin American utopianism, however, is more strongly disputed. This prejudice has to do with another, more general one: the widespread assumption that Iberian cultures in general — chiefly the Spanish- and Portuguese-speaking countries on both sides of the Atlantic — are little inclined towards utopianism. The mainstream literary canon, which extolls certain works written in English, French, Italian and German as key utopias, has tended to neglect the significance of Hispanic and Portuguese utopianism, both in Europe and America. According to this stereotype, based on unfounded assumptions, Iberian cultures are little given to boldness of imagination or creative freedom and are more inclined towards pragmatism and tradition. To express it in terms of well-known literary figures taken from a Spanish work, these would be societies made up of Sancho Panzas, not Don Quixotes.

This stereotype comes from a need to justify the relative economic backwardness of Iberian countries and to blame their victims for it. It has been further reinforced by asymmetrical scientific developments, in which the spotlight has shone on works that feed the rich Anglo-Saxon, French, Italian and German utopian traditions, but left the treasures of Spanish and Portuguese utopianism buried away in semi-darkness. In this instance, the difference in the perceived relevance of utopianism is more in the eye of the beholder than in the reality observed and could be turned in the opposite direction just as soon as a dedicated line of research pointed out, not only the wealth of utopian literary productions in Latin America, but also the many other forms of utopia, such as political and social movements, community experiences, approaches to urban planning, artistic and cultural trends, institutions and practices. An example of the former — utopian literary productions from Latin America — is examined in Chapter 11, and examples of the latter — active utopias in the form of social experiences and movements — in Chapters 12 and 13.

At this point we arrive at one of the central problems in any study whose subject is utopias, namely, which definition of utopia we are going to work with, what is included and what excluded. The boundaries of the term are blurred because it has been used to refer to quite different things with very different connotations during its history. The first chapter in the book deals, to some extent, with the historical fluctuations in the concept of *utopia* and, more specifically, its uses in the Spanish language. We shall dispense here with the political use of the concept, which has generally been as a weapon to discredit proposals made by opponents and dismiss them as impossible dreams. Our interest lies in those analytical uses of *utopia* that are helpful in academic contexts for pinning down a real phenomenon and accounting for its evolution, modalities, characteristics, causes and consequences. At the end of the twentieth century, these uses seemed to be important enough to give rise to the formation of the new scientific field of Utopian Studies, defined, precisely, by its object of study, utopia. This implies accepting that it is a recognizable object and, on this matter, there is academic consensus.

It is important to emphasize that the institutionalization of Utopian Studies has developed around two poles — one North American, the other European — to the extent that there are two functioning scientific societies, one American and another European, with two sets of periodic conferences where developments are exchanged and an academic community is constituted.[8] Latin America does not really belong to either of these two areas and an equivalent process of defining the scientific field has not yet emerged in the region. As in the social sciences as a whole, Latin America has been relegated to a marginal role in the academic study of utopias. Its scholars occasionally participate in European or North American activities and maintain communication with both worlds, but Latin America continues to appear more as the object than the subject of Utopian Studies.

In its origins, this interdisciplinary academic field had a markedly philological stamp, which still persists. Among the classic works, the central thread of the history of utopias was made up of literary works belonging to the genre initiated by Thomas More's *Utopia* and all its later variants, down to works of science fiction in the present day.[9] This tendency to identify utopia with a literary genre has been widely challenged and overcome because the indifference to the content of literary works considered utopian led to loss of focus. To put it another way, the type of proposal that shaped individual imaginary societies became irrelevant.

Nowadays it is generally accepted that a history of utopia cannot confine itself to ideas expressed in literary form, let alone a single literary genre. Utopia is not only text, it is also experience and action. Certain social movements — revolutions, for example — even when they are not inspired by utopian texts, are responses to an undeniably utopian stimulus, because of their tendency to imagine an ideal future that contrasts with the limitations of the present. Certain community experiences, ranging from the model communities of utopian socialists and anarchists in the nineteenth century to the hippy and eco-pacifist communes in the twentieth, are also responses to the utopian mechanism, which consists of thinking up ideal ways of constituting society and putting them into practice directly, with or without a canonical text to define the undertaking.

The concept of *utopia* that applies at present in Utopian Studies is therefore a plural one, and extends beyond the purely literary. It includes, at the very least, community experiences and social movements.[10] After expanding the boundaries of the definition, there then remains the problem of defining which movements and experiences can be considered truly utopian for their boldness in imagining a better future that is radically different from the present. The field of Utopian Studies continues to be a lively academic space, with many intersecting polemics, and the work of definition is still incomplete, leaving its frontiers open to further research and theoretical discussion.

With respect to Latin America particularly, expanding the concept of utopia to include the political and social spheres, as well as the merely literary, simultaneously means giving it greater prominence. While the tradition of utopian writings in Spanish and Portuguese remains relatively hidden, the abundance of community experiences and social movements in the history of Latin America that can be considered utopian is glaringly obvious. Hence, the region is sure to have a central place in any developmental scenario of Utopian Studies that is not determined by cultural prejudices or nationalistic distortions.

The opening chapter by Juan Pro (Universidad Autónoma de Madrid), entitled "Utopia in the Spanish Language: The Origin of a Word, the History of an Idea", makes reference to the performative capacity of language by focusing on the different ways that the term *utopía* has historically been used in Spanish. This review of its uses in journalistic, literary, and political texts shows the interrelationship between semantic change and the political and social changes taking place in each period. Juan Pro also shows that, in spite of the dominant trends, there has always been some margin for interpretation. *Utopía*, together with its derivatives, was a polysemous word, which made it possible for the speaker or writer to emphasize particular aspects of its meaning, whether the negative connotation of extreme idealism, bordering on fantasy, or the positive one of creative imagination necessary to drive progress in a particular direction. Placing this chapter at the beginning of the book serves to draw attention to three things: first of all, the role of language in the construction of historical reality, particularly when discussing cultural realities mediated through language use, such as this one on the creation of utopias. The second draws attention to the multiple — even contradictory — definitions of concepts like utopia as they are used in the various academic today, and also within the particular historical spatio-temporal framework that is being studied. Lastly, attention is also drawn to the fact that recourse to utopian strategies has been cyclical, waxing and waning in popularity, lest we fall into the temptation of confusing the most recent of these cycles, or any other, with a definitive long-term trend. Bearing in mind these caveats about the concept and the ways in which utopia has been used, we then go on to study a different grammar that needs to be considered: the grammar of memory.

The second chapter, "How to Do Things With Utopias: Stories, Memory and Resistance in Paraguay", is the work of Marisa González de Oleaga (Universidad Nacional de Educación a Distancia, Spain). She focuses her attention on five of the many utopian colonies that were established in Paraguay in the nineteenth and twentieth centuries and analyses the way in which these experiences have been passed down to the present day, resulting in their memory being 'disabled' for present-day social movements. The text is a reflection on the constraints imposed by literary genre on the transmission of utopian experiments and calls for new forms of writing and representation. Whereas the first chapter alerts us to the linguistic dimension of the construction of utopia, the second one raises the question of the paradoxes of historical memory. Thus, by triggering the tension that exists between history and memory in relation to complex cultural objects such as utopias, the question of the tension between past and present, which is intrinsic to this book, is also raised.

Leaving behind the "grammar of utopia" considered in the first two chapters, Chapter 3 is the first in a chronological sequence covering a series of historical milestones that take utopia from the beginnings of modernity in the colonial period to present-day Latin America. The text "Vasco de Quiroga rewrites *Utopia*", by Geraldo Witeze Junior (Instituto Federal de Goiás, Brazil), analyses the early attempt by the bishop of Michoacán to address the issues and abuses of Spanish colonization in Mexico by using an alternative model of *pueblos-hospitales* for the indigenous population. Vasco de Quiroga developed these communitarian experiences, inspired by Thomas More's Utopia, in the 1530s, and wrote a description of them as a model for harmonious coexistence for the whole of Spanish America. This is therefore one of the earliest examples of Hispano-American utopianism, in which three fundamental characteristics can be noted: first, the tendency to accompany the production of texts — like those written by Vasco de Quiroga — with the practical implementation of their ideas in the field; secondly, it is a re-reading from the American continent of ideas originating in other places in Europe (in this case, England) expressed from a Hispanic point of view; and thirdly, the emergence of the indigenous question, which is central to Quiroga's project and cannot be avoided in any forward-looking approach for Latin America.

In Chapter 4, "Where is Columbus's Helmsman Taking Us?: *The City of the Sun* by Tommaso Campanella as a Utopia Critical of the Iberian Empires", Carlos Berriel (Universidade Estadual de Campinas, Brazil) shows us that the classic seventeenth-century utopia, *The City of the Sun* (1602) contains a criticism of the Hispanic Monarchy and the way it governed its European kingdoms — in the Iberian peninsula and Italy, where Campanella was writing — and also those in America. From the time of Philip II, the monarchs of the House of Austria had incorporated the Portuguese Crown and its dominions in Brazil into their Monarchy. When we talk about a criticism of the Hispanic Monarchy therefore, we are referring, in this case, to a political formation that encompassed the whole of what we today call Latin America, plus the whole of the Iberian peninsula and other territories in Europe, Asia, Africa and

Oceania. It was a genuine world power with aspirations to a Universal Monarchy. *The City of the Sun* used utopian philosophy as an instrument to launch an early, sharply critical attack on that Monarchy's methods of governance: absolutism, wars, fanaticism, aversion to modern science, in short, everything that, from Campanella's point of view, was leading to ruin and backwardness in the dominions of the Hispano-Portuguese empire. Starting from the premise that *The City of the Sun* is one of the key works in the literary canon of the utopian genre, the important aspect of this chapter is that it highlights two fundamental ideas: first, the fact that all utopian content involves implicit or explicit criticism of something existing in the historical moment, and second, the centrality of the Latin American world in any history of utopias, even when discussing works written in Italy, as in this one by Campanella (or in England, such as More's, for that matter).

Chapter 5, "Utopian Imagination Across the Atlantic: Chile in the 1820s", by Carlos Ferrera (Universidad Autónoma de Madrid) takes us to nineteenth-century revolutionary South America. From the very start of its independent existence, Chile, perhaps because of its remoteness and relative inaccessibility, was seen as a prime example of utopian space, among conservatives, as well as liberals and socialists. This view was rooted in eighteenth-century philosophical Geography and proponents of it used moral criteria to define territories — often without visiting them — as places of barbarism, but also as utopias. Based on this, the Atlantic became a space for the channelling of ideas, projects and people that permeated the culture of the Creole elites. Independence stimulated experimentalism by triggering processes for constructing utopias with a view to creating harmonious republics of virtuous citizens.

In Chapter 6, "Cabet's Utopia, from Minorca to Argentina: Bartolomé Victory y Suárez", Horacio Tarcus (CeDInCI — Universidad Nacional de San Martín — CONICET, Buenos Aires, Argentina) provides a somewhat later example of these transatlantic utopian flows in the specific field of what came to be known as *utopian socialism* or *romantic socialism*. He considers the figure of Bartolomé Victory (1833–97), a Spanish printer, publisher and publicist close to the ideas of Cabet, who emigrated to Argentina after the defeat of the revolutionary movements of 1854–56. Here we see the importance of political exile among Europeans in generating utopian expectations of America, and also as a vehicle for the circulation of utopian ideas in search of environments where they could be disseminated and implemented. Victory was one of those prophets of advanced ideas who, when contemporaries accused them of being utopians, responded by proudly defending the term:

> utopia has recognized the human personality and declared it sacred; utopia allows us to cross thousands of leagues in just a few hours; utopia enables us to converse in thirteen minutes from one world to another; utopia has changed the face of laws, customs, and even religions themselves.[11]

In short, Victory argued in favour of an idea that would become common in the second half of the nineteenth century: the utopias of today are the realities of tomorrow. The idea that utopia is a purely historical concept that should always be approached from a historical perspective is one of the key ideas that this book seeks to establish. There are no intrinsically utopian ideas. New ideas surface in a specific historical context, and it is in context that they are conceived of as utopian. Only later, after time has passed, can it be demonstrated whether or not that idea was realizable. History is a succession of utopias, some of which come to fruition, while others fall by the wayside as mere possibilities.

The following two chapters also have the same theme of transatlantic utopian socialism and its development in Latin America. In Chapter 7 "The Utopia of the 'Latin Race': Michel Chevalier, Victor Considerant and Public Debate in Spain Concerning the Intervention in Mexico, 1861–67", Nere Basabe (Universidad Autónoma de Madrid) considers the joint French, British and Spanish intervention against the government of Juárez in Mexico and the subsequent Mexican Empire under Maximilian I as framed in utopian concepts and proposals. European utopian socialists viewed this neo-colonial expedition as an opportunity to implement some of their proposals in America; there were, however, proposals of a different kind, such as those linked to the concept of the "Latin race", a utopia of intercontinental unity and critical redemption that contrasted with the modern tendencies imposed by the Anglo-Saxon and Germanic nations to the north. This chapter raises the interesting question of the relationship between utopias and power. As many of the utopias that are commonly studied arose from contestation against established power, those who formulated utopias have always sought the help of political power in order to implement them, and quite a number of historical experiences of the use of power can be found that were driven by more or less explicit utopian ideals. In Latin America alone, a whole host of political experiments can be identified that started off as ideas that contemporaries regarded as utopian. Nere Basabe mentions the Second Mexican Empire of Maximilian of Habsburg because it is connected to the idea of the "Latin race", although others mention José Gaspar Rodríguez de Francia, Dr Francia, in Paraguay (1814–40), to whom Fourier himself wrote, seeking help in the creation of a model phalanstery. In the twentieth century, we could point to the utopian content at the heart of the Cuban Revolution (1953–59) or the Sandinista Revolution in Nicaragua (1978–79).

The Chapter 8, "Rhodakanaty in Mexico", by Carlos Illades (Universidad Autónoma Metropolitana, Mexico), shares the same stage: Mexico in the middle decades of the nineteenth century. Plotino Rhodakanaty was a cosmopolitan European of Greek origin, who not only set in motion a socialist utopia in Mexico using elements taken from Fourier, Proudhon and the whole of the previous socialist tradition, but also actively disseminated his own version of utopia by contributing to the revolutionary mobilization of the peasants and the emergence of the workers' movement in Mexico. Rhodakanaty's version of utopia was an original one, however much it harks back to known precedents and models. His aim was the creation of sovereign federated bodies, which, when added together, would make up society. In the long term, the State would have to be suppressed and replaced by a social pact endowed with political functions, although this pact could be revoked at any time.

While we catch glimpses of an anarchist utopia in Rhodakanaty, this is fully developed in Giovanni Rossi's project, which is the subject of Chapter 9: "The Cecilia Colony: Echoes of an Amorous Utopia in the Libertarian Press", by Laura Fernández Cordero (CeDInCI — Universidad Nacional de San Martín — CONICET, Argentina). This Italian anarchist developed the Cecilia Colony experiment in Brazil between 1890 and 1894. It involved putting into practice an 'amorphist', radically anarchist utopia, a society without institutionalized authority in which absolute freedom prevailed, extending even to emotional relationships between men and women. The chapter analyses the intensive publicity given to that Latin American experiment with a view to making it a universal model.

The topics treated in Chapters 8 and 9 introduce us to the problems that utopias face when translating theory into practice. In both cases, this step involved a shift in space from one place to another — from Europe to America — that seemed to be comparable to time travel, since it made it possible to realize something that had never been achieved before on the Old Continent. The theme, therefore, is the journey to America as a journey in time.

Both the "Greek", Plotino Rhodakanaty, and the Italian, Giovanni Rossi, were non-conformists who wanted to go further than simply set out their utopias in theoretical terms and who managed to make that leap onto American soil, in Mexico and Brazil, respectively. The two experiments enjoyed a measure of success, Rhodakanaty's because of his ability to mobilize the peasants, and Rossi's because he was able to set up a human community that survived for three years living according to anarchist principles. Nonetheless, both ended badly. The 1869 peasant revolt, inspired by Rhodakanaty's teaching, ended in a bloodbath, suppressed by the Mexican government, while Rossi's small community, the Cecilia Colony, had to be abandoned in 1894 after being sacked the previous year by Brazilian government troops. What stands out in the first case is the way that social groups can appropriate particular utopian discourses for their own purposes and launch into violent actions in order to turn them into reality, so running considerable risks; the second experiment — as its founder recognized — revealed how difficult it was for participants in those communitarian experiments to overcome the system of values and patterns of behaviour instilled earlier by the dominant social order. The outcomes of both experiments raise the issue of the destiny of utopias, which conservative historiography considers by definition to be doomed to failure. The meaning of that failure can be interpreted in different ways, in as much as such episodes have continued to be remembered as models down to the present day, and not only continued to inspire those who took part in them, but also later generations of activists and researchers.

Ana Sabau (University of Michigan) is the author of Chapter 10: "Technologies of the Afterlife: Spiritualism and Social Imagination in Nineteenth-Century Mexico". In this chapter, she addresses an issue that is crucial to the history of utopianism, namely, the tendency to explore alternative forms of knowledge and unconventional belief systems, such as spiritualism, alongside the new possibilities of social and political organization. Between the middle and end of the nineteenth century, Mexico, like many other countries, witnessed the rekindling of spiritualist doctrine. A great number of men and women became devotees of this new worldview somewhere between a science and religion, attending spiritualist clubs and societies that spread across the country. Even though such practices seem bizarre today — maybe even damaging the reputations of those who promoted them, along with their ideas about social change — spiritualist circles were inspired by a strong common impetus for freedom. The publications of these societies show that, at the same time that spiritualism was imagining communion between the worlds of the living and the dead, it was also trying to promote a freer, more equal society. Followers of spiritualism were committed to freeing themselves from conventions that would constrain the ambitions of emancipatory thought, and that utopian commitment included shedding the conventionalisms of inherited religion, as well as the new impositions of positivist science. The link between utopia and alternative spirituality has been underlined many times, starting at least as far back as Thomas Müntzer and the peasant uprising in Germany, in 1524–25.[12] In Latin America, thanks to the repeated revolutionary readings of the Christianity of the Scriptures, this journey has been particularly intense and continuous. We could refer again to the evangelical inspiration of the Jesuit communities in Paraguay or the *pueblos-hospitales* of Bishop Quiroga in Mexico, but much closer in time, there is a whole history to be told about the utopian content of Liberation Theology and the impact that it has had in Latin America in the twentieth and twenty-first centuries.[13]

The title of Chapter 11 by Andrew Ginger (University of Birmingham), "Universopolis: The Universal in a Place and Time", makes reference to an essay by the Mexican, José Vasconcelos, *La raza cósmica* (The Cosmic Race) (1925), in which he conceived the Latin American populations as a "fifth race", a "race of bronze" or a "cosmic race", an agglomeration of all the races in the world, ready to usher in a universal era of humanity. He envisioned the utopia of a new civilization, Universopolis, constituted of knowledge and values from around the world. Andrew Ginger presents this philosophical-literary utopia as the epitome of universalism, arguing that universalism is one of the quintessential manifestations of utopia. The universal constitutes literally a no-place, as well as a no-time, precisely because it escapes the confines of geographical and historical contextualization. Universopolis was a utopia that set out to break down the barriers that marginalized the misfits and those on the fringes, giving them a prominent role in the development of the world. Today, when national identities, borders and clashes of civilizations threaten irreconcilable conflicts everywhere, the idea of universalism reflected in the work by Vasconcelos acquires fresh currency. His idea of linking universalism to a particular geographical and cultural space like Latin America is not without interest. In essence, each of the American countries is a conglomerate made up of migrants of diverse origin and native populations that have survived under very difficult circumstances. Miscegenation and cultural syncretism are characteristic features of Latin America, which is why it is better positioned than other regions to offer the world a possibility of overcoming racial and national conflict by pursuing the utopia of a humanity without borders. In view of the fact that the rejection of immigrants and refugees has reached such a height in Europe and North America that it challenges the framework of democratic coexistence, might it not, therefore, be important to explore this utopia?

Each of the final two chapters of the book takes a look at utopian approaches active in the Latin America of the twenty-first century. In Chapter 12, Dario Azzellini (Cornell University) discusses "The Commune in Venezuela: A Utopian Prefiguration" as the utopian impulse latent in the "Bolivarian Revolution" in Venezuela, activated by means of key institutions, such as communal councils and communes. These are forms of local self-government, aimed at creating a new geometry of power. Although precedents can be found in the history of revolutionary movements in Europe, the communes are also firmly anchored in Latin American traditions — the experiences of resistance among the indigenous peoples and the Afro-Americans, for example — or those reflected in Simón Bolívar's writings on toparchy (a system of local self-government). They also constitute new utopian proposals, characteristic of their time, that offer a response to globalization.

Finally, in Chapter 13, "Walking towards Utopia: Experiences from Argentina", Marina Sitrin (State University of New York, Binghamton) reminds us of the continued vigour of initiatives resistant to the hegemonic order in present-day Latin

America, initiatives that are opposed to the seemingly unstoppable logic of capitalism and globalization. By means of various recent expressions of social defiance in Argentina, Marina Sitrin shows that the vital utopian spark that drives them, as well as the world views and human relationships that lie behind them, are undoubtedly constitutive elements of operational utopias.

The point of these final chapters is to show that history continues, even after the critical moments of 1989–91, which some interpreted as signalling the "end of History".[14] Although the crisis in the Communist bloc — exemplified in Europe by the fall of the Berlin Wall and the disappearance of the Soviet Union — was important, it could not be taken to mean the end of the historical process of humankind, and even less so in Latin America, where the persistence of the Castro regime in Cuba reminds us that the rhythms and tempos there are different. History has not ended with the victory of capitalism and the establishment of a unipolar — imperial — order in which the only social, economic and political model is the one offered by the United States. There have been conflicts since, and there will continue to be conflicts, dissatisfaction with the existing order, innovation and change, the driving forces of History.

As we try to show in this book, utopias are powerful driving forces of that historical dynamism. The yearning to achieve a better world, whether it turns out to hit the mark or to be totally misguided, continues to prefigure the future and to set the direction for building a country that is very different from the present and past versions that we are already familiar with. These final two chapters demonstrate that, in the twenty-first century, there are still some who dare to dream of alternative systems to capitalism, and to do so from one of the peripheral regions of the world, which Latin America continues to be. The communes in Venezuela and the occupation of factories in Argentina are two good examples. The fact that utopian approaches of this type should flourish and proliferate on Latin American soil cannot in any way be regarded as the product of chance. This part of the world is blighted by extreme poverty and inequality, appalling shortages that contrast with the abundance of natural resources, and governments that are especially antisocial, repressive and corrupt. Utopia germinates spontaneously in these conditions, in its attempt to escape a reality that is both stubborn and cruel, raising its voice to demand what ought to be obvious, gearing itself up time and time again to protest against some unjust order. It all germinates in the form of utopia because this seems to be the very language of Latin America, the language of a boundless imagination, flowing from the cultural richness of a continent that still has much to offer the world.[15]

Latin American utopianism — which is easy to recognize in the Early Modern centuries, in the experiences of nineteenth-century utopian socialists and the revolutionary guerrillas of the twentieth century — has continued down to the present day and manifests itself in various ways, some of which appear in these last two chapters of the book. Utopia is not only a thing of the past, but also has a place in the present. This sets up other kinds of polemics, because it turns the utopian object into something much more dangerous and challenging than if it were placed in a museum showcase as an object of contemplation, an archaeological object, a relic of the past. There is no shortage of writings — in Latin America especially — that talk about the end of utopias from the comforting standpoint that this form of imagining the future had its moment but has now passed. According to this approach, utopia would be a primitive way of dealing with uncertainty about the future by imagining fanciful solutions to real problems. The repeated failures that mark the history of utopias have created the widespread conviction that experiments of this type end up causing more problems and suffering than the situations they set out to overcome.

This is certainly a conservative interpretation, closely inspired by the collapse of the Soviet Union in 1991 and the subsequent discrediting of what had been one of the grand utopias of the twentieth century: communism. Perhaps, with its downfall, communism caused the collapse of other traditions of alternative thinking, even the very drive for social transformation itself, giving rise to a moment of hegemony of the so-called *pensée unique* or 'single way of thinking'. Capitalism, the class society, the nation state, the established world order; it all seemed for a moment unshakeable, discouraging any kind of critical thought or dreams of an alternative. For a short space of time, perhaps, the only meaning of *utopia* was once again that discardable fantasy of dreaming of ideal worlds that can never come true. Nonetheless, it did not take long for alternative thinking to start up again and the movements of contestation, which had never ceased, often nurtured fairly well-defined utopias of the future. Not even the annihilation of the communist utopia and the complete transformation of the world that it brought in its wake managed to put an end to the production of utopias; they simply changed their form and content, as they always had. Different utopias spring up in every age, expressing the concerns, hopes, and conditions of possibility of their times. This historical dimension is one of the most interesting aspects of the study of utopias. It allows us to read these singular phenomena for certain dimensions of their time — such as their expectations for the future — which are perhaps less immediately obvious when analysing other kinds of sources.

Utopias that have long been a fundamental ingredient in the historical trajectory of the region, are far from disappearing from present-day Latin America. What can be detected, however, is the existence of historical cycles in the growth and decline of utopias, both in Latin America and the rest of the world. This is an interesting challenge for Utopian Studies, because the reasons why a utopian outlook is discredited at certain times but appeals at others has to do with cultural frameworks and changing socio-political circumstances, which is why investigating it could be so illuminating. In the same way that we can establish an origin for utopia and associate it with the beginnings of modernity, so the possibility of the complete disappearance of utopian proposals in the future could also be considered. Utopia could disappear at some point, although it has not happened yet. Utopian cycles of growth and decline, which have always existed, should not be confused with the complete disappearance of the phenomenon, the premature announcement of which seems to be more a case of conservative wishful thinking.

The twenty-first century did not begin with the end of utopia, but rather with the upsurge of a new utopian cycle. Myriad political, social, artistic and intellectual movements confirm this, with particular intensity in Latin America. In short, utopia is a necessity in our societies, a response to their deficiencies. Perhaps this is why the utopian impulse is so strong in Latin American countries, where the deficiencies are so appallingly obvious in contrast to the expectations and possibilities. In such conditions, imagining alternative futures is not such a wild idea; what is a wild idea is to take for granted that the future will be nothing more than the mere continuation of what we already know.

Notes

1 Ernst Bloch, *The Principle of Hope* (Oxford: Basil Blackwell, 1986).

2 Thomas More, *De optimo reipublicae statu, deque nova insula Vtopiae* (Leuven: Dirk Martens, 1516). The academic commemoration of this fifth centenary, which was celebrated throughout the world, also reached Latin America, with a symposium organized by the Universidad de la República (Uruguay), *Utopías: pasado y presente: A 500 años de Utopía de Tomás Moro* (27–28 October 2016).

3 See especially the essay that Alfonso Reyes began in 1924, "No hay tal lugar", in *Obras completas* (Mexico: Fondo de Cultura Económica, 1960), 11: 336–390.

4 An example of the cultural studies approach, with particular attention to an analysis of literary works is the book by Kim Beauchesne and Alessandra Santos, eds., *The Utopian Impulse in Latin America* (New York: Palgrave Macmillan, 2011). A more recent book, Kim Beauchesne and Alessandra Santos, eds., *Performing Utopias in the Contemporary Americas* (New York: Palgrave Macmillan, 2017), concentrates on the study of present-day protest movements and actions, extending the focus to the Americas as a whole.

5 An early reflection on the utopian meaning of such expectations in the myths and legends of the discovery of America was made by Alfonso Reyes in another of his essays in 1942: "Última Tule", in *Obras completas* (Mexico: Fondo de Cultura Económica, 1960), 11: 11–155.

6 This change is usually considered to have occurred after the publication of the work by Louis-Sébastien Mercier, *L'an deux mille quatre cent quarante, Rêve s'il en fût jamais* (London, 1772).

7 Pierre-Luc Abramson, *Las utopías sociales en América Latina en el siglo XIX* (Mexico: Fondo de Cultura Económica, 1999), p. 217.

8 The academic institutionalization of Utopian Studies in the United States came about in the form of the Society for Utopian Studies (SUS), founded in 1975. Since 1987, it has published the journal, *Utopian Studies*. The equivalent organization in Europe is the Utopian Studies Society, established in 1988.

9 Lyman Tower Sargent, *British and American Utopian Literature 1516–1975: An Annotated Bibliography* (Boston: G.K. Hall, 1979); Gregory Claeys and Lyman Tower Sargent, eds., *The Utopia Reader* (New York: New York University Press, 1999).

10 There is an up-to-date discussion of academic definitions of utopia in Ruth Levitas, *The Concept of Utopia* (Berne: Peter Lang, 2010); or, from the Latin American perspective, in Graciela Fernández, *Utopía: contribución al estudio del concepto* (Mar del Plata, Argentina: Suárez, 2005).

11 B. Victory y Suárez, "La verdad social", *El Artesano* 11, 9 May 1863, pp. 1–2. Reproduced by the author as an appendix in *El comunismo de Esteban Cabet* (Buenos Aires: Imprenta Central B. Victory y Suárez Editor, 1864), pp. 152–57: "la utopía ha hecho consagrar el reconocimiento de la personalidad humana; la utopía nos hace cruzar millares de leguas en pocas horas; la utopía nos hace conversar en trece minutos desde uno a otro mundo; la utopía ha cambiado la faz de las leyes, de las costumbres y hasta de las mismas religiones".

12 A reading in utopian terms in the 1921 study by Ernst Bloch, *Thomas Müntzer als Theologe der Revolution* (Frankfurt: Suhrkamp, 1977); also in Karl Mannheim, *Ideology and Utopia* (Oxford–New York: Routledge, 2013), pp.190–97, originally published in 1936.

13 Tom Moylan, "Mission Impossible? Liberation Theology and Utopian Praxis", *Utopian Studies* 3 (1991): 20–30.

14 Francis Fukuyama, "The End of History", *The National Interest* (Summer 1989): pp. 3–18.

15 The vitality of the utopian impulse in the context of social and political struggles in present-day Latin America can be traced in works such as Eugene Gogol and Latin American Colleagues, *Utopia and the Dialectic in Latin American Liberation* (Leyden: Brill, 2015); or Camilo A. Pérez-Bustillo and Karla V. Hernández Mares, *Human Rights, Hegemony, and Utopia in Latin America: Poverty, Forced Migration and Resistance in Mexico and Colombia* (Leyden: Brill, 2016).

1 Utopia in the Spanish Language: The Origin of a Word, the History of an Idea

Juan Pro

The capacity of language to reflect complex historical processes and also to have an effect on the changes that determine those processes is particularly evident in certain complex polysemic concepts, such as *utopia*, which has charted a course through five centuries of Western culture since it was first invented by Thomas More in his 1516 work of the same name.[1] The process by which the fictional toponym of *Utopia* became a common noun in the Spanish language was a long and tortuous one. While the uses of the term have gradually changed over time in response to transformations in the social, political and cultural environments, the same uses have, at the same time, also affected the social and cultural changes themselves, showing the capacity of language to construct reality.

Using the press, non-periodical print publications and dictionaries of the Spanish language as the principal sources, this study explores the history of the uses of the term *utopia* in Spanish language texts on both sides of the Atlantic, from its origins until that moment in the nineteenth century when it became part of the political battlefield. My text ends when the word was officially recognized in dictionaries in the 1850s and 1860s. The hypothesis that is explored here is that the word *utopia* was a fundamental tool that paved the way for utopian thought and the activism that derived from it, both in peninsular Spain and in Latin America. In short, the semantic field of utopia was also a social and political battlefield during the nineteenth century.

Thomas More's island in the Hispanic Atlantic

It took a hundred and twenty years before Thomas More's *Utopia*, originally written in Latin, was published in a Spanish translation. The first translation was actually done in America in the sixteenth century, when the Bishop of Michoacán, Vasco de Quiroga, translated at least the first book — and possibly the entire work — into Spanish. He sent the translation to his friend, Juan Bernal Díaz de Luco, in the hope that he would persuade the Council of the Indies to have it published; in the end, however, it was never printed and was subsequently lost.[2] There were no more Spanish translations until Medinilla's in 1637.[3]

Two years before that date, Francisco de Quevedo had translated a passage, which he inserted in Spanish into his political lampoon, *Carta a Luis XIII* (Letter to Louis XIII).[4] The translated passage, incidentally, was the one in which the supposed traveller, Raphael Hythloday, says that it is pointless for philosophers to try and counsel kings, since, if the latter are surrounded by ignorant, self-serving courtiers, they will be given bad advice and be persuaded [against their better judgment] to go to war anyway. This dilemma goes back to the very origins of utopian thought, even before More coined the term in the Renaissance, since it refers to Plato's failed attempt to counsel the tyrant of Syracuse.

It was probably the success of Quevedo's work, with seven editions in two years, that contributed to the launching of Medinilla's Spanish version of More's classic; indeed, that edition appeared with some "Preliminaries" written by Francisco de Quevedo. Even then, however, the Spanish version of *Utopia* was incomplete, since Medinilla opted to publish only the second book and left the first one out of his translation.[5] There were two further editions of that translation before the Spanish monarchy collapsed in 1808.[6] No new Spanish translations of Thomas More's *Utopia* appeared until the twentieth century.[7]

For a long time, therefore, *Utopia* was a Latin word of Greek origin that circulated in Spain and Spanish America only among certain educated elites, which included Quevedo, who seems to have had a copy of the 1548 Louvain edition at his disposal.[8] For all of them, *Utopia* was a learned word that was written (in Spanish) without the diacritical mark that it has nowadays, and with a capital letter, since it was a proper noun. In his *Información en Derecho* (1535), Bishop Vasco de Quiroga comprehensively analysed More's *Utopia* with a view to proposing its application to the *pueblos-hospitales* that he founded for the Indians in Michoacán.[9]

After that, we come across a host of references during the sixteenth, seventeenth and eighteenth centuries, generally written with a capital letter and no accent; some examples are Andrés de Poza (in his work *De la antigua lengua, poblaciones y comarcas de las Españas*, 1587), Fray Juan Márquez (in *El gobernador cristiano*, 1612), Francisco de Cabrera (in the *Discurso de los tvfos, copetes y calvas* by Bartolomé Jiménez Patón, 1639), Juan de Solórzano y Pereira (*Política indiana*, 1648), Benito Jerónimo Feijoo (*Teatro crítico universal*, 1726) and so on.[10] The newspapers of the Ancien Régime also contained mentions of Utopia.[11] The word did not, however, appear in Francisco Sobrino's dictionary, which was the standard lexical reference work for educated Spanish speakers from the time the Bourbons were established on the Spanish throne in the early eighteenth century.[12]

Throughout the ancien regime, whenever the word *utopía* appears, it is in connection with Thomas More's work or his imaginary island. When the French invasion (1808) and the Cortes assembled at Cadiz (1810) completely transformed the social and political landscape of the Spanish Monarchy in the first two decades of the nineteenth century, the term *utopía* appeared more frequently, but continued to be used in the same way as in the earlier period.[13]

On utopia in the revolution

In the nineteenth century — not before Riego's revolt (1820) — *utopía* began to make a tentative appearance as a common noun, which transferred the characteristics of

More's project, by a process of metonymy, to other proposals that could be called by the same name. The first time it appeared in this form was in 1820, in the translation of a French text, an indicator of the many words — and practices — that made their way into Spanish politics at that time from its influential French neighbour. The text reproduced passages from the plea by Dupin, the legal counsel for the defence in the trial of abbé Pradt, in a case brought against him because of a work that he had published about electoral law: "Aunque esta idea fuera una exageración está llena de grandeza; y aunque fuera una *utopia,* inspira por su sola elevación, meditación y respeto" (Even if this idea were an exaggeration, it is full of grandeur; and even if it were a *utopia,* it is inspiring for its uplifting quality, meditation and respect alone).[14] Shortly afterwards, it was used in the same sense in *El Censor* of Madrid, in an article in 1821:

> Having already settled as an incontrovertible principle that even anarchy could be considered a kind of government, and that its beginning is lost in the mists of time, it is obvious that we consider the others as mere babes-in-arms with respect to that utopia, with a few centuries here or there not making any difference whatsoever.[15]

Here utopia, no longer in italics, was associated with anarchy very early on. In 1822, the Madrid newspaper, *El Universal,* published a brief article critical of another one on law and order in the newspaper, *El Tribuno.* With reference to the implicit contradiction of the *Tribuno* criticizing the lack of security on the roads while at the same time rejecting the use of the National Militia or public law enforcement by the Government, it commented with heavy sarcasm: "To tell the truth, this discovery by the Tribunes [of the people] is remarkable, and they would perform a most important service for the nation by publishing the details of this extraordinarily ingenious *utopia.*"[16]

As can be seen, at that time the word generally appeared without the written accent on the *í* (its correct spelling later), underlining the fact that it came from another language. It was also italicized, indicating that it was transcribed directly from the Greek. When the common noun, *utopía* entered the Spanish lexicon, therefore, it was associated with the liberal revolution and the influence of France. It formed part of the new language with which the change of political regime between 1808 and 1840 was implemented.[17]

At the end of that decade, when the revolutionary experience seemed have been left behind and the restoration of absolutism was prevailing, the word *utopía* appeared in a poem by the Romantic writer, Bretón de los Herreros, taking advantage of the greater freedom offered by the poetic genre to use language in unusual or unfamiliar ways. His lines portray the Spanish worker of his time as more pragmatic than his French counterpart, and his mind less contaminated by utopias:

> The artisan here, without that imbroglio
> That inflames the man from Lutetia [i.e. Paris] and turns him into a fanatic,
> Makes sure of his daily pittance, and his noodle
> Does not seethe with so much horrendous or foolish utopia.[18]

For Bretón, then, utopias were horrendous or foolish and the consequences of them were to create confusion, inflame people and turn them into fanatics. A politically conservative viewpoint, certainly, but linguistically revolutionary, because of the innovative way the word was used. In that context of absolutist hegemony and timid reformism that accompanied the last period of Fernando VII's reign, utopia prospered only as a derogatory label, as in this memorial by Olavarría in 1833:

> It is but a short step from a system to a utopia, and speaking of the wonders worked by industrial power, let us avoid lapsing into the defect common to so many philosophers and legislators who have educated man exclusively for the land, against the manifest design of Providence, which created the land for man.[19]

The way was open for what would then become the most common use of the term utopia, namely, something deplorable, on account of the perverse ignorance of the constraints of reality. Within a short period of time of each other, two Madrid newspapers used the term *utopía*, first, in 1828, to discredit economic policies in favour of free trade and then, in 1844, to discredit protectionist ones. *Utopia* was used to criticize free trade policies by reproducing a speech that Friedrich List gave in Philadelphia while travelling around the United States, which leads us to think that the concept was also entering the lexicon of Spanish speakers through translations from the English: "Mr Say is very popular in France. As far as his theory of *laissez-faire* is concerned — a very naïve utopia — it is well known that it is the basis of the pseudo-cosmopolitan system of the Adam Smith school."[20] But equally, *utopía* was used to criticize protectionist policies, particularly by attacking customs protectionism in the United States: "The United States, apparently abandoning one of the main conditions of its prosperity, is increasing its import duties, to give its manufactures some illusory protection. This is the utopia that has been created."[21]

Reactionaries, traditionalists and conservatives of every stripe would load this word with negative connotations over the centuries, until it became a synonym for a childish dream or a capricious fancy. This pejorative meaning of the concept of *utopía* has most certainly been the dominant one in the Spanish language in Spain and America in all periods, and should be taken as the backdrop against which to observe subtle historical changes in meaning that have taken place over time. These changes allow us to observe the cycles of appreciation and contempt for the concept, which reflect the historical cycles of appreciation and contempt for utopian ideals.

It seems that *utopía* started to be used as a common noun somewhat later in Latin America than in Spain, since, until the 1830s, it is only found in the traditional form of the colonial era, that is, capitalized and alluding directly to Thomas More's work. In 1830, Andrés Bello made early use of the term to refer to a literary genre, giving the name *utopía* (in lower case) to a similar type of work to the one that Thomas More had written in 1516.[22]

It is in 1836 when we come across the modern form of *utopía* as a common noun, detached from its specific origin in More's work. It is found in Mexico, in an article that practically gives a definition of the concept, in connection with Eduardo Turreau de Linieres' project:

> Citizen Eduardo Turreau de Linieres' huge project, aimed at establishing a college of arts and sciences for a thousand young people, a house of education for two hundred girls, and a hospice where a thousand respectable old people can find comfortable shelter; all this founded on the association of many individuals contributing a small sum of 18 rs. a month

> until it totals 270 ps., which will be recognized as capital and earn an income in perpetuity; the project, we say, colossal in its dimensions, born of such modest sacrifices, and presented to a country to which the spirit of association, [which is] the vital principle of the greatest European establishments, is still quite foreign, was looked upon by many of our compatriots as a utopia; the dream of a well-meaning man, full of attractive ideas, but impossible to implement.[23]

The same publication used the term a year later, now with a written accent, to refer once again to a project that nobody denied was something positive, desirable and laudable, but which was assumed nevertheless to be quite unworkable, even if — as in this case — such an assumption could well turn out to be mistaken and a way somehow be found — here, through industrial development — to bring the utopia into being:

> The unity of industry is such, that the slightest modification that it undergoes, whatever the circumstance, is immediately communicated everywhere with the speed of an electric current. It is also true that barriers disappear before it, and that it alone could achieve the famous utopia of universal society.[24]

The concept can also be found in a speech by the president of the Republic of Guatemala, José Francisco Morazán, at the closure of the sessions of the ninth legislature, on 31 August 1836, in a passage defending the institution of trial by jury.[25]

A concept travelling across time (and the political spectrum)

In his study on the history of the concept of *utopia* in Spain, Juan Francisco Fuentes discerns three different strands in the way the term was used in the nineteenth century. The first was a tale or account set in an imaginary place with satirical or moralizing purposes; the second was synonymous with a pipe dream that was unworkable, used pejoratively to refer to proposals of social, economic or political change; and the third was a global alternative, revolutionary in nature, to the established order. He suggests that, although these three uses of the concept combine with and overlap each other, there is, nonetheless, a certain chronological order. The first one corresponds to the early years of the nineteenth century, continuing the trends of the eighteenth, whereas the second is specific to the period of transition between the revolution and the liberal post-revolution from the 1830s onwards. The third corresponds to the period in the 1860s when utopian socialism and revolutionary movements of socialist or anarchist tendency burst onto the scene. [26] His assessment holds up, although the question is a little more complex.

The term *utopía* throughout the 1830s and 1840s was used with no other meaning than some ideal that was impossible or very difficult to achieve. This usage became the standard one, since it is found in publications spanning the entire ideological and sociological spectrum of the time. The process of standardization involved writing the word with its proper stress by placing an accent on the *i,* something that began to be done occasionally in the 1830s, although it existed alongside the unaccented version until the beginning of the twentieth century.[27]

All ideological currents then used *utopía* in the negative sense of the quintessential innocence that chasing impossible dreams tended to imply, although with some key nuances in the application of the concept. The progressive press generally put a positive gloss on the word and acknowledged the innovative nature of utopia, since, even if the objectives pursued were beyond reach, they were always laudable and heading in the right direction. *El Eco del Comercio* (The Commercial Echo), for example, when it gave an account of Olavarría's 1833 memorial, quoted earlier, said that:

> The author supposes that his project is as portentous as the means to execute it is simple; but at the same time he knows that it is but a short step from a system to a utopia. It is certainly the case that, in this century, in which only the positive is appreciated, it is right and well-founded for nations to be wary of all systems and theories.[28]

The sense in which the term is used is clearly the opposite of the positive sense and associated with the realm of the imagination and the unworkable. These reservations, nonetheless, were framed within the context of a generally positive evaluation of the project mentioned, insofar as it proposed ideas about progress, advanced ideas with which the progressives sympathized: putting an end to vagrancy, correcting the Spanish temperament, improving the condition of the working classes, introducing technological innovations in all branches of production, introducing exotic crops and improving the welfare of the population.

The reactionary press, on the other hand, which also took the concept of *utopía* as equivalent to an unworkable fantasy, did so with no sympathy whatsoever. For example, *El Amigo de la religión y de los hombres* (The Friend of Religion and Men), in an article that defended "la sublime verdad de la suprema soberanía de la Iglesia sobre la cristiandad" (the sublime truth of the supreme sovereignty of the Church over Christianity), wondered:

> [Can it be that] so many wise bishops, so many illustrious princes of the Church like Bernard, Thomas Aquinas, Antoninus Pius, Raymond of Penyafort, Bonaventure, Gregory the Great, Gregory VII, Pius II, who have fought to defend that basis of civilization have endured so many hardships and laboured so tirelessly for the sole purpose of recreating their fantasy with the simulacrum of a utopia? No.[29]

Meanwhile, in the liberal-conservative press, the term appeared more as a synonym for something that could never be realized, without bothering to consider either the positive assessment of the progressive press or the negative one of the reactionary press. This can be seen in the newspapers edited by Andrés Borrego in the 1830s, for example, in *El Español* (The Spaniard) of Madrid, referring to the disturbances taking place in Malaga at the time:

> Perhaps the government took for self-interest, the frankness, good faith and trust with which its friends — patriots as interested as itself in not letting the programme of 14 September be reduced to a *utopia* — sought to join it in order to prevent one of the principal cities in Spain from setting the sorry example that we are lamenting today, and we shall not do the government the injustice of believing it is not lamenting with us.[30]

Or in *El Correo nacional* (The National Post): "this, in the language of experience, is a utopia, an impossibility."[31] If anything, in this ideological context, even more nuances could be pointed out, since the most combative and conservative part of the moderate press tended to use the concept in the same negative sense as the reactionaries, in other words, as something not only unworkable, but intrinsically wrong. So, for example: "The forlorn utopia that Rousseau invented, which the constituent assembly of France tried to put into effect, made executive power the slave of the legislature."[32]

So, *utopía* gradually came to be associated with a few favourite targets that tended to assimilate the dimension of unattainability that the concept implied. The first of these was peace. In 1836, Larra inaugurated the realistic assumption that it was as tempting as it was impracticable:

> Merciless law of nature: either devour or be devoured. Peoples and individuals, either victims or executioners. And even in peace, a fanciful utopia still not achieved in the continuous struggle of beings, even in peace, peoples devour, just as calm waters carve out their course more surely, although with less of a roar than the torrent.[33]

Other ambitious objectives of revolutionary politics were immediately branded utopian. These included national sovereignty, universal suffrage, democratic monarchy, reconciliation between liberals and absolutists after the civil war, the unity of the various parties into which liberalism had splintered, the creation of a party of the centre, the unification of Germany, the abolition of slavery, communism (understood as property that was fully held in common), the distribution of land ownership and restricting private property. With respect to the latter, Alcalá Galiano referred to those who looked for alternatives to that fundamental principle of bourgeois society as *utopistas* (utopists), thereby initiating a semantic family branching out from the root *utopía*.[34]

Soon after, attempts were also made to stop people thinking about democracy by dismissing it as a utopia, as can be seen in the following speech in 1842 in Mexico:

> Let the events of eighteen years speak for themselves, they will persuade us more eloquently that social utopia, which congress is inclining towards, is a piece of fiction for a respectable majority of the nation untutored in the knowledge of their duty and rights; knowledge that can only be inculcated through study, and the experience of a long political career. Since the unfortunate fall of the hero of Iguala [Agustín de Iturbide, 1824], a big mistake has been made in applying pure democracy — which carries within it the seed of disunity, turbulence and factions — to the Mexican people.[35]

In Spain, the republican form of government, which was the banner of the struggle for democracy, was also denounced as utopian.[36] Not surprisingly, the Democratic Party felt obliged to publicly reject the accusation:

> Our fellow democrats will understand that democracy is not a utopian dream or pious hope, as our enemies have tried to imply, but a party already organized for the struggle in the sphere of reality and practice, now ripe for power, with victory now in sight.[37]

From the 1840s, the term was associated with revolutionary, radical or extreme positions with respect to the dominant political spectrum, and used in particular to target socialist proposals and experiments. An article in the *Revista de España* in 1869, for example, warned against the socialist ideologies that were beginning to spread, describing them as utopian: The shameful scenes taking place in Andalusia are the product of the bizarre combination of ancient customs and laws and delusional modern utopias."[38] And in another article later, it went on:

> In the name, no longer of the bizarre utopias of socialism that deranged intellectuals seek to raise to the category of a scientific system, but [basking] in the warmth of the enthusiasm that promises of the most brutish socialism cannot help but sow among the masses, a few Deputies come to the new Assembly with the inevitable mission of forgetting the commitments in the name of which they have acquired such an honourable investiture or of defending solutions against which solemn protest will rise up from the very depths of their souls.[39]

Utopia was associated with the likes of Owen, Blanqui, Cabet, Fourier, Proudhon, Saint-Simon, Considerant and Barbès,[40] and the word appeared without fail in countless critical essays published in condemnation of socialism.[41]

With this negative meaning, the term became part of the arsenal that helped the language of conservatism to impose its hegemony after the post-revolutionary turn that accompanied the consolidation of state institutions in all countries from the middle decades of the nineteenth century. The word *utopia* was used overwhelmingly to disqualify and discredit; it would be regarded as something illusory and nonsensical, something that evoked, in that particular age of pragmatism and materialism, the revolutionary period that they were trying to bring to an end and hence, clearly reprehensible. It was always used in opposition to realism, so that, on occasions, it was associated with contemplation, as if it were a literary genre of pure fantasy and only useful for aesthetic delight,[42] or else with extreme ideological positions or doctrinal rigours that made it impossible to apply in practice.[43] The term *utopía* began to appear collocated with adjectives such as *ridícula* (ridiculous), *irrealizable* (unworkable), *despreciable* (contemptible), *insensata* (senseless), *imperdonable* (unforgivable), *vana* (futile), *impracticable* (impracticable), *loca* (mad), *absurda* (absurd), *funesta* (disastrous), *ruinosa* (ruinous), *descabellada* (hare-brained) and so forth.[44]

It is true that the term *utopía* is also associated at that time with adjectives with apparently positive meanings, such as *bella* (beautiful), *bellísima* (most beautiful), *feliz* (happy), *brillante* (brilliant), *admirable* (admirable), *seductora* (seductive), *galana* (elegant), *lindísima* (lovely), *apetecible* (appealing), *halagüeña* (promising), *generosa* (generous), *suprema* (supreme), *magnífica* (magnificent), *deslumbrante* (dazzling), and even *posible* (possible) and *realizable* (feasible).[45] In general though, these were either found in texts that sought to create ironic or sarcastic effects, or deliberate journalistic oxymorons in which the generally negative appraisal of the term did not change. The negative connotation held sway. To say, therefore, that someone was "poco utopista" (not very utopian), in the sense of "no se apasiona por las concepciones poéticas de la fantasía" (he is not passionate about poetic notions of fantasy) — as Miguel Luis Amunátegui said of Bernardo O'Higgins — was praise, praise that could be applied to a nation as a whole, since he went on to add that "bajo este aspecto, puede decirse que era muy chileno" (in this respect, it may be said that he was quite Chilean).[46]

Furthermore, from the middle of the century, the idea arose in Latin America that the transfer to the new continent of ideas and approaches that had been conceived in Europe for societies of a different kind was turning out to be utopian. This is how José Mármol's use of the word *utopía* appears, for example, in the 1850s, in what might be regarded as the first Argentinian novel, *Amalia*:

> The history of the world's social revolutions is the most perfect treatise on logic: such effects must follow on from such causes. And the great upheaval that the monarchist principle underwent here, an improvised republic in which there was neither enlightenment nor virtues to preserve it, and the sudden implantation of civilized ideas and habits among peoples accustomed to the comfortable inertia of ignorance, was a magnificent but impractical utopia, which barbarism would demolish, until a more thorough lesson in the same school of public misfortunes created a generation to raise it up and put it into practice, such a thing was bound to happen, and that, unfortunately, is what did happen.[47]

This perception of utopia as something essentially negative would culminate in the kind of approach found in *El Clamor Público*:

> Is there any chance anyone who will be able to vouch for the fact that everything that seems to us today to be the very pinnacle of freedom as far as the institutions are concerned will not, in ten thousand years time, be an actual utopia, an aberration of the understanding, a bondage typical of our ignorance, the child of our social concerns, the fruit of our selfishness?[48]

The paragraph, nonetheless, contained an argument that would be relevant later to future evaluations of utopia, namely, that the boundaries of utopia shift in every age in response to the cultural frameworks and enabling conditions that change with each generation. The journalist in 1850 was using the same argument in a conservative sense, suggesting that what seems reasonable today could well turn out to be utopia tomorrow, equating utopia with an "aberration of the understanding" or "bondage of ignorance", "social concerns" (in other words, prejudices), or "selfishness". The opposite argument would be useful to others later to rehabilitate the concept of *utopía*, when, in the light of experience, it was realized that what looks like a utopia at one point in time may not be one when the apparently unachievable objectives have actually been achieved or are within reach.

The vindication of utopia

In the mid-nineteenth century, *utopía* was already so common in texts written in Spanish that a Madrid magazine ventured to emphasize how odd it was that, despite being used so often, it could not yet be found in the Dictionary of the Spanish Royal Academy.[49] It was relatively common to express the feeling of living in utopian times, as was said with reference to political developments in France after 1848: "Since the February revolution till now, there has been no novelty that has not been tried out, no utopia that has not been followed, no absolute principle that has not had its cohorts of followers. The sea of ideas has risen to such a level that it has flooded the world."[50]

The situation reached such a pitch in those middle decades of the nineteenth century that the liberal newspaper, *La Iberia*, devoted an editorial to the concept of *utopia* in 1854, condemning the self-serving ways in which it was being used. The article was quite a manifesto. It rejected the idea that the conservatives had such foreknowledge of what would be possible or impossible in the future that they were able to discredit every plan for progress, every emancipatory or progressive idea as utopian (since that was the main use to which the term was being put until then). The author of that article had no other meaning in mind than the commonest one in circulation at the time, namely an imaginary project impossible to put into practice. Nevertheless, brushing aside this type of attack and vindicating every idea that the reactionaries were dismissing as utopian initiated a new shift in the way the word was used; those who were on the receiving end of the insult proudly adopted the word as their own and went on to invest it with more favourable overtones, precisely because it encapsulated everything that the reactionaries had tried to block:

> There are men of diminished intelligence and faint of heart for whom anything that embraces a horizon any wider than the limited circle of their faculties and opinions is illusory and fanciful. For them, humanity must crawl around in eternal infancy. . . . In the absence of reasons with which to oppose the doctrines that terrify them so much, in the absence of material means to halt the flight — which grows more powerful every day — of the principles against which they rail in vain, absolutists of every stripe and school have invented a word, which — to judge by its frequency and, especially, the aplomb with which they utter it — must be a real talisman for them: that word is *utopia*.
>
> Immense indeed must be the power of that word when it is invoked with so much faith and when it is raised, so to speak, as if it were a veritable barrier, before every generous aspiration, every progressive thought, every humanitarian idea, in order to block their path, or at least hinder their progress.
>
> Try arguing that the human species is not a flock whose fate it is to be forever shepherded by those who, judging it to be so, have arrogated to themselves the privileged role of *shepherds*, and you will immediately hear them cry: *utopia*! Say that peoples are not the patrimony of certain classes, families or individuals, that the institutions that rule them should be the product of their will, that national sovereignty is the only fount of justice and law, that indefinite progress is a providential law, and it will not be long before the word *utopia* will be ringing in your ears. In short, support those ideas that so gratify the mind, that help form such a lofty Concept of man and of the designs of the Creator, ideas that enlarge the spaces of moral and material creation, that make this arid present enjoyable, and awaken in us the comforting intuitions of a brilliant future and this world of sweet hopes, magnificent prophecies, presentiments worthy of human intelligence will doubtless be opposed and emphatically waved away with that wretched word.[51]

This veritable manifesto in defence of utopia went on to appeal to an argument that was destined to have a long life, namely, that ideas denounced as utopian in the present may simply be the precursors of what, in future years, will be a fully accepted reality. In the end, the same argument became a weapon to be used against the conservatives' resistance to change:

> Many ordinary souls before you regarded as utopias, and even as signs of madness, the immense ideas which, with the passing of the centuries, would later shine as glorious truths. How many of these, of every kind, were not described as fanciful, out of stupidity and alarmed interests! And yet they triumphed, because the inevitable future of the truth is to triumph; and the petty-minded creatures who relate everything to themselves, who see everything through themselves, had to confess their short-sightedness and the shameful inaccuracy of their opinions.

> What looked like a wild dream yesterday, will be a reality tomorrow; what is condemned today as impracticable, will shortly, perhaps in the not too distant future, be a protective institution; what, in our turbulent days, flickers as uncertainly as a breath of lukewarm air, will tomorrow be giving off torrents of light and heat. The measure of the mysterious line separating the possible from the impossible has still not been taken, and perhaps it never will be, because it is probably one of the Creator's great secrets; refrain, then, so-called politicians, from constantly repeating any word whose meaning is known only to Him . . .
>
> Do not believe that everything that is baptized with this undefined and indefinable term is a utopia; you yourselves, on more than one occasion, have seen what you thought was impossible to achieve when you were young has now been achieved. Who can determine what will, within in the space of a few years, have similarly passed into the sphere of reality, crossed over the present hazy boundary of the problem, and the axiom of theory and institution? . . .
>
> And believe us, if there are real utopias in the world, they are yours, when, despite what is happening in front of your eyes, you still think of yourselves as destined to play some role in this colossal drama.[52]

After that, it seems that there was a certain amount of public discussion about the concept. Not long afterwards, it once more occupied the pages of the democratic daily, *La Europa*, which also devoted an article to the word *utopía*. The paper rejected once more the negative sense in which it was generally used and called upon all progressives to eliminate that particular meaning of it from their vocabulary:

> For Charles V, steam power was a dream; for Napoleon an explainable theory. For both a utopia. . . . So, utopia, in the figurative sense used by the conservatives when they want to establish that something cannot be done, is a meaningless phrase. Utopia used in this sense is not Spanish. Progressives! Erase it from your dictionary.[53]

As the article pointed out, technological progress was demonstrating how what was previously impossible had, in a very short period of time, become possible, so that it was imperative to use the same reasoning in the realm of political and social ideas. Indeed, from the middle of the nineteenth century, the idea of *utopia* started to be associated with technological change and the way in which it could be used to transform life. Hence, as a result of the observation of technological progress, the concept gradually became decontaminated as it started to shed the negative connotations that it had accumulated in the previous decades. For example, the enthusiasm for the Great Exhibition of 1851 in London gave rise to some of these comments, which linked the great social utopias to modern industry: "Universal peace has ceased to be a utopia since industry took the first steps towards its achievement by bringing peoples of different races, religion and customs together in a common centre."[54]

More widespread use of the concept meant that it appeared in the 1850s in a few privately published Spanish language dictionaries, such as those by Eduardo Chao and Rafael Baralt.[55] In Chao's dictionary, *utopia* was defined as "any idea that is acceptable in theory, although impossible to achieve,"[56] whereas in Baralt's it was a "system that is principally based on fevered imaginings with no real basis in history, customs or human propensities, although made with the best of intentions to improve the condition of man and the state of peoples." This second one is particularly significant because it comes from a *Diccionario de galicismos* (Dictionary of Gallicisms), which means that it explicitly recognized utopia as "one of the words, expressions and phrases of the French language that [had] been introduced into modern Spanish speech" in the first half of the nineteenth century, not because it was a French word, but because its customary use as a common noun and its value as a political concept came from France.[57]

In Spain, the concept became firmly rooted following the triumph of the coalition of liberal unionists, progressives and democrats that led the Glorious Revolution in September, 1868. The hour of utopias had arrived at last, at a time when the republican groups and the socialist and workers' movements were finally free to openly engage in their activities. The concept even began to turn up in official Government documents, thus denoting its now irrevocable entry into Spanish political vocabulary.[58]

It was this Revolution, then, that created the conditions that made it possible to overcome the reluctance of the conservative Spanish Royal Academy to add concepts such as *utopía*, which was already in common use in the Spanish language, to its *Dictionary*, and indeed the term was introduced into the eleventh edition of the *Dictionary* in 1869. Since then, it has appeared as a noun in every edition, without the definition ever failing to stress the unworkable nature of projects called utopian: "noun fem. A plan, project, system or doctrine that sounds pleasing in theory, but is impossible to put into practice" (f. Plan, proyecto, sistema ó doctrina que halaga en teoría, pero cuya práctica es imposible). One would have to wait until the 1992 edition of the Academy's *Diccionario* to find the added detail that its nature as something unworkable referred only to the moment of the project's formulation, so leaving the door open for its possible realization at some later date: "An optimistic plan, project, doctrine or system that seems to be unworkable at the time of its formulation."[59] Academy dictionaries have clearly lagged well behind linguistic practice in Spanish-speaking countries, because this particular nuance in the way the term was used had appeared at least one hundred and fifty years before it was officially reflected in dictionary. The same can be said about the appearance of words derived from *utopía*, such as the adjectives *utópico* and *utopista* (*utopic* and *utopian*), which did not enter the Dictionary until the 1884 edition, even though we know that both were in common use long before that,[60] not to mention the noun, *utopismo* (*utopianism*), which denotes "Tendencia a la utopía" (A tendency to utopia), which did not appear until the twenty-second edition, in 2001.

Although the Spanish Royal Academy Dictionary does not create language, but confines itself to giving an account of it, its effect on the way words are used should not be underestimated. Many native users of Spanish, including professionals, such as journalists and writers, resolve lexical queries by consulting the Royal Academy Dictionary. Hence, the inclusion of these terms, after four decades of trying, certainly had the effect of legitimizing them, which helped consolidate the presence of the concept of utopia in the Spanish language.

At that stage of the nineteenth century, there were still no significant differences in the ways the term *utopía* and its derivatives were used in peninsular Spain and Spanish-speaking America. *Utopía* appeared with similar meanings in Spain — in the writings of Benito Pérez Galdós, José Echegaray, Joaquín Costa, Leopoldo Alas, also known as *Clarín*, Emilia Pardo Bazán, Marcelino Menéndez Pelayo, Miguel de Unamuno and Juan Valera — as in texts by Spanish American writers, such as *Recuerdos de provincia* by Domingo Faustino Sarmiento (1850), the *Apuntamientos para la historia de la Nueva Granada* by José María Samper (1853), a Cuban novel like *Ambarina* by Virginia Auber Noya (1858), the *Excursión a los indios ranqueles* by Lucio Mansilla (1870), Adolfo Alsina's Report on the Argentinian frontier (1877), or the River Plate pioneers of the novel, such as José Mármol, Eduardo Acevedo or Manuel Podestá.[61] The reason, on both sides of the Spanish Atlantic, may have been to do with the common influx of translations of works in other languages — from Latin to English and French — in which the term *utopía* had taken root earlier. It is indicative that the Ecuadorian writer, Juan Montalvo, used the term *utopistas* (utopians) in 1882, precisely in a text in which he called upon Spanish intellectuals to translate the great European authors into Spanish and to stop sending minor Catholic devotional works to America:

> If you are freethinkers, translate Laplace, Littré; if kindly utopians, Flammarion, Delaage; if avowed heretics, Renan, Peyrat. . . . If you want French authors in everything and for everything, there is an illustrious throng of historians, orators, scientists, philosophers, and even novelists, great novelists, like the author of *René*, the author of *Obermann*, the author of *Corinne ou l'Italie*. Translate the *Encyclopaedia* for us, for God's sake, translate it for us, you Spaniards who are so fond of and partial to Rousseau, Diderot, d'Alembert, Grimin and other shining stars in the great constellation of the eighteenth century, whose pole star, the helix of hell, is François-Marie Arouet, converted into Voltaire by the grace and work of the devil. But those little books, those novelettes, the little saints and little religious pictures that the bookshops are stuffed with. . . .[62]

Conclusion

The concept of *utopía*, which was linked to modernity from the start, became a normal feature of the Spanish language during the period of the revolutions on both sides of the Atlantic Ocean, and, in large measure, as a result of the influence of French. Between 1820 and 1850, it received a bad press, because it was first and foremost a concept of the conservatives, aimed at discrediting progressive political approaches that were more imaginative in nature or broke with the existing regime. It was in the 1850s and 1860s, in the heat of the new revolutionary cycle initiated in 1848, that the term was recuperated with a positive value, to express the capacity for imagining new, better worlds to fight for. Rehabilitation was undoubtedly driven by the speed of technological progress, and even took utopia into the pages of the dictionaries.

This review of the conceptual uses of *utopía* in the intellectual history of the Spanish language up to 1870 clearly shows that it was not only a linguistic history, but a manifestation of the performative capacity of utopian language. The war of words to establish the language of utopia and surround it with positive values was a tool used by intellectuals in particular historical situations to critique and transform reality.

The men and women of the nineteenth century witnessed such great changes, in both the technological and political spheres, that the idea took hold that they were living at a time when principles that they had until recently considered to be impossible to achieve or illusory were now becoming reality (always going hand in hand with the progress that technological innovations were bringing about). Corroboration of that provided the permanent impetus to rehabilitate the concept of *utopía* and surround it with positive connotations, in spite of the conventional definitions.

Notes

1 Thomas More, *De optimo reipublicae statu, deque nova insula Vtopiae* (Leuven: Dirk Martens, 1516).
2 Silvio Arturo Zavala, *Recuerdo de Vasco de Quiroga* (Mexico: Porrúa, 1965), pp. 161–165.
3 *Utopia de Thomas Moro, traducida de latin en castellano por Don Geronimo Antonio de Medinilla i Porres* (Cordoba: Salvador de Cea, 1637).
4 Francisco de Quevedo, *Carta al serenissimo muy alto, y muy poderoso Luis XIII, Rey christianissimo de Francia . . . en razon de las nefandas acciones . . . que cometio contra el derecho diuino... en la villa de Tillimon en Flandes Mos de Xatillon vgonote; con el exercito descomulgado de franceses herejes* (Zaragoza: Hospital Real y General de Nuestra Señora de Gracia, 1635).
5 Francisco López Estrada, "Quevedo y la Utopía de Tomás Moro", in *Actas del Segundo Congreso Internacional de Hispanistas*, compiled by Norbert Polussen and Jaime Sánchez Romeralo (Nijmegen, Instituto Español de la Universidad de Nimega, 1967), pp. 403–409.
6 The second edition in Madrid by Pantaleón Aznar in 1790; the third, also in Madrid, by Don Mateo Repullés in 1805.
7 Thomas More, *Utopía (el estado perfecto). Traducción, prólogo y notas de Ramón Esquerra; va añadido un juicio crítico sobre Utopía y su autor por D. Francisco de Quevedo Villegas* (Barcelona: Apolo, 1937). There was, however, an earlier translation into Catalan: Thomas Morus, *Utopía, traducció catalana... precedida d'un comentari sobre l'autor y l'llibre per J. Pin y Soler* (Barcelona: Massó, Casas et C^{a}, 1912).
8 López Estrada, "Quevedo y la Utopía de Tomás Moro", p. 405.
9 "Información en Derecho del licenciado Rojas sobre algunas provisiones del Consejo de Indias" (1535), in *Colección de documentos inéditos relativos al descubrimiento, conquista y colonización de las posesiones españolas en América y Oceanía, sacados de los Archivos del Reino y muy especialmente, del de Indias*, edited by Joaquín F. Pacheco, Francisco de Cárdenas and Luis Torres de Mendoza, vol. 10 (Madrid: Imp. de M. Bernaldo de Quirós, 1868), pp. 333–525.
10 References are taken from the Corpus of the Nuevo Diccionario Histórico del Español (Real Academia Española), where there are 2,036 allusions to the term *utopía* in 892 documents. Available online at http://web.frl.es/CNDHE/org/publico/pages/consulta/entradaCompleja.view (accessed 14 July 2016).
11 *Memorial literario instructivo y curioso de la Corte de Madrid*, December 1790, p. 63; *Diario de Madrid*, 9 February 1791, p. 3, 7 March 1793, p. 1, 5 February 1806, p. 2; *Nuevas efemérides de España históricas y literarias*, 5 July 1805, p. 2; *Memorial literario o Biblioteca periódica de ciencias, literatura y artes*, 10 February 1808, p. 10. The instances in journalistic texts are taken from the Hemeroteca Digital (Biblioteca Digital Hispana), available online at http://www.bne.es/es/Catalogos/HemerotecaDigital/ (accessed in July 2016).
12 Francisco Sobrino, *Diccionario nuevo de las lenguas española y francesa* (Brussels: Francisco Foppens, 1705). *Utopía* does not appear in this first edition, or in the later ones of 1734, 1751, 1760, 1769 and 1775.
13 *El Revisor político*, 17 August 1811, p. 1; *Almacén de frutos literarios inéditos de nuestros mejores autores antiguos y modernos*, 1813, p. 10; *Crónica científica y literaria*, 31 December 1819, p. 8; *Miscelánea de comercio, política y literatura*, 13 June 1820, p. 2.
14 *El Constitucional, o sea, Crónica científica y literaria* (Madrid), 25 September 1820, p. 2. The fact that it appears in italics in the original shows that its figurative use was still not very common.
15 "Sobre el despotismo", 8 September 1821, pp. 282–83: "Habiendo sentado ya como principio inconcuso que hasta la anarquía se podía considerar como un género de gobierno, y que su principio se oscurece en la noche de los tiempos, claro es que consideramos á los demás unos mozalbetes respecto de aquella utopia, sin que influyan para nada algunos siglos mas ó menos."
16 10 April 1822, p. 4: "A la verdad que este descubrimiento de los Tribunos es admirable, y

que harían un servicio importantísimo a la nación en publicar los pormenores de esta ingeniosísima *utopia.*"

17 A change about which the following, among others, have given an account: María Paz Battaner, *Vocabulario político-social en España, 1868–1873* (Madrid: Real Academia Española, 1977), María Cruz Seoane, *El primer lenguaje constitucional español: Las Cortes de Cádiz* (Madrid: Moneda y Crédito, 1968) and Eduardo García de Enterría, *La lengua de los derechos: la formación del Derecho Público europeo tras la Revolución Francesa* (Madrid: Alianza Editorial, 1994).

18 Manuel Bretón de los Herreros, *La desvergüenza: poema joco-serio* (Madrid: Mellado, 1856): "El artesano aquí, sin esa embrolla / Que exalta y fanatiza al de Lutecia, / Su pitanza asegura, y no en su cholla / Hierve tanta utopía horrible o necia."

19 Juan de Olavarría, *Memoria dirigida a S. M. sobre el medio de mejorar la condición física y moral del pueblo español* (Madrid: Fundación Banco Exterior, 1988): "De un sistema á una utopía el trecho es corto, y hablando de las maravillas que obra la potencia industrial, evitemos el caer en el defecto comun á tantos filósofos y legisladores que esclusivamente han educado al hombre para la tierra, contra el fin bien manifiesto de la Providencia, que crió la tierra para el hombre."

20 *Correo literario y mercantil,* 1 December 1828, p. 1: "Mr. Say es muy popular en Francia. En cuanto a su teoría de *dejar hacer,* utopia muy inocente, es sabido que es la base del sistema pseudocosmopolita de la escuela de Adam Smith."

21 *La Esperanza,*17 October 1844, p. 4: "Los Estados-Unidos, renunciando al parecer á una de las condiciones principales de su prosperidad, aumentan los derechos de importación, para dar una protección quimérica á sus manufacturas. Esta es la utopia que se han formado."

22 Andrés Bello, "La oración inaugural del curso de oratoria del Liceo de Chile de José Joaquín de Mora", in *Obra literaria,* edited by Pedro Grases (Caracas: Ayacucho, 1985): "El único fiador de la moderación de Ciro es Jenofonte en una obra que el mismo Jenofonte parece haber querido que se mirase como una utopía o novela política, pues la contradice abiertamente cuando escribe como historiador." [The only guarantor of Cyrus's moderation was Xenophon, in a work that Xenophon himself seems to have wished to be looked upon as a utopia, or political novel, for he openly contradicts it when he writes as a historian.]

23 *Diario del Gobierno de la República Mexicana,* 2 January 1836, p. 3: "El gigantesco proyecto del ciudadano Eduardo Turreau de Linieres, dirigido á establecer un colegio de artes y ciencias para mil jóvenes, una casa de educación para doscientas niñas, y un hospicio en que mil ancianos respetables encuentren cómodo asilo; todo esto fundado en la asociación de muchos individuos que contribuyan con la pequeña cantidad de 18 rs. al mes hasta completar 270 ps. que serán reconocidos como un capital, y ganarán una renta perpetua; un proyecto, decimos, colosal en sus dimensiones, nacido de tan módicos sacrificios, y presentado en un pais á donde el espíritu de asociacion, principio vital de los mayores establecimientos europeos es todavia extrangero, fué mirado por muchos de nuestros compatriotas como una utopia; como el sueño de un hombre de bien, lleno de ideas agradables, pero imposibles en la ejecución."

24 "Es tal la unidad del hecho industrial, que la mas leve alteracion que sufra en cualesquiera circunstancias, se comunica en el acto por do quiera, con la rapidez del fluido eléctrico. Es verdad también que ante él desaparecen las barreras, y que él solo podría realizar la famosa utopía de la sociedad universal." *Diario del Gobierno de la República Mexicana,* 19 January 1837, p. 4. In the same vein, *ibid.*, 29 January 1838, p. 3: "Una sociedad sin crímenes es una utopia" [A society without crime is a utopia]; also *ibid,* 8 November 1838, p. 3: "La igualdad absoluta de condiciones y fortunas es una utopia, es un delirio de la excéntrica Mrs. Francés Wrigth [sic] que divierte en los teatros, y carece de efecto en la escena del mundo" [Absolute equality of conditions and fortunes is a utopia, it is a delusion of the

eccentric Mrs Francis Wright who is entertaining in the theatres, and has no effect on the world stage].

25 "Mas no pudiendo ser una utopia la que se nos presentaba" [But it was not possible for the one that was being presented to be a utopia], reproduced in the *Diario del Gobierno de la República Mexicana*, 7 January 1837, p. 2.

26 Juan Francisco Fuentes, "Utopía", *Diccionario político y social del siglo XIX español*, edited by Javier Fernández Sebastián and Juan Francisco Fuentes (Madrid: Alianza Editorial, 2002), pp. 685–688.

27 The progressive Madrid daily, *El Eco del Comercio*, 15 March 1835, p. 4, was one of the first to Hispanicize the term: "cuando las autoridades se hacen superiores á recelos tan vanos como peligrosos, la unión entre los liberales españoles no es una quimera ni una utopía" [when the authorities rise above such futile yet dangerous misgivings, union between Spanish liberals is not a pipe-dream or a utopia].

28 *El Eco del Comercio*, 15 June 1834, p. 2: "El autor supone que tan portentoso es su proyecto como sencillo el medio de ejecutarle; pero conoce al mismo tiempo que de un sistema á una utopia el trecho es corto. Ciertamente que en este siglo, en que solo se aprecia lo positivo, es justa y fundada la prevención de los pueblos contra todos los sistemas y teorías."

29 *El Amigo de la religión y de los hombres*, 1836, pp. 27–28: "¿Un Bernardo, un Tomas de Aquino, un Antonino, un Raimundo de Peñafort, un Buenaventura, un Gregorio el grande, un Gregorio VII, un Pio II, tantos sabios obispos, tantos ilustres principes de la Iglesia, que han combatido en defensa de esa base de la civilización, no han empleado tantas fatigas y esfuerzos sino con el objeto de recrear su fantasía con el simulacro de una utopia? No."

30 *El Español*, 26 February 1836, p. 3: "Quizás el gobierno tuvo por interesada la franqueza, la buena fe y la confianza con que amigos suyos, patriotas tan interesados como él mismo en que el programa de 14 de setiembre no se quedase reducido á una *utopia*, buscaron á unirse á él para evitar que una de las principales ciudades de España diese el triste ejemplo que hoy lamentamos, y que no haremos al gobierno la injuria de creer que no lamenta con nosotros."

31 *El Correo nacional*, 20 February 1838, p. 1: "esto en el lenguaje de la esperiencia es una utopia, un imposible."

32 *El Guardia nacional* of Barcelona, 31 May 1838, p. 1: "La triste utopia que inventó Rousseau, y que quiso llevar á efecto la asamblea constituyente de Francia, hizo al poder ejecutivo siervo del legislativo."

33 Mariano José de Larra, "Horas de invierno" (1836), in *Fígaro. Colección de artículos dramáticos, literarios y de costumbres* (Barcelona: Crítica, 2000), p. 607: "Ley implacable de la naturaleza: o devorar, o ser devorado. Pueblos e individuos, o víctimas o verdugos. Y hasta en la paz, quimérica utopía no realizada todavía en la continua lucha de los seres, hasta en la paz devoran los pueblos, como el agua mansa socava su cauce, con más seguridad, si no con tanto estruendo como el torrente."

34 Antonio Alcalá Galiano, *Lecciones de Derecho Político* (1843–1844) (Madrid: Centro de Estudios Constitucionales, 1984): "Entre los reformadores de la sociedad en estos nuestros tiempos, entre los llamados utopistas está recibiendo el dogma de la propiedad individual duros embates, con el intento de substituirle una como propiedad común, entendida o repartida, según los diferentes formadores de planes para variar la sociedad de diversos modos" [Among the reformers of society in our times, among the so-called utopians, the dogma of individual property is receiving a severe pounding, with an attempt to replace it with one of common property, understood or distributed, according to the different devisers of plans so as to vary society in diverse ways].

35 "Discurso pronunciado por D. Francisco E. de Castro, en la junta de autoridades, empleados y vecinos, que tuvo lugar el dia 19 del corriente mes de Diciembre" [A speech delivered by D. Francisco E. de Castro, at the board meeting of authorities, employees and local residents, which took place on the 19th of this month of December] in 1842 in Tamaulipas

and published by *El Tiempo* in Santa Anna de Tamaulipas and reproduced by the *Diario del Gobierno de la República Mexicana*, 13 January 1843, p. 3: "Hablen los hechos de diez y ocho años, ellos nos persuadirán con mas elocuencia que la utopia social, á que propende el congreso, es una novela para una respetable mayoría de la nación que no está aleccionada en la ciencia de su deber y sus derechos; ciencia que solo puede inculcarse por el estudio y esperiencia de una larga carrera política. Desde la caida infortunada del héroe de Iguala, ha habido un gran error en aplicar al pueblo mexicano la democracia pura que en sí misma lleva el germen de la desunión, de la turbulencia y de las facciones."

36 The conservative Madrid newspaper *El Castellano*, 26 February 1842, p. 1 associated *utopía* with republicanism in an article in which it commented on the rumours of a possible armed insurrection: "Los republicanos esperan ganar mas que ningún otro partido, no solo porque los progresistas se desacreditan completamente y dan muestras ostensibles de inhabilidad para gobernar, sino porque llegado el momento de una nueva insurreccion, y aun sin mas que los rumores de ella, conseguirían establecer juntas en las provincias, acusar de ineptitud y aunque sea de traición á los ministros, formar su deseada junta central y acabar de una vez con las instituciones todas, dando principio á la realización de su desatinada utopía" [The Republicans expect to win more than any other party, not only because the Progressives are completely discredited and show clear signs that they are unfit to govern, but also because, once a new insurrection takes place, and even with mere rumours of one, they would succeed in establishing juntas in the provinces, accusing the ministers, even if it is treason to do so, of ineptitude, setting up the central junta that they have longed for and putting an end to all institutions by initiating the implementation of their hare-brained utopia].

37 "Manifiesto del comité central del Partido Democrático" in Madrid, 15 March 1865, reproduced in Miguel Artola, *Partidos y programas políticos, 1808–1936. II: Manifiestos y programas políticos* (Madrid: Alianza Editorial, 1991), p. 77: "Nuestros correligionarios comprenderán que no es la democracia el sueño utópico o la esperanza insensata, como han querido suponer nuestros enemigos, sino el partido organizado ya para la lucha en la esfera de la realidad y de la práctica, maduro ya para el poder, apercibido ya a la victoria."

38 *Revista de España*, vol. 6 (1869): 135: "Las vergonzosas escenas que tienen lugar en Andalucía, son producto de la extravagante combinación de costumbres y leyes antiguas y de delirantes utopías modernas."

39 *Revista de España*, vol. 6 (1869): 291: "En nombre, no ya de las extravagantes utopías del socialismo que locas inteligencias quieren elevar á la categoría de sistemas científicos, sino al calor del entusiasmo que no pueden menos de sembrar en las masas promesas del más brutal socialismo, vienen á la nueva Asamblea algunos Diputados con la inevitable misión de olvidar los compromisos en cuyo nombre han adquirido tan honrosa investidura ó defender soluciones contra las cuales se levantará en el fondo de su propio ánimo solemne protesta."

40 With Robert Owen in the conservative Barcelona periodical, *El Museo de familias* ("Los socialistas ingleses", 1841, p. 326). With Blanqui in *El Clamor Público*, 25 July1848. With Cabet in *El Balear* of Palma de Mallorca, 21 September1848; *Diario Constitucional de Palma*, 22 February1849; *El Heraldo*, 27 February1850 and *El Nuevo Observador* of Madrid, 18 June 1852. With Fourier in *El Heraldo*, 27 February 1850 and *El Nuevo Observador*, 18 June 1852. With Proudhon, Saint-Simon and Considerant in *El Nuevo Observador*, 18 June 1852. With Barbès in *El Clamor Público*, 25 July 1848 and *El Nuevo Observador*, 18 June 1852.

41 As in the book by Nicomedes Pastor Díaz, *Los problemas del socialismo: lecciones pronunciadas en el Ateneo de Madrid en el curso de 1848 a 1849* (Madrid: Imp. de Manuel Tello, 1867).

42 As did Joaquín Francisco Pacheco in his *Lecciones de Derecho Político* of 1845 (Madrid: Centro de Estudios constitucionales, 1984, p. 109): "El propósito de este curso no es el de examinar utopías, que dejamos intactas a los que gusten dedicarse a su contemplación.

Debemos tomar por nuestra parte el mundo como es, como Dios le hizo, como nos lo muestra la historia de cuatro mil años, y la observación de su estado presente" [The purpose of this course is not to examine utopias, which we leave untouched for those who like to spend their time contemplating them. We must, for our part, take the world as it is, as God made it, as four thousand years of history, as well as observation of its present state, have shown us that it is].

43 Nicomedes Pastor Díaz used it in this other sense in the Cortes in 1845 in his speech on the return of property to the clergy (reproduced in *Discursos*, Barcelona: Fundación Caja Madrid-Anthropos, 1996, pp. 339–340): "Nuestro cargo es asimismo práctico: nosotros no estamos autorizados para sostener utopías, sino cosas hacederas: no hemos venido aquí a representar nuestras exigencias de sistema, ni aun la pureza ortodoxa de nuestras doctrinas" [Our task is likewise practical; we are not authorized to uphold utopias, but things that are feasible; we have not come here to represent our requirements for a system or even the orthodox purity of our doctrines].

44 *Ridícula* in *El Espectador* of Madrid, 9 March 1845; *El Eco del Comercio*, 11 March 1845; and *El Clamor Público*, 5 September 1852. *Irrealizable* in *El Eco del Comercio*, 19 September 1845; *El Castellano*, 5 December 1845; *El Pensamiento de la Nación*, 10 June and 1 July 1846; *El Clamor público*, 4 December 1847, 9 March 1850 and 4 April 1850; *El Católico*, 20 December 1847; *La Época*, 1 May 1849; *El Heraldo*, 21 July 1849; *Revista mensual de agricultura*, 1850; *La Ilustración*, 9 February 1850 and 26 May 1850; *La España*, 26 March 1850; *La Nación*, 4 April 1850; *El Áncora* of Barcelona, 3 August 1850. *Despreciable* in *El Espectador*, 18 October 1845. *Insensata* in *El Español*, 20 December 1846; *El Heraldo*, 18 February 1849. *Imperdonable* in *El Clamor público*, 23 May 1847 and 26 January 1849. *Vana* in *Revista barcelonesa*, 4 July 1847. *Impracticable* in *La Esperanza*, 29 July 1847; *El Heraldo*, 7 January 1849. *Loca* in *Boletín de Medicina, Cirugía y Farmacia*, 1 April 1849. *Absurda* in *La Nación*, 11 June 1850. *Funesta* in *El Clamor Público*, 28 July 1852. *Ruinosa* in *El Áncora* of Barcelona, 10 October 1852. *Descabellada* in *La Esperanza*, 4 September 1854.

45 *Bella* in *El Español*, 16 November 1845; *El Popular*, 11 April 1848; *El Heraldo*, 2 August 1850. *Bellísima* in *El Castellano*, 26 May 1845; *La Esperanza*, 5 July 1852. *Feliz* in *El Espectador*, 26 July 1845; *La Nación*, 4 August 1849. *Brillante* in *El Eco del Comercio*, 22 August 1845. *Admirable* in *El Eco del Comercio*, 7 September 1845 and 11 July 1846. *Seductora* in *El Castellano*, 4 March 1846; *El Heraldo*, 12 October 1848. *Galana* in *El Pensamiento de la Nación*, 1 July 1846. *Lindísima* in *El Popular*, 3 September 1847. *Apetecible* in *El Español*, 9 October 1847. *Halagüeña* in the *Boletín de Medicina, Cirugía y Farmacia*, 16 January 1848. *Generosa* in *La Patria*, 12 March 1850. *Suprema* in *El Áncora* of Barcelona, 13 May 1850. *Magnífica* in *La Esperanza*, 27 June 1850. *Deslumbrante* in *El Áncora* of Barcelona, 23 June 1852. *Posible* in *La Esperanza*, 4 October 1845. *Realizable* in *Semanario pintoresco español*, 25 April 1847.

46 Miguel Luis Amunátegui Aldunate, *La dictadura de O'Higgins* (1853), in *Biblioteca Virtual Miguel de Cervantes*, Universidad de Alicante, 2002.

47 José Mármol, *Amalia* (1851–55) (Madrid: Cátedra, 2000), pp. 581–82: "La historia de las revoluciones sociales en el mundo es el tratado de lógica más perfecto: a tales causas han de suceder tales efectos. Y el gran trastorno que sufría aquí el principio monárquico; la improvisación de una república, donde no había ni ilustración ni virtudes para conservarla; y la plantificación repentina de ideas y de hábitos civilizados, en pueblos acostumbrados a la cómoda inercia de la ignorancia, eran una utopía magnífica pero impracticable, con la cual la barbarie daría en tierra; hasta que una enseñanza más prolija, en la escuela misma de las desgracias públicas, crease una generación que la levantase y la pusiese en práctica tal cosa debía suceder; y así ha sucedido, por desgracia."

48 *El Clamor Público*, 3 September 1850: "¿Acaso podrá nadie responder de que todo cuanto nos parece hoy el colmo de la libertad en punto á instituciones no sea dentro de cien siglos una verdadera utopía, una aberración del entendimiento, una servidumbre propia de

nuestra ignorancia, hija de nuestras preocupaciones sociales, fruto de nuestro egoísmo?". A very similar expression, probably by the same author, in the editorial in *El Clamor Público* of 15 June 1851, p. 1: "Lo que se elogia hoy como la mas sublime concepción, se condena mañana como una utopia ridicula" [What is praised today as the most sublime of conceptions is condemned tomorrow as a ridiculous utopia].

49 *El Laberinto*, 1 April 1845, p. 173: "El hombre feliz, que es la especie mas *utópica* (de *utopia*, con permiso del Diccionario) que se conoce en el humano enjambre" [The happy man, the most *utopian* species (from *utopia*, with the permission of the Dictionary) that is known in the human swarm.]

50 *El Heraldo*, 12 August 1849, p. 1: "Desde la revolución de febrero acá no ha habido novedad que no se haya ensayado, utopia que no se haya seguido, principio absoluto que no haya tenido sus secuaces. El mar de las ideas ha salido de madre, y ha inundado al mundo."

51 *La Iberia*, 21 September 1854, p. 1: "Hombres hay de menguada inteligencia y asustadizo corazón para quienes es ilusorio y quimérico todo aquello que abarca un horizonte más vasto que el limitado círculo de sus facultades y opiniones. Para ellos la humanidad debe arrastrarse en eterna infancia (...). Á falta de razones que oponer a las doctrinas que tanto les horripilan, á falta de medios materiales con que detener el vuelo, de dia en dia más poderoso, de los principios contra que en vano declaman, los absolutistas de todos los matices y todas las escuelas, han inventado una palabra, que, á juzgar por la frecuencia, y sobre todo por el aplomo con que la pronuncian, debe ser para ellos un verdadero talismán: esta es la palabra *utopia*.

Inmenso, en efecto, debe ser el poder de ese vocablo, cuando con tanta fé se le invoca, y cuando haciendo de él un verdadero valladar, se levanta, por decirlo asi, ante toda aspiración generosa, ante todo pensamiento de progreso, ante toda idea humanitaria, para cerrarles el paso , ó por lo menos dificultar su marcha.

Sustentad que la especie humana no es un rebaño destinado á ser siempre conducido por los que, conceptuándola tal, se han reservado el ventajoso papel de *pastores*, y les oiréis al punto esclamar: *¡utopia!* Decid qué los pueblos no son patrimonio de determinadas clases, familias ó individuos; que las instituciones que los rijan deben ser producto de su voluntad; que la soberanía nacional es la única fuente de la justicia y el derecho; que el progreso indefinido es una ley providencial, y no tardará en resonar en vuestro oido la voz *utopia*. Apoyad, en fin, esas ideas que tanto halagan la mente, que tan alto Concepto hacen formar del hombre y de las miras del Criador; ideas que agigantan los espacios de la creación moral y material, que nos hacen agradable este árido presente, despertando en nosotros las consoladoras intuiciones de un brillante porvenir; y á este mundo de dulces esperanzas, de magníficas adivinaciones, de presentimientos dignos de la inteligencia humana, sé os opondrá, sin duda, con enfático ademan aquella asendereada palabra."

52 *Ibid.*: "Muchas almas vulgares juzgaron antes que vosotros utopias, y aun rasgos de demencia, las gigantescas concepciones que, andando los siglos, debian brillar como gloriosas verdades. ¡Cuántas de estas, en todo género, no fueron calificadas de quimeras por la estupidez y los intereses alarmados! Y sin embargo, triunfaron, porque el triunfo es el porvenir infalible de la verdad; y los seres mezquinos que todo lo refieren á sí, que todo lo miran á través de sí mismos, hubieron de confesar la pequeñez de sus miras y la vergonzosa inexactitud de sus juicios.

Lo que ayer parecía un delirio, será mañana un hecho; lo que hoy se condena como impracticable, será dentro de un plazo, tal vez no lejano, una institución protectora; lo que en nuestros turbulentos dias brilla indeciso y tímido como una tibia exhalacion, despedirá mañana torrentes de luz y calor. La línea misteriosa que separa lo posible de lo imposible no ha sido medida aun, y quizá no lo será jamás, porque es acaso uno de los grandes secretos del Hacedor; guardaos, pues, pretendidos políticos, de repetir á todas horas una palabra cuya acepción solo de él es conocida (...).

No creáis utopias todo lo que con esta no definida é indefinible calificación se bautiza;

vosotros mismos habéis visto realizado lo que en vuestra juventud juzgasteis irrealizable, mas de una vez; ¿quién puede fijar lo que en el discurso de algunos años habrá pasado asimismo á la esfera de los hechos, salvado el hoy oscuro confin del problema, y el axioma de la teoría y la institución? (…).

Y creednos: si hay en el mundo verdaderas utopias, son las vuestras, cuando, á despecho dé lo que pasa á vuestra vista, os considerais aun llamados á representar algún papel en ese drama colosal."

53 *La Europa*, 29 October 1854, reproduced by *El Genio de la Libertad*, Barcelona, 3 November1854, p. 2: "Para Carlos V fué el vapor un sueño; para Napoleón una esplicable teoría. Para ambos una utopia (…). Utopia, pues, en la acepción figurada con que los conservadores quieren verificar lo irrealizable, es una frase sin sentido. En tal concepto utopia no es castellano. ¡Progresistas! borradla de vuestro diccionario."

54 *La Nación*, 23 October 1850, p. 2: "La paz general deja de ser una utopia, desde que la industria dé los primeros pasos hacia su realización, reuniendo en un centro común á pueblos de razas, religión y costumbres diferentes."

55 Cited by Juan Francisco Fuentes: "Utopía", in *Diccionario político y social del siglo XIX español*, edited by Javier Fernández Sebastián and Juan Francisco Fuentes (Madrid: Alianza Editorial, 2002), pp. 685–88.

56 Eduardo Chao, ed., *Diccionario enciclopédico de la lengua española: con todas las vozes, frases, refranes y locuciones usadas en España y las Américas españolas, en el lenguaje común antiguo y moderno; las de ciencias, artes y oficios; las notables de historia, de biografía, de mitolojía y geografía universal, y todas las particulares de las provincias españolas y americanas* 2 vols. (Madrid: Imprenta y Librería de Gaspar y Roig, 1853–1855): "cualquier idea que es aceptable en teoría, aunque imposible de realizar."

57 Rafael María de Baralt, *Diccionario de galicismos: o sea de las voces, locuciones y frases de la lengua francesa que se han introducido en el habla castellana moderna* (Madrid: Imprenta Nacional, 1855): "sistema que principalmente se funda en lucubraciones sin fundamento real en la historia, en las costumbres, o en las propensiones humanas, si bien formadas con el buen deseo de mejorar la condición del hombre y el estado de los pueblos." And "una de las voces, locuciones y frases de la lengua francesa que se han introducido en el habla castellana moderna." Even as late as 1874, Baralt still regarded the term utopia as a neologism, although he considered its use to be advisable because of Thomas More's work.

58 The minister of Internal Development of the provisional Government that emerged from the Revolution, Manuel Ruiz Zorrilla, used it in the Explanatory Memorandum to the Decree of 28 January 1869 (*Gaceta de Madrid*, no. 30, 30 January 1869, p. 1). At the Ministry of Overseas, Adelardo López de Ayala also used it in the Explanatory Memorandum to another Decree, on 30 January 1869 (*Gaceta de Madrid*, no. 32, 1 February 1869, p. 1).

59 Real Academia Española, *Diccionario usual de la lengua española* (Madrid, 1992): "Plan, proyecto, doctrina o sistema optimista que aparece como irrealizable en el momento de su formulación." This definition was retained until the 2001 edition.

60 Real Academia Española, *Diccionario usual de la lengua castellana*, 12th ed. (Madrid, 1884): *Utópico*, with its obvious meaning of "adj. Que halaga en teoría, pero que es irrealizable" [Adj. That which is gratifying in theory, but which is unworkable], simplified in the 1925 edition to "adj. Perteneciente o relativo a la utopía." [Adj. Belonging or relative to utopia], a definition that was kept until 2001. *Utopista*: " Que forma utopías ó es dado á ellas. Ú. m. c. s." [Adj. That which forms utopias or is given over to them. Used more as a noun], also until 2001.

61 José María Samper, *Apuntamientos para la historia política y social de la Nueva Granada* (Bogota: Imprenta del Neo-Granadino, 1853), p. 526; Virginia Auber Noya, *Ambarina* (1858), edited by Isabel Ruiz Apilánez (Santiago de Compostela: Xunta de Galicia, Consellería da Presidencia e Administración Pública, 1989): Lucio Victorio Mansilla, *Una*

excursión a los indios Ranqueles (1870), edited by Saúl Sosnowski (Caracas: Ayacucho, 1957), pp. 195, 276, 382; Adolfo Alsina, *La nueva línea de fronteras. Memoria especial del Ministro de la Guerra* (Buenos Aires: Imprenta del Porvenir, 1877), p. 51; Eduardo Acevedo Díaz, *Brenda* (1886) (Alicante: Biblioteca Virtual Miguel de Cervantes, 2002); Manuel T. Podestá, *Irresponsable. Recuerdos de la Universidad* (1889) (Buenos Aires: Biblioteca Clarín, 2003).

62 Juan Montalvo: *Siete tratados* (Besançon: Imprenta de José Jacquin, 1882), pp. 381–82: "Si sois librepensadores, traducid á Laplace, Littré: si amables utopistas, á Flammarion, Delaage: si herejes declarados, á Renan, Peyrat (...). Si en todo y para todo quereis, autores franceses, ahí están en ilustre muchedumbre, historiadores, oradores, científicos, filósofos, y hasta novelistas, grandes novelistas, como el autor de René, el de Obermann, el de Corina. Traducidnos la Enciclopedia, por Dios, traducídnosla, vosotros que sois, oh españoles, tan amigos y partidarios de Rousseau, Diderot, d'Alembert, Grimin y más puntos luminosos de la gran constelacion del siglo décimoctavo, cuya estrella polar, el hélice del infierno, es Francisco Maria Arouet, convertido en Voltaire por obra y gracia del demonio. Pero esos libritos, esas novelitas, esos santitos, esas estampitas de que están atestadas las librerías . . ."

2 How to Do Things With Utopias: Stories, Memory and Resistance in Paraguay

Marisa González de Oleaga

"En la suela de los zapatos está la verdad de la historia."
[The truth of history is found on the soles of shoes]
J. Consiglio, *Hospital Posadas*

"Truer in the sense of poetic or emotional truth."
J. M. Coetzee and A. Kurtz, *The Good Story: Exchanges on Truth, Fiction and Psychotherapy*

"C'est agir en vaincus: (...) ne pas voir l'espace — fût-il interstitiel, intermittent, nomade, (...) — des ouvertures, des possibles, des lueurs, des malgré tout."
[It means acting like the defeated; (...) not seeing the space — whether it be interstitial, intermittent, nomadic, (...) — within the openings, what is possible, the flashes, the *nevertheless, the in-spite-of-everythings*]."
G. Didi-Huberman, *La survivance des lucioles* [1]

The West has an abundant literature on the subject of utopias. These exercises have traditionally concentrated on what was called "utopian thought" and "literary utopias", a genre of writing with its own conventions and rules. Actual cases of utopias that were set up at some point in contemporary history seem to have generated less interest, possibly because the very definition of utopia — as a "no-place" or an "ideal-place" — appears to preclude any project carried to completion, which, for that very reason, would cease to be utopian. How to do things with utopias is a proposal that is developed at length in *En primera persona: Testimonios desde la utopía*,[2] an analysis of the limitations that the literary genre imposed on the representation and circulation of the memory of utopian/dystopian experiments, and which calls for new forms of writing and representation.

A clotted memory

Paraguay is a country that is unknown in Mediterranean Europe. Hemmed in by Brazil, Argentina and Bolivia, little is known of its existence or its history. Whereas the imaginaries of the other nations bordering it are associated with emblematic objects and historical events (*caipirinha* and slavery; *mate* and the last military dictatorship; *coca* and the first indigenous government of Evo Morales) or are linked to more or less stock scenes associated with music and dance (the samba and the bossa nova; the tango and the Andean flute), nothing seems to represent Paraguay on the European side of the ocean. Perhaps the two most significant landmark historical events associated with the country are the Jesuit Indian settlements (popularized by Hollywood) and the harsh dictatorship of Alfredo Stroessner.

Nevertheless, the nineteenth and twentieth centuries were quite prolific in historical events in Paraguay. Not just dramatic ones, like the War of the Triple Alliance (1864–1870) or the Chaco War (1932–1935), to cite two episodes of major importance for the population; it was also the setting for alternative social projects. In spite of being marginalized — or perhaps because of it — Paraguay proved to be fertile ground for religious and political utopias and the occasional dystopian experiment. These included anarchist and theosophical colonies, Mennonite and Hutterite communities, socialist enclaves and even a racist Aryan colony founded by Elisabeth Förster-Nietzsche, the sister of the famous German philosopher. These were veritable social, political and economic laboratories on the margins of the State and the market with a clear vocation for communitarianism and self-management.

From the end of the nineteenth century until the middle of the twentieth, at the same time as a new political order was being organized throughout Latin America, utopian communities proliferated in a number of countries in the area. In Paraguay, the government's "racial whitening" policies offered incentives (in the shape of land grants, tax reductions, religious and linguistic freedom, autonomy in education and security, exemption from military service) that led immigrant contingents to consider the country as a place to settle. The Paraguayan authorities looked upon the members of these groups — mostly European — as disciplined, dedicated workers who were going to "infect" the local population with their industriousness, and make a crucial contribution to launching the economy of the nation. For the newcomers — quite a few of them, political exiles — Paraguay was that empty fertile space where they could build a better society.

However, none of these goals was realized. The arrival of immigrants with alternative projects did not provide the kind of stimulus that the "fathers of the nation" desired, while the founding of self-managing colonies did not extend any further than their own symbolic, physical limits, and most of them had very little impact on the local populations. Stranger still was the complete absence of contact between these communities;[3] as if they had become prisoners of the very metaphor that had given rise to them, these political and religious utopias remained islands in the national landscape for decades; and what was even worse, the memory of them fared no better. In a country not known for its writing, a good deal has been written about the utopias in Paraguay. But in spite of the abundant bibliography on each of these experiments (articles, monographs, web pages, documentaries and other audio-visual devices), memory of them — the copious accounts of their experiences — has failed to circulate and the image of these ventures that has become consolidated is of failed, non-repeatable experiments. Little or nothing of that enormous symbolic capital appears in the form of tradition, trace or reference in the new social movements or in the discourse of progressive organizations. They are known to have existed, but the memory of them has not been inscribed within the framework of alternative social models, but portrayed rather as experiences of foreign settlers doomed to failure. It is not uncommon to hear that these ventures were "cosas de gringos" (gringo affairs), a statement accompanied by the twofold sentiment of envy (of what others had or have) and impotence (for what they will never have because they are Paraguayans).

But how do we explain and understand the fate that the memory of these experiments suffered? In the first place, political historiography must accept its share of responsibility. Concerned in recent decades about the creation of the State and building citizenship, it neglected those other phenomena that ran parallel to the organization of the liberal order, by stressing — more by omission than commission — that they were expendable. The questioning of the role of the State that accompanied the neoliberal wave unleashed in Latin America as a whole-in the 1990s drove historians to dig deeper into the past to recreate the different institutional genealogies, although they forgot about or scorned these other forms of collective organization. In the second place, the utopian/dystopian character of these ventures also worked against them. The very notion of utopia as an unworkable project, or a venture that sooner or later would degenerate into its opposite, dystopia, hampered the transmission of these memories.[4]

And, last but not least, and this is the subject of the present chapter, it is likely that if, despite the relative abundance of stories about these utopian experiences, the memory of those experiments has not been passed down, perhaps there is something about the form or the narrative structure of those stories that hindered or contributed to their failure. It would be a question, then, of failed or faulty transmission that prevented some experiences of the past from becoming part of the knowledge of the present.

How to do things with utopias[5]

In Paraguay, the memory of the utopian/dystopian colonies has not surfaced in the present day. But, then, why should it have done? After all, is it not possible that if there is no echo of those experiences in the present, it is perhaps because they are not useful or have little to offer today? Why attempt to make it easier for society to appropriate other people's experiences? Why refer to the aesthetic responsibilities of the narrators rather than to the truth of their stories as a fundamental element in transforming other people's experiences into knowledge of one's own? For what reason and to what end do we do things with utopias?

In the first place, utopias/dystopias — like other alternative historical processes – can offer *possibilities*. In a globalized world where homogeneity rules and difference is reduced to a format or a brand, historical accounts — those narratives that give an account of "what was" — can serve as spaces of alterity, places in which "the different" and "difference" can be intuited or represented, and utopian experiments — which arose as alternatives in the interstices of State power — turn out to be a clear and encouraging example, in this respect. To be able to incorporate that difference into one's own imaginary, however, requires the ability to conceive of those transformations as possible, to think that there is nothing natural or irremediable in any historical situation, past or present. And that imagination that constructs new landscapes draws nourishment from different sources, one of which may be the historical account, the narration of scenarios that are no less luminous for being non-repeatable.[6] The effectiveness of those historical accounts lies not so much in the fact that they are instruments of sociological engineering (which tell us how social life works and how we could alter it) as in their poetic potential to inspire (which clears the way to the imagining of other presents).[7] If socio-political difference was possible in the past, why should it not be so in the present and the future? It is the discontinuity that difference (of utopias/dystopias in this case) opens up in our expectations of the past, which enables us to imagine that same disparity (other alternative forms of organization) in the present or the future. That historical imagination that wonders about other possibilities is a condition, although not the only one, of every transforming action.

Secondly, and following on from the previous argument, transformative actions require, besides the idea of possibility — change is possible — a certain conviction on the part of the subjects about their own capacity, because somebody may know that other worlds are possible but not consider that they are competent to take part in those changes. This is the case of public opinion in Paraguay, which usually considers successful utopian experiments, such as the Mennonite colonies, as instances that do not concern them or from which nothing can be learnt because "they are (the product) of another culture". These are all formulas that seek to explain how other people triumph over natural and economic adversity, but at the same time, shut down any possibility of experimental appropriation for the locals. The complex processes of identification and identity intervene in this construction of qualified subjects, both individual and collective, and, in this transition, a leading role is played by transmission via stories, because these are what create identities. As Stuart Hall points out, identities "[…] are the names we give to the different ways we are positioned by, and position ourselves within, the narratives of the past".[8] *Narratives enable identifications*, which is the key to actions that promote reasoned changes and help create imaginaries of resistance. Transmission is a human and social need; social and individual identities are not created from nothing, as something inaugural, but form a fabric, a weave made with the appropriation of some narrative threads from the past. As Jacques Hassoun points out, in all transmission of memory of the past, an attempt is made to reconcile what is received with desire.[9] To put it another way, one takes from the weave of the past that which enables desire in the present to be recognized, sustained and relaunched. Thanks to this transmission, the past is symbolized, and assigned a value and a place, which enables some identifications to be made, certain features, certain threads, to be appropriated, which leads to the symbolization or resignification of the present. Lack of transmission in the present, or silence about the past, makes it difficult for subjects to take part in social life. The historical account, the story about the memory of utopias/dystopias may be the scenario, or point of departure from which to realize that passage, where that transmission can take place.

In the third and final place, if every transformative action requires the idea of possibility, and a subject or subjects convinced of their capacity to bring about change, it is also worth enquiring into the relationship between the two elements, because the transmission of memory is neither repetition, nor rupture, but elaboration. And in the process of elaboration, the dystopian possibility that lurks within any human project serves as a limit, and a warning. It is not a matter of "copying" (assuming such a thing were possible) or of wanting to imitate the ways of life of utopian/dystopian communities of a hundred years ago. While there is a level of practice that can be reproduced (forms of cultivation, the governability strategies of certain communities such as the Hutterite-Bruderhof), nevertheless what is transmitted is not only information that aims at a sort of sociological engineering, but *inspiring images or emotions* that enable the subjects to identify themselves and recognize their desire. In transmission, what is at stake is not the truth of the facts, but the subjective truth of that which was important to those who preceded us (desire) and with which we can identify and use to empower ourselves in the present.

Hence enquiring about the reasons for the failed transmission of these alternative memories is not a trivial matter. It represents a major resource in the building of collective imaginaries of resistance and the creation of traditions (communitarian and self-managing) that are understood as "a version of the past which is intended to connect with and ratify the present."[10]

The structure of the stories

We counted five colonies with an abundance of written or audio-visual material about each one of them, regularly present in the national media: anniversaries, distinctions, events, cultural activities. These colonies were: Nueva Germania [New Germany]; Puerto Bertoni / Colonia Guillermo Tell [Port Bertoni /William Tell Colony]; Nueva Australia [New Australia] / Cosme Colony; the Mennonite Colonies of the Paraguayan Chaco; and the Hutterite Colonies of Primavera. It is not ignorance of their existence or their history that has prevented their memory from being inscribed within the common imaginary, but rather the particular way in which these records have been worked into collective stories. The memory of those communities has undergone a process of depoliticization and the message that circulates does not account for their status as alternative projects to the State and the market, but their lack of continuity in time and their foreignness.

How can the "obliteration" of this enormous symbolic capital be explained? How can this be understood, particularly among the ranks of progressive social movements looking for references and traditions to appeal to? I have already pointed out the way in which the circulation of their memories was hampered by the fact that the experiments were regarded as utopias/dystopias, so clotting their transmission. But there was something else, and it might be supposed that that remainder must be inscribed somehow into the way that these experiences have been told whenever the histories of these communities have been transmitted in the form of stories. But what is that narrative structure like?

Five utopian/dystopian colonies and a variety of stories in different formats: political and religious communities that had features in common (despite their many differences), a collective ideal that transformed them into living communities and some form of communal property. There are written, photographic and audio-visual narratives of all of them. The choice of the corpus calls for two preliminary clarifications.[11] Firstly, some explanation is needed for the inclusion of a dystopia (Nueva Germania) among the set of utopian colonies; and secondly, why such disparate stories have been included in the same set. The five colonies can all be considered utopian projects in the traditional sense of the term, as proposals for creating "another place" or "an ideal place." They were all regarded as utopias by their founders and first colonists, and as dystopias by many of their members throughout their history. In the case of Nueva Germania, an instance of an Aryan racist colony, I find it difficult not to describe it as a dystopia from the start after the aberrations that that particular aspiration took in twentieth-century European history. Nonetheless, as far as this study is concerned, it is not so much the way in which their utopian or dystopian nature is defined that is of interest here — a task that is always complicated and partial — but their status as alternative projects to liberal order. It is their very alterity that justifies the inclusion of such disparate ventures in one and the same group. As for the diverse nature of the stories that form the corpus — written in different languages by authors from different disciplines — this is a virtue rather than an obstacle; in spite of these differences, we are able to talk of a common literary genre that would be imposed on the various disciplinary mandates, a sort of metanarrative that would condition the narrative structure of the stories about utopias/dystopias.

In all cases, the predominant literary genre is the travelogue, a mode of writing that can encompass quite diverse forms and in which a variety of discursive strategies intersect.[12] In the case that concerns us here, two elements have been included: the traveller and the route he took, in a three part movement in which they narrate the journey of the historical protagonists, incorporate the narrator's encounter with that experience, and try to repeat this exercise with the reader, who accompanies the protagonist and the narrator on that journey.[13] The stories about utopias/dystopias in Paraguay do not set out to describe what each community was like in an objective way — as a sociologist would, for example —, but are exercises in discovery of other forms of life organized from the standpoint of an autobiography. These are not "scientific" accounts as some sociological currents or other social disciplines might set out to be, but they are not autobiographical accounts in the strictest sense either, as in the case of an extraordinary individual retelling his adventures. The genre we are considering is in tension somewhere between these two extremes.

They all share some common features and one reaction that is repeated. What is surprising about this corpus is the type of emotion that the stories arouse: fascination. It is a very particular emotion, one triggered by something exceptional and non-repeatable that the protagonists in the colonies talked about in their relationship with nature,[14] which is retold by the narrators of those experiences that others lived at first hand, and is then referred to by the literary critics or reviewers of those narratives.[15] Here are some examples:

> Works such as the great falls of Guayrá are not described; they are marvelled at. Scenes of nature inspire the poet: the Guayrá is one of those that leave him speechless (...) and the roar silences thunder, in the stunning solitude of the jungle — all of that stifles words, confounds reason and subjects the heart to the most diverse and mixed of feelings.[16]

> The most fascinating of all to me was the unwritten story of New Germany, the racist colony Elisabeth helped to found in the middle of South America over a century ago.[17]

> A sparkling idea, and its realization . . . yields vivid travel writing and information of a ghostly but fascinating sort.[18]

> The book Mr Macintyre has written about Elisabeth Nietzsche is a fascinating, provocative and highly eccentric volume that is part biography, part travelogue, part detective story.[19]

I could reproduce other quotes in which the word "fascination" appears or where some characteristic typical of this emotion is described. And this is just as true of the protagonists of the utopian/dystopian experiences — Bertoni's relationship with nature in the example given above — as of the ones who related those experiences second-hand — Macintyre on the Nueva Germania experiment — and of the critics — Steiner and Kakutani — who reviewed the accounts. As the dictionary of the Spanish Royal Academy indicates, *fascinación*, or fascination, is synonymous with delusion or hallucination; it also defines the word as "irresistible attraction". Fascination is a singular passion, an irresistible attraction that blocks any possibility of exchange or appropriation, locks the stories in like clots of meaning. To feel fascinated is to be captivated, enraptured, absorbed by the object (Freud, 2006).

In theory, we do not know what produces fascination but we can certainly try to find something common to all the stories. It seems plausible to suggest that if they all produce fascination it is because they contain some similar ingredient. If there is a common effect it may be that there is a common cause. If we confine ourselves only to the content of the accounts, there is a constant in the narrative structure of all of them — that which is traditionally called "subject matter" or "argument" — and in the way that that content is organized. Regardless of the variety of detail in the actions or the number of characters that perform them, all these accounts have a common three-part structure that can be summed up as follows. The subject — individual or collective — the hero of the tale abandons the comforts of a modern city to set out on an adventure in an unknown country, represented by a mysterious nature — at times bountiful, at others terrifying — and succeeds, in spite of, or because of his suffering, in taming that savage nature and redeeming it for the good of humanity. His is a journey of no return, a one-way journey because even if he returns to his country of origin, he will no longer be the same, since he pays for his resolve with his life or his identity. In this drama, in which the subject/hero decides to exchange the comfort of the familiar for the unknown, there is a reward that he will never enjoy, a legacy that is passed down to all mortals: a fascinating intellectual work or significant moral values, the strength of someone who dared to go beyond what was known. But it also conceals a threat: the possibility that that legacy will be lost for ever, due to human apathy, neglect or indifference. It is a threat that the author(s) of the story/stories warn(s) about and seek(s) to invoke with narrations that are a way of making the hero return to his place of origin and so allow him to rest in peace.

All the accounts analysed share a constant three-part structure: *separation*; *initiation*; and *return*, somewhat similar to the folktales analysed by Vladimir Propp.[20] In spite of the differences between the stories, these three functions are common to all the texts. The protagonists set off leaving family and friends behind, and begin an adventure in territories unknown, where they arrive, more often than not, by chance rather than as a result of deliberate choice. This was the case of Bertoni who, having had his fill of the decadence of modern urban life, sought out that ideal place, first in the northwest of Argentina, then in the port that bears his name in Paraguay. Or the case of William Lane, who left Australia after the failure of the 1891 shearers' strike, convinced that his socialist ideals could only be realized in some country in South America, first in Argentina, then in neighbouring Paraguay. Elisabeth Nietzsche, the driving force behind Nueva Germania, would also leave her native country to follow her husband, Dr. Bernhard Förster and establish their project of racial purity in Paraguay, that *terra incognita*. Much the same can be said of the Hutterite, later Hutterian, Brethren of the Primavera Colonies who, in exile in Great Britain during the Second World War, sought refuge for their way of life and Anabaptist beliefs in Paraguay. Finally, the Mennonites also left several countries (Russia, Canada and Mexico) persecuted by the new laws of the States that were being constructed, to find safe haven in the "new paradise" of Chaco — a dense "spiny forest" — where they could live according to their religious and cultural traditions.

In all the cases analysed, the stories in which these experiences are related single out this first function or stage, and do so by using it to trigger the narration. They could have indicated other functions or made a different aspect the centre of the narration. There is nothing self-evident about selecting the move away from the country or place of origin as the initial element of the story. In Bertoni's case, he could have used some other aspect as the organizing principle of the account, for example, the omnipresence of his mother (in a somewhat peculiar family structure), who abandoned her husband to accompany her son's family to the "promised land".

Likewise, the second stage — *initiation and difficult task* — is present in every one of the five cases studied. Separation gives rise to initiation, a new path to be taken and, in this new beginning, the protagonists will rely on the help of a supernatural being (God in the case of the religious colonies) or the collaboration of powerful entities (such as nature or an overarching ideology, in the case of the other communities). Nothing will be easy, the stories tell us, and the heroes are not going to achieve their objectives straightaway. They will pay a price, *no gains without pains*, and that price will be paid in the form of a difficult task. The supernatural beings that help and guide them, or the natural and political entities, which, without being supernatural, have enormous powers, put the heroes to the test and these tests mark a turning point in the story. The narrations about New Australia depict William Lane as obsessed with socialist ideals, intent on building a *Workingman's Paradise*, and subjected to tests from which he did not emerge unscathed: having to accept the gulf between his ideals and human and social reality. The tales about the Port Bertoni / William Tell Colony speak of a hero naively confident in the powers of nature, an ambiguous nature: exceptionally fertile, but which will put him to the test more than once. The death of his little daughter, Inés, and the loss of his plant collection — the work of many months lost as the result of a flood in the first settlement in Missions — seem to be part of the price Bertoni had to pay to sustain his ideals. Something similar is recorded in the story about Nueva Germania. The ideal of a racially pure community was what drove the Förster-Nietzsches to Paraguay, and they would do their utmost, or at least try, in spite of the tests to which husband and wife were put: the accusations of the colonists, Bernhard Förster's suicide in San Bernardino, and the desertion of the peasants. The Mennonites and Hutterites, who shared many distinguishing characteristics, relied — according to their accounts — on the help of God, who steered their lives but also placed a number of tests in their path towards achieving his kingdom on earth, namely, the death of hundreds of new arrivals from disease and malnutrition, the dreadful, extreme climatic conditions as well as the internal and leadership crises that threatened their survival.

Separation, initiation and return are the three functions common to all the accounts; the return stage does not necessarily have to be physical but it will always be symbolic. Moisés Bertoni died in 1929, in Foz de Iguazú, and rests "en una plenitud misteriosa bajo este majestuoso árbol de ciprés" (in mysterious plenitude beneath this majestic cypress tree), as it states on the sign accompanying his grave (placed there by the Itaipú Hydroelectric Company) in the Port that bears his name. He never returned to the canton of his birth; nonetheless, the story arranged another kind of return for him, according to which his death — his return to nature — enables the gift of his legacy — his wisdom, his courage — to be bequeathed to the whole of humanity. The Mennonites did not return to their places of origin either (that place being much more difficult to determine given the centuries-old diaspora of the community); nevertheless, the stories that are told about them emphasise their symbolic return, namely, the attainment of their objectives, their ability to maintain their identity and traditions and, at the same time, achieve great economic success. The Hutterites of Paraguay moved on again in 1960, after what came to be known as the "Great Crisis", which was basically a crisis of leadership. They left the Primavera colonies and settled in the state of New York. The stories of this event and the later development of the community also record this symbolic return and the legacy that it represented for the protagonists and the new members.

The stories of the colonies founded by William Lane and Elisabeth Nietzsche turn out to be rather different. Lane, in his time, was accused of being authoritarian and purist — which was what caused his departure from New Australia to found a parallel community, Cosme Colony — while Elisabeth Nietzsche was accused of having embezzled the funds of the colonists. Both returned (physically) to their places of origin, but the accounts seem to say that they did not succeed in passing the tests imposed by their ideologies, or the constraints of reality. Even so, such is the power and the force of this three-part narrative structure in the development of the story, that there is a sort of return and gift. Lane returned to his own country and the New Australia and Cosme Colony would end up being dissolved as socialist communities shortly afterwards, and their lands divided into lots and distributed among the settlers. The stories, however, insist on picking up on the legacy of their memory; it was a failure, but it was those first colonists and their socialist leader who laid the necessary foundations that enabled a figure like the anthropologist León Cadogan to be born and grow up in that community. Cadogan became a specialist in the Mbyá-Guaraní and Aché Indians, and was acknowledged by Claude Levi-Strauss as the foremost authority on those cultures and a staunch defender of indigenous rights in Paraguay.[21] The accounts acknowledge that Lane failed to the extent that he did not manage to keep the socialist colony going, but see the birth of Cadogan — who, during his lifetime, was more than critical of the extinct socialist project — as a pretext for the story to follow the three-part structure.

The stories about Nueva Germania are even more eloquent. This was an experiment that is not in tune with present-day sensibilities. Elizabeth Nietzsche went back to Germany and decades later struck up an excellent relationship with the Nazi party and its leader. The narrative structure is imposed on the content to such an extent that the author of the story himself feels compelled to seek, and indeed finds, a kind of symbolic return and legacy, by stressing the dark side of the experiment, which would lead years later to concentration camps and gas chambers for millions of people. Nevertheless, his fascination is such that Elisabeth is seen as a woman of enormous energy, "of extraordinary courage, character [...] and *chutzpah*".[22] There is no doubt that the author of the account, Ben Macintyre, feels no sympathy for or has any ideological affinity with Elisabeth Nietzsche and, yet, he cannot help being drawn to her and has to look for some reason (her fighting spirit, her capacity for transgressing limits in order to go beyond what is known) to justify his fascination. And he finds it in this lady's personal qualities, a sort of decontextualized personal legacy.

Apart from the three-part structure — so reminiscent of the folktales and fairy stories of oral folklore that Propp studied — there is another interesting ingredient: the binary oppositions that shore up the stories. I remember a newspaper feature that I read in Foz de Iguazú in 1991, which was when I first became aware of Port Bertoni. Its front-page headline said: "Bertoni: A *Museum* in the *Jungle*" and the title of the article: "A *Swiss Genius* in the Upper *Paraná River Jungle*". These polarities, or binary oppositions, turn up in the majority of the titles of the works cited. By way of example, take *Paradise Mislaid: In Search of the Australian Tribe of Paraguay*, a work by Anne Whitehead, who plays with this opposition of a *paradise* that is found, but then mislaid or has gone astray (has become hell?), and a *tribe* of *white* Australians in Paraguay. *Strangers Become Neighbors* is the title of the book by Redekop about the relationship between the Mennonites and the indigenous peoples. Once again, oppositions: the *strange* versus the *familiar*. Macintyre's account of Nueva Germania, *Forgotten Fatherland*, recreates that duality with a title that alludes to the loss of memory (of the fatherland, the origin, the place of the father). One of the works by Baratti and Candolfi about Port Bertoni repeats this pattern: *Vida y obra del sabio Bertoni: Moisés Santiago Bertoni (1857–1929): un naturalista suizo en Paraguay* (The Life and Work of the Learned Bertoni: Moisés Santiago Bertoni (1857–1929): A Swiss Naturalist in Paraguay) reinforcing a twofold opposition. On the one hand, the one forged between his birthplace, *modern Switzerland*, and his new homeland, *unknown and wild Paraguay*; on the other, in this latter country, Bertoni's status as a *learned man*. All these oppositions are inherent in mythical tales — those tales about marvellous deeds and supernatural beings — whose structure was described by Levi-Strauss. Among these oppositions, the culture-nature opposition is the one that is characteristic of the myth of the hero, exactly as shown in the research of the mythographer Joseph Campbell, and also in that of Hugo Bauzá.[23] The hero is, more than anything, a transgressor; a being in permanent conflict between two worlds, a mediator between the civilized and the wild, between order and disorder; he is a being capable of going beyond the limits imposed on mere mortals.

The three-part narrative structure — the functions of Propp's fairy tales — and the binary oppositions are the two common elements that appear in every story about the utopian/dystopian communities in Paraguay. And these two constants seem to be associated with the fascination that was, as we saw, the prevailing emotion in the reception of the stories. We can say, then, that we are in the presence of mythical stories about heroic figures that follow the structure of fairy tales. But there is something else.

The paratexts

Those common elements of the narrative structure with the power to fascinate us — the three-part functions and binary oppositions — are not restricted to the accounts themselves but can also be observed in the paratexts, those visual elements that turn a story into a book.[24] In almost every case, the paratexts, those discourses of transition/transaction that serve as pragmatic devices anticipating the structure of the book, seem to match each other. I refer here to the book covers, a sort of letter of introduction to the book, which draws in the reader and directs the type of reading or interpretation. The book covers of the cases studied here have characteristics in common. Exotic motifs predominate, whether they are photos or illustrations; jungles, Indians, and animals on the one hand, and modest buildings and white settlers on the other. The semantic oppositions that are so characteristic of the texts have also migrated to the covers, suggesting that we are not looking at academic books but at stories that are more like the fables or folk tales that Propp spoke about. Apart from the way the covers "exoticize" the text, an abundance of maps and photos in the paratexts is another common feature in all the accounts that serves to differentiate them from others.

The presence of maps is a constant and their inclusion does not seem to be due to any need to be informative. In most cases, the maps are generic, contribute no information whatsoever, or are so basic — such as Paraguay outlined on the continent as a whole — that they seem to be there for other purposes. If the studies were about a colony in Sheffield instead of being about Paraguay, nobody would include the outline of Great Britain on a map of Europe as a guide, unless the book was written for young children, when they would either include other types of map (showing the use of space, distribution of the population, communication networks) or would dispense with school maps altogether. However, in the works analysed, these simple maps appear, reinforcing the idea of remoteness, the uncharted place of the territories of utopia/dystopia. Hence, Paraguay appears inscribed as a place so remote that the mere representation of its outline is sufficient to give it substance, to make it emerge from the shadows. Remoteness and empty space; the outline of the country is marked and then the site where the colony settled, as if the colonists had arrived in *terra nula*, an empty uninhabited place, and, incidentally, naturally adopting that colonial perspective that makes the original peoples invisible.

All the works have a large number of photos, divided between historical photographs of the colonies, those showing the activities that were carried out in those enclaves, and other photographs — and they are always there — in which the author(s) of the story is/are seen decades later with the descendants of the settlers and posing at emblematic sites in the colonies. In principle, there is no reason to include this material, not even to support the story with pictures, unless the purpose is to guide the reading in some way. The inclusion of period photos looks like an attempt to anchor the story in the real world, a way of tying that "exoticized" tale — so close to the narrative structure of fairy tales — to "what really happened". Reproducing these pictures and including the narrator in the scene form part of a strategy for increasing the mimetic capacity of the story; historical photographs in which the author of the story invariably appears indicating two time periods: the past and the present. This inscription bears the obvious meaning of authorization, branding, the "I was there" that constitutes the hallmark of scientific ethnography.

Paratexts also follow the three-part structure that we saw in the narrations: *separation*, *initiation*, and *return*. The maps and photographs exoticize, *separate* the object of the story and summon the reader to an unknown space, deliberately made strange. Exotic, but translatable; the past is brought up to date in a present that rests on the author's point of view. Once the reader has been summoned, he may find in the reading of the text a form of *initiation* into knowledge, the possibility of understanding something that, in principle, is presented as alien. And just as the hero mediated between two worlds, so the author — serving as a bridge — mediates between the alien and the familiar, transforming the strange into the readable, intelligible, making the reader *return* to the known.

Transmission

Everything seems to indicate that something about the narrative structure — shaped in a similar fashion to folktales — and the exoticizing paratexts have influenced the circulation of the stories about utopias/dystopias in a negative way; it is as if the most contemporary narrative versions of those experiments were dragged down by the genre markers associated with the literary utopias/dystopias; a genre close to the folktales and fairy stories of European folklore that structured the content by mythologizing it, heightening the aspects of those experiences that were exceptional and non-repeatable. If we accept this argument, it begs another question; in what other way could stories about utopias/dystopias be written that would enable those experiences to be actively and critically appropriated?

Let us reconsider some of the characteristics of the stories about utopias/dystopias in Paraguay. We saw that one of the markers of the way those stories were narrated was the inclusion of the author in the story, both in the narrative structure and the paratexts, which is what I shall call the *autobiographical character*. *Reflexivity and autobiography*: the inclusion of the author in the story is an interesting and necessary marker. This inclusion can make it easier for the memories of the communities to circulate, but it is not without its problems. In recent decades, it has been common to find an authorial presence in the social sciences. It occurs in some disciplines — like ethnography — more than others, but it is becoming increasingly common for the narrator to appear as one more character in the story, an attempt to debunk the positivist fallacy of the objectivity of the scientific account. Forming part of the narration that gives an account of a phenomenon is to uncover the inherently biased, perspective-laden nature of any interpretation. Thinking and giving an account of the position from which one is speaking (one among many) implies considering the relationship with that which is the object of study and analysis. The result is a shift from the idea of truth — my story represents what happened — to the idea of responsibility — my story is, at best, one of the possible, interested, partial views of what was observed, and for which I must answer. It is in this sense that the appearance of the autobiographical element as the marker of the stories of utopias/dystopias in Paraguay seems to me to be interesting.

However, instead of resolving the problem, the inclusion of the author in the scene simply creates other problems. The inclusion of the author in the story may make one think of a centred subject, with a fixed identity, master of his word, an autonomous, rational, unified subject, one of the great inventions — along with the subject/object split — of modernity. Feminists, poststructuralists and deconstructionists have all justifiably attacked this assumption. And I say justifiably because the debates about the crisis of the subject are political debates that concern our daily lives, and the possibility of thinking about alternative ways of life. After all, this modern subject — regarded as rational, self-centred and autonomous — is the cause and effect of modern forms of domination. Without going so far as to take up the most radical positions, in which the subject is a mere effect of discourse, how does one include oneself in the story while accepting that all processes of identification are unstable and relational, and that the postmodern individual is being created and transformed in and through the story?

Let us recall what was said about the inscription of the subject in the stories about utopias/dystopias in Paraguay; this is a marker of authority, it is about the need to endow the narration with a certain realism threatened by the exoticization of the object. This authorial function and its anchorage are a long way from the relativist, perspectivist, decentred function that the inclusion of the postmodern individual seems to bring with it. Nevertheless, the idea of a journey, which runs through these stories as a metaphor, fits in with the idea of a narrator who is created and transformed in the course of the narration. It is a narrative journey from which one emerges at the end different from the way one was at the beginning, and can generate a similar effect in the reader. Hayden White and Roland Barthes have analysed an alternative mode of writing, applied to certain historical accounts and testimonies about traumatic memories, such as those of the Holocaust.[25] This is the so-called "middle voice", a voice halfway between the active and the passive voice.[26] The subject is, at one and the same time, subject and object of the action. In the case that concerns us here, the employment of this pronominal form would mean starting from a premise that would be difficult for the empiricists to digest, namely, that when we narrate, we are not giving an account of something external to the story itself, but rather the story generates a relationship, an encounter, an exchange with that which we are studying. It is not a question of narrating a prior experience; the story is the experience.[27]

The inscription of this type of *subject in process* ought to go for a choral, polyphonic story in which the author included in the account would show his other facets, his contradictions, his losses; in other words, a chorality that does not appeal to other voices, but other voices of *his own*, a plural-singular voice, because a story about an alternative experiment or about a utopian/dystopian colony is, ultimately, an encounter, an exchange, a friction between the person writing and the traces of past experiences. It would be as well, then, to think of how to inscribe oneself into an account of utopias/dystopias, how to demolish the fallacy of the centred subject without at the same time eliminating all possibility of thinking about another type of subject (or individual, if you prefer). A subject in process would be unable to transmit stable, finished knowledge; however, if there is no communicable knowledge, what might the relationship of the social actors — in other words, those subjects who are going to appropriate the alternative experiences of the past — be with the stories about the historical utopias/dystopias in Paraguay?

Representation or evocation/friction: representation is the characteristic expression of the social sciences in modernity. Returning that knowledge of the past to presence is what historiography and other related disciplines have done for more than a century. But what aspect of historical experience — in this case utopias/dystopias — may be useful today? For after all, is not the knowledge that we can extract from the past firmly gripped between irrelevance and exceptionality? What I mean is, if there were some more or less specific knowledge derived from the utopian/dystopian historical experiences, it would be so general and so obvious that it would not require looking into specifically. Pointing out that utopian/dystopian colonies have problems maintaining themselves at some point during their history (problems of leadership, conflict of interests and so on) does not seem to be such a far-reaching conclusion that it would deserve our attention. Pointing out that religious communities seem to last longer — because they have stricter rules — does not seem to be a great discovery either. An analysis of each particular case would enable us to reach more finely calibrated conclusions but what would be the point of such findings if their conditions of possibility cannot be repeated? What I mean by this is that what is general to all the communities is too obvious and what is peculiar to each one cannot be repeated. What then can be taken from those experiences that would be of use today? Representing — in other words, returning what was and what took place to presence — would mean accepting that there is some stability in the signification of events, which is questionable at the very least. Can we account for the significance of utopias/dystopias once and for all? But even if we were capable of doing so, what use would it have? "What was", in its most diverse modalities, is over, but "that thing that was" contains one facet that has not really been explored: "what might have been and was not". So, representation, based on the similitude or mimesis between the story and what happened, is replaced by another operation, evocation,[28] a movement founded on difference, on what did not take place. "What might have been" are those other possibilities that were discarded, or not possible because they belonged to other codes of signification and value.[29]

If, as I have pointed out, "what was" is unstable, why should "what might have been and was not" be any more so? The intention is not to replace one representation with another, but to subvert the very idea of representation by means of friction. Friction is a kind of dialogue in which creation takes place, not appropriation. When there is friction between a reader and a text, this movement sparks images, ideas, scenarios, landscapes that were not there before. Friction indicates returning its many presents to the past, at the same time as it converts the present into historical material, relativizing it and putting it into perspective.[30]

The irony of the story and binary oppositions: with a decentred subject who evokes and is in friction with the past, the resulting story ought to be quite different from the traditional one. This new subject, since it is constantly shifting, would opt for an ironic text that would mark the doubt in every assertion, reject single significations and suggest other possibilities. One of the most characteristic features of the narrations about utopias/dystopias in Paraguay were the binary oppositions, those formulas — together with other ingredients — that generate fascination and tend to perpetuate the legacy of the experiments as myths. How can that structure be altered? By introducing polyphony and chorality. Faced with the nature versus culture opposition, it is not a question of inverting the value of either of the two terms but of opening up play to other voices, in other words, changing the duality into multiplicity. This polarity that keeps appearing in contemporary stories could perhaps be altered and denatured if other voices are summoned, that of the Mbyá-Guaraní, for example, that are far from conceiving that relationship as a duality; introducing other codes of signification and value is to denature one's own. The same could be said of the fate of the colonists, of that heroic version that the stories introduce. If those binary oppositions fracture, almost certainly, one of the most important ingredients of the myth of the hero as mediating between two worlds will collapse. Even at the risk of being cacophonous, it would be necessary to superimpose different stories about that fate (of the protagonists, the native witnesses, the contemporaries in their places of origin), to change the tempos and modes of narrating (the corollary of including other voices will be to deploy these different tempos and stories with different structures), to experiment ... with knowledge that is open, in process, which knows that it does not know and shows its lack of knowledge. Experimenting and opening ourselves up to experimentation because:

> If one can stop looking at the past and start listening to it, one might hear echoes of a new conversation; the task of the critic would be to lead speakers and listeners unaware of each other's existence to talk to one another. The job of the critic would be to maintain the ability to be surprised at how the conversation goes, and to communicate that sense of surprise to the other people, because a life infused with surprise is better than a life that is not.[31]

Acknowledgement

This chapter forms part of project HAR2012-31212, financed by the Spanish Ministry of Economy and Innovation.

Notes

1 Jorge Consiglio, *Hospital Posadas* (Buenos Aires: Eterna Cadencia Editora, 2015), p. 55; J. M. Coetzee and Arabella Kurtz, *The Good Story: Exchanges on Truth, Fiction and Psychotherapy* (New York: Viking Random House, 2015), p. 11; George Didi-Huberman, *La survivance des lucioles* (Paris: Minuit, 2009), p. 36.

2 Marisa González de Oleaga, ed., *En primera persona. Testimonios desde la utopía* (Barcelona: NED/Gedisa, 2013).

3 There are two cases that show the almost complete indifference of these communities: the projected anarchist colony, William Tell, and the Cecilia colony on the one hand, and the Anabaptist colonies of Friesland and Primavera, on the other. In spite of their ideological and geographical proximity, contact was practically non-existent or very sporadic. See Danilo Baratti and Patricia Candolfi, *L'Arca di Mosè. Biografia epistolare de Mosè Bertoni* (Bellinzona: Casagrande, 1994), pp. 39 and 44; Yaacov Oved, *The Witness of the Brothers: A History of the Bruderhof* (New Brunswick: Transaction, 1996), pp. 109 and 118.

4 Already in the nineteenth century, in the work of Friedrich Engels, *Socialism: Utopian and Scientific* (1880), so-called utopian socialism was discredited for advocating a model that was impossible to realize, and was described as a mere step along the road to formulating scientific socialism. Edward Forster, Yevgeny Zamyatin, Karl Popper, Aldous Huxley, George Orwell are only a small sample of the authors whose literary works and essays sealed the fate of the concept. If to this, we add that a significant number of these ventures did not last, the hypothesis of utopia as a daydream, or a project doomed to failure, was reinforced. A detailed development of the critiques of utopia can be found in Rafael Sánchez-Mateos, *De la ruina a la utopía: una constelación menor. Potencias estético-políticas de la infancia* (Unpublished PhD diss., Universidad Nacional de Educación a Distancia, 2015), pp. 290–318. Daniel W. Hollis III, *Utopian Movements* (Santa Barbara, California: ABC-CLIO, 1998), p. xv; Krishan Kumar, *Utopia and Anti-utopia in Modern Times* (London: Blackwell, 1991); Richard C. S. Trahair, *Utopias and Utopians* (Connecticut: Greenwood Press, 1999), p. xii.

5 Both the title of this chapter and this epigraph are indebted to the classic work by John L. Austin, *How to Do Things with Words*, 2nd ed. (Cambridge, MA: Harvard University Press, 1975).

6 Experiences are non-repeatable as such, although some repetition is necessary in the gesture of creating and constructing ways of life that represent alternatives to the hegemonic way. If this were not so, if there were not some repetition in the gesture we would not be talking about utopias and we would not be able to compare them. Repetition, however, does not mean copy but rather re-creation, in the sense of something new meeting the old. Jacques Derrida, "Firma, acontecimiento y contexto", in *Márgenes de la filosofía* (Madrid: Cátedra,

1989), pp. 347–372 and by the same author, *Acts of Literature* (New York: Routledge, 1992).

7 Michèle Petit, *El arte de la lectura en tiempos de crisis* (Mexico: Océano, 2009) and Greg Dening, *Readings/Writings* (Victoria: Melbourne University Press, 1998).

8 Stuart Hall, "Cultural Identity and Cinematic Representation", *Framework: The Journal of Cinema & Media* 36 (1989): p. 69.

9 Jacques Hassoun, *Los contrabandistas de la memoria* (Buenos Aires: Ediciones de La Flor, 1996), p. 29.

10 Raymond Williams, *Marxism and Literature* (Oxford: Oxford University Press, 1977), p. 115.

11 The following is a list of the colonies and the stories used: Nueva Germania (San Pedro), an Aryan racist colony founded in 1887 by Bernhard Förster and Elisabeth Nietzsche; descendants of the original colonists are still living in the colony; Ben Macintyre, *Forgotten Fatherland: The Search for Elisabeth Nietzsche* (New York: Harper Perennial, 1993); Puerto Bertoni or Colonia Guillermo Tell (Alto Paraná), an anarchist colony project transformed into a family colony, founded in 1893 by the Swiss Moisés Bertoni; its last inhabitant died in 1929; Danilo Baratti and Patricia Candolfi, *L'Arca di Mosè. Biografia epistolare de Mosè Bertoni* (Bellinzona: Casagrande, 1994), and by the same authors, *Vida y obra del sabio Bertoni, Moisés Santiago Bertoni (1857–1929). Un naturalista suizo en Paraguay* (Asunción: Helvetas, 1999); Colonia Nueva Australia and Cosme Colony (Caaguazú), a socialist colony founded in 1893; in May 1894, William Lane and other colonists broke away and founded Cosme Colony, seventy kilometres from the first settlement; Lane left the community in 1899, which continued until 1905; Gavin Souter, *A Peculiar People: William Lane's Australian Utopians in Paraguay* (Queensland: University of Queensland Press, 1991 [1968]); Anne Whitehead, *Paradise Mislaid: In Search of the Australian Tribe of Paraguay* (Queensland: University of Queensland Press, 1997); Anabaptist Mennonite Colonies (Menno, Fernhein and Neuland in the Department of Boquerón, Chaco Paraguayo) founded in 1927, 1930, 1947 to the present day; Calvin Redekop, *Strangers Become Neighbors:Mennonite and Indigenous Relations in the Paraguayan Chaco* (Ontario: Herald Press, 1980); Peter and Elfreida Dyck, *Up From the Rubble: The Epic Rescue of Thousands War-Ravaged Mennonite Refugees* (Pennsylvania: Herald Press, 1991); the Hutterite Primavera Colonies (Caazapá), founded in 1945; in 1960, they moved to the United States; Yaacov Oved, *The Witness of the Brothers: A History of Bruderhof* (New Brunswick: Transaction Publishers, 1996); Bob and Shirley Wagoner, *Community in Paraguay: A Visit to the Brotherhood* (Farmington: Plough and Hutterian Brethren, 1991).

12 The following are research studies of undeniable interest: Diana Salcines de Delás, *La literatura de viajes: una encrucijada de textos* (Unpublished PhD diss., Universidad Complutense de Madrid, 1996); G. Percy Adams, *Travel Literature Through the Ages: An Anthology* (New York: Garland, 1988); Eric Leeds, *Shores of Discovery: How Expeditionaries Have Constructed the World* (New York: Harper Collins, 1995); John Needham, *The Departure Lounge: Travel and Literature in the Postmodern World* (Manchester: Carcanet Press, 1999); Mary Louis Pratt, *Imperial Eyes, Travel Writing and Transculturation* (New York: Routledge, 1992); Christian Kupchik, "Elogio de la fuga. En busca de la identidad perdida", unpublished manuscript, 2005.

13 Northrop Frye, "The Journey as Metaphor", in *Myth and Metaphor: Selected Essays* (Charlottesville–London: University Press of Virginia, 1990).

14 Stephen Greenblatt, "Resonance and Wonder", in *Exhibiting Cultures: The Poetics and Politics of Museum Display*, edited by Ivan Karp and Steven D. Lavine (Washington and London: Smithsonian Institution Press, 1991), pp. 42–56, defines *wonder* precisely, as a way of displaying an object so that it fascinates, and contrasts that with *resonance*, which has the power to evoke a much more productive reception.

15 Frye, "The Journey as a Metaphor".

16 Moisés Bertoni in Baratti and Candolfi, *L'Arca di Mosè*, p. 129. "Obras como el gran salto del Guayrá no se describen: se admiran. Las escenas de la naturaleza inspiran al poeta: el Guayrá es de aquellas que lo enmudecen (...) y el estruendo que enmudece al trueno, en la soledad imponente de la selva -todo aquello ahoga la palabra, confunde a la razón y somete al corazón a los más diversos y encontrados sentimientos."

17 Macintyre, *Forgotten Fatherland*, pp. xi–xii.

18 George Steiner, "Review of Forgotten Fatherland written by Ben Macintyre", *The New Yorker*, 19 October 1992, p. 122.

19 Michiko Kakutani, "Books of the Times; On the Trail of the Other Nietzsche", *The New York Times*, 16 October 1992.

20 Vladimir Propp, *Morfología del cuento* (Madrid: Akal, 1998).

21 Richard Arens, ed., *Genocide in Paraguay* (Pennsylvania: Temple University Press, 1976).

22 Macintyre, *Forgotten Fatherland*, p. xii.

23 Claude Levi-Strauss, "La estructura de los mitos", in *Antropología Estructural* (Barcelona: Paidós, 1995). Joseph Campbell, *El héroe de las mil caras: psicoanálisis del mito* (Madrid: Fondo de Cultura Económica, 1998) and Joseph Campbell and Bill Moyers, *El poder del mito* (Buenos Aires: Emecé, 1988). Hugo F. Bauzá, *El mito del héroe. Morfología y semántica de la figura heroica* (Buenos Aires: Fondo de Cultura Económica, 1998).

24 The paratextual elements are those "devices and conventions, both within and outside the book that form part of the complex mediation between book, author, publisher and reader: titles, forewords, epigraphs and publishers' jacket copy are part of the public and private history of the book", in Gérard Genette, *Paratexts: Threshold of Interpretation* (Cambridge: Cambridge University Press, 1997).

25 Hayden White, "Historical Emplotment and the Problem of Truth", in *Probing the Limits of Representation*, edited by Saul Friedlander (Cambridge: Cambridge University Press, 1992), pp. 37–53. Roland Barthes, "Escribir, ¿un verbo transitivo?", in *El susurro del lenguaje* (Barcelona: Paidós, 1994), pp. 23–34.

26 This is a formula characteristic of Greek, Sanskrit and Indo-Persian that was lost when the Greeks began to use a vocabulary associated with the idea of the will and began to think of the agent as the source of all action. Barthes, "Escribir, ¿un verbo transitivo?", pp. 23–33; White, "Historical Emplotment".

27 Martin Jay, *Songs of Experience: Modern American and European Variations on a Universal Theme* (Berkeley-Los Angeles: University of California Press, 2005). We have an interesting corpus of experiments in this respect: Alun Munslow and Robert Rosenstone, eds., *Experiments in Rethinking History* (London: Routledge, 2004) and Art Spiegelman, *Maus* (New York: Pantheon Books, 1973).

28 José Ramón San Miguel, "La evocación. Un análisis fenomenológico", *Catoblepas* 35 (Jan. 2005): 10.

29 Elizabeth D. Ermarth, "The Closed Space of Choice: A Manifesto on the Future of History", in *Manifestos for History*, edited by Keith Jenkins, Sue Morgan and Alun Munslow (London: Routledge, 2007), pp. 50–66.

30 Greg Dening, "Performing cross-culturally", in *Manifestos for History* eds. Jenkins et al.; Hayden White, "Afterword. Manifesto Time", in *Ibid.*, p. 225.

31 Greil Marcus, *Lipstick Traces: A Secret History of the Twentieth Century* (Cambridge, MA: Harvard University Press, 1989), p. 23.

3 Vasco de Quiroga Rewrites *Utopia*

Geraldo Witeze Junior

This chapter examines the Utopian project that Vasco de Quiroga (1478?–1565) developed in New Spain from 1532 onwards, which resulted in two *pueblos-hospitales*: one in Michoacán and the other close to Mexico City. We discuss some of Quiroga's texts and argue that his project is a rewriting of *Utopia* by Thomas More (1478–1535). The intention is to show the close relationship between Quiroga's texts and More's, and the many similarities, but also the differences between the two, especially on issues such as slavery and the death penalty.

Thomas More published his book in 1516, a work that has enjoyed great success down to the present day. Vasco de Quiroga set off for America not long after that date — in 1531 — to take up a post as judge in the second *Audiencia* (Assize Court) of New Spain, after the first one had been dissolved. In 1532, while he was visiting the region of Michoacán, he founded a community, which he referred to as a *pueblo-hospital*, for the Indians to live in. He returned to Mexico City, where he organized another similar *pueblo*. He saw these two *pueblos* as a model to be followed in the Spanish dominions and a way of solving the problems resulting from the Spanish Conquest.

Fintan B. Warren has recounted the history of these *pueblos* and I shall not repeat it here.[1] The idea of bringing the Indians together in groups to instruct them in Christian doctrine was not invented by Vasco de Quiroga, although there was one major feature that was unique to his project: its very close link with *Utopia*. It was associated with the work that inaugurated and lent its name to the utopian literary genre, rather than with utopian thought in its widest sense, rooted in Plato's *Republic*. Quiroga interpreted the book more as a proposal for America than as a critique of English society or of Spain's colonial system.

Quiroga did not manage to extend his project to the whole of New Spain. Nevertheless, those *pueblos* survived until the end of the nineteenth century, which gives some idea of how important they were. Perhaps because they were set up on the periphery of the world, the story of these *pueblos* tends not to occupy a prominent place in histories of utopias, especially in Anglo-Saxon circles. This is a gap that needs to be filled, as some authors have already tried to do,[2] because in fact Vasco de Quiroga answers the question "what is Utopia?" with some precision. We do not necessarily have to agree with the answer he gave, but his contribution helps clarify the meaning of Utopia, and since his reading is such an early one, it is of even greater interest.

The concept of pueblos-hospitales

Vasco de Quiroga probably read *Utopia* after he had already arrived in America. His friend, Juan de Zumárraga (1468–1548), who was the first bishop of Mexico, owned a copy of the 1518 edition, published by Johann Froben (1460–1527). Silvio Zavala explains that Zumárraga lent the volume to Quiroga, and the idea that More's book would be useful for improving the organization of the *pueblos-hospitales* in Mexico and Michoacán was the result.[3]

The decision to build *pueblos* for the Indians did not come from *Utopia*, but the project was carried through as a result of reading it. During his very first year in New Spain, the judge wrote a letter to the Council of the Indies, the supreme governing body of Spain's colonies in America, in which he stated that it was both important and necessary for the Indians — who had been scattered throughout the territories since the conquest — to be grouped together in *pueblos*. Quiroga saw it as the only way to guarantee their survival, enable them to be evangelized and ensure the continuity of the colony, and he clung to this idea until the end of his life. The *pueblos* would be "ordered along civilized, law-abiding lines and with good holy Catholic ordinances",[4] ordinances that Quiroga himself was responsible for writing later.

As was the case with other Europeans who knew the New World, Quiroga held the Indians in high regard. He stressed their intellectual capacities, especially their oratorical skills, but also praised their morality. Their good qualities were akin to the Christian virtues of humility, obedience, poverty and contempt for the world. They were the ideal men to undertake the renewal of the Church that so many humanists were dreaming of.

Most of the Spanish settlers were the opposite of the Indians. Motivated by gain and greed, they did not possess the virtues necessary to be part of Quiroga's utopian project. They were not categorically excluded from it, but they were not suitable for it either. Nor did the settlers readily accept the project, since it went against their own interests, which was to make slaves of the natives. Their stance was essentially anti-utopian.

The Indians were the main characters in that utopia in New Spain. In his letter to the Council of the Indies in 1531, Quiroga described the Indians as "a clean slate and very soft wax".[5] Later, in 1535, in his *Información en derecho* (Information on Law), he said that the Indian was "teachable and made of wax for all things good".[6] They were like the apostles of the early Church and could help revive it. What was missing was for them to learn about the Gospel and acquire a few good habits, although this course of action would be made easier if the idea of the *pueblos-hospitales* became a reality.

This did not mean that the Indians were a kind of "good savage". Far from it, they were humans, and even with some negative characteristics: they did not have good governance, they continued with their age-old religions, they were habitual drunkards, and so on.[7] Even so, they were better than the Spanish settlers, more willing in spirit, more amenable to correction.

There was a multitude of indigenous ethnic groups and we can almost never be certain which one of them Quiroga was referring to. In the territory he was operating in, there were the Mexicas, the Purhépechas or Tarascans,[8] and the fearsome Chichimecas. He does not specify their customs and he refers to Indians in generic terms, a mental construct based on his daily experience of the natives. It would be an uphill task to find out about the Indians from what Quiroga wrote about them, and this should be taken into account. Nonetheless, from what we know of his life and his movements in New Spain, it may be assumed that he had more contact with the Mexicas in the early days when he was a judge in Mexico City, and that, after 1536, when he was bishop of Michoacán, he was probably closer to the Purhépechas. The hospital close to Mexico City was still under his jurisdiction.

The Indians themselves would have to build the *pueblos* that Quiroga was proposing, since, as he said, "it is all for them, their children, their descendants and relatives, and for the common good of everyone; where the orphans and the poor of the territory round about must be gathered together and indoctrinated in and taught the things of our holy faith".[9] The major question of the exploitation of the indigenous workforce was not a concern in this case, since the Indians themselves would enjoy the fruits of their labour. Quiroga was not interested in exploiting Indian labour, as the settlers were wont to do, and even invested his own money in the construction of the *pueblos*, a factor that critics of the project should bear in mind.

These *pueblos* were the opposite of "the hellish chaos of the mines", where the Indians toiled until they died of exhaustion, starvation or disease.[10] They offered the Indians salvation on two levels: physical salvation, an opportunity for them to live and work for themselves, and also spiritual salvation, through knowledge of the Gospel. For Quiroga, the two were inseparable.

Fulfilling the will of God was a very important aspect of his philosophy. He considered that the only legitimate reason for the European presence in America was to preach the Gospel to the natives; everything else was superfluous. He believed that God intervened in history and that his own projects were in accordance with the divine plan. This also strengthened his line of argument, since the Catholic Monarchy would not go against the will of God; a hypothesis that proved to be a risky one.

Interpreting *Utopia*

Utopia emerged as a way of organizing the *pueblos*, once the project was already underway. Quiroga talks of many features being taken from More's book and adapted according to need. Although Quiroga's interpretation forms part of the history of readings of Thomas More's work, he intended his project to serve useful social and political purposes, rather than form part of a literary debate or a humanist game.

Quiroga described the republic of Utopia so as to "talk about it and to state that it was something seen, tried and tested, because, if it was not tried out once, it seems that it could not be believed".[11] Imagination and literary creativity helped conceive possibilities for society, as though many of its problems had already been resolved. In this way, one could see not only what the desired new world would be like, but also the new man, the key element for achieving it.

It was like a dream, but it was more than that: "anyone who has experienced it, does not question it".[12] In other words, for Vasco de Quiroga, the meaning of the experiment in America was to confirm that utopia was a practical possibility, not only in the future, but in the here and now. Rather than a critique, it was a sort of prophecy. The interesting thing is that Quiroga was not interpreting the New World on the basis of his reading of More, but the reverse; Quiroga read *Utopia* based on his own experience of daily life among the Indians.

Thomas More knew of Amerigo Vespucci's accounts of the New World and drew on them for inspiration for his own book. For some authors such as Frank Lestringant, *Utopia* did not depend on America for its existence, which is open to discussion,[13] although it can at least be stated that without America, *Utopia* would have been a very different book. Trying to pinpoint the geographical location of More's island is a joke, as the etymology of the word attests. Quiroga was not concerned about this; he thought that if a vision of the New World as simple as More's could produce such a good description, living there could enable one to dream of so much more.

So, he did not doubt that More knew "what these most simple peoples of this New World were like", adding that they seemed to him to be "just like those of the people of gold in that first Golden Age".[14] According to this interpretation, *Utopia* not only depended on America but was actually talking about it.

Quiroga saw a likeness between the Indians and the people of the Golden Age and projected his understanding onto More, most probably because his edition of *Utopia* also contained the *Saturnalia* by Lucian of Samosata (125–180?), translated into Latin by More himself. Indeed, Quiroga quoted Lucian in his *Información en derecho* when he compared the life of the Indians with the Golden Age. It seemed obvious to him that More, being familiar with the customs of the Indians and the description of the Golden Age in Lucian, had made a connection between the two things.

Quiroga referred to More as a "distinguished man and of a more than human genius", which qualified him to describe the best state of a commonwealth, very useful for building a new world. He went further and declared that the book was not merely a human work, but was "inspired by the Holy Spirit".[15] God was working through More to remedy the problems in New Spain.

This link with the Holy Spirit would certainly not have displeased More, who was a devout Christian, and even lived with the Carthusians for four years, but it was far removed from his thinking as expressed in the letters that preceded *Utopia*, and even in the text itself. One example was the sentence almost at the end of the book where More said that he did not subscribe to all the institutions that Raphael Hythloday described, in spite of the fact that he admired them.

This did not appear to matter to Quiroga; his interpretation was in keeping with the tone of a letter by Guillaume Budé, who treated the account of Utopia as a historical reality. It seems that this letter was supposed to be interpreted in a metaphorical sense, although Quiroga chose to interpret it literally. *Utopia* emerged then as the set of "regulations and the best state of the commonwealth" suited to the qualities of the Indians and it was then a question of making sure that there were "enough of them [the Indians], for them to be able to keep and maintain themselves, and to introduce them to the faith and a mixed *policía*, which is the only thing they lack".[16] Thus, the evangelizing plan and More's book were inextricably linked.

Christian virtues came naturally to the Indians, but, nonetheless, they did not know what good government was. The political aspect was a key part of Quiroga's thinking. One of the paradoxes of his utopia was that the Indians were as virtuous as the Apostles and yet, at the same time, barbarians who were strangers to acceptable political forms.

Quiroga thought that it would be easier to teach those virtuous Indians Christianity and politics than to incline the hearts of the Spaniards towards the Christian virtues.

Furthermore, More was not just anybody, but a "great Greek scholar and a major expert, with a good deal of authority".[17] His words were worthy of consideration and respect. If he had translated Lucian and taken an interest in the men of the Golden Age, if the Indians were like them and he had written a book to solve the problems of the American peoples, this should be appreciated.

Quiroga insisted that "for such people, such a well-wrought state of a commonwealth was both in their interests and necessary and that their preservation could only be assured in this [commonwealth] and in no other, for all the reasons mentioned".[18] It may have been that he had noticed, while living among the Indians, that they were accustomed to quite a disciplined kind of life and that those ancient societies had strict standards of behaviour, so that the ordered life described in *Utopia* would suit them much more than it would the Europeans. The Europeans were already developing a kind of individualism that made it difficult to establish a life in which communal interests were favoured over individual wishes.[19] The modern world was moving in the direction of strengthening individuality, while Quiroga's utopia was heading in the opposite direction.

Utopia was much more than a good project and a good idea; it was the only way to preserve the Indians, convert them to Christianity and teach them good government. So, just as Fray Bartolomé de Las Casas (1474–1566) entitled his work *Del único modo de atraer a todos los pueblos a la verdadera religión* (On the Only Way to Attract All Peoples to the True Religion), here we have the only way of preserving the Indians. This rhetorical device demonstrates that Vasco de Quiroga was involved in the disputes about the fate of America that were being decided in the Council of the Indies. In fact, he sent his *Información en derecho* to his friend, Juan Bernal Díaz de Luco (1495–1556), who held a seat on the Council and was in a position to exert influence on any decisions that it took.[20] The question of what was to be done with the New World was one of overriding importance and the response would affect the lives of many people. At that time, there were many responses and still dreams of a happy outcome for the indigenous population. The history that we know, however, is a terrible tragedy, although some tried to prevent it from happening.

Since it was important in debate to anticipate objections in order to rebut them, the first one that Quiroga envisaged was that "a human, civilizing *policía* of such perfection could not be preserved, if they were not all good, something which seems impossible".[21] He recognized, and never doubted, that the Indians were human, and it was consistent, therefore, also to assume that they had a fallen nature as a result of original sin. Since they were human, they were not perfect. Why introduce a *policía* suited only to perfect men? It would not be practicable in the real world unless men themselves were perfect.

Quiroga's response points to something that he had always talked about: the combination of faith and politics that he called *policía mixta*, or a mixed *policía*. In order to be successful, it was essential to cut out "the roots of all discord and anxiety, all lust, greed, idleness and misspent time". This was how temporal and spiritual affairs would be ordered and would make it possible to implement the best state of the commonwealth. Of course, there were no perfect men and they all needed to be perfected by God, so that "with the introduction of peace and justice, everyone in it [the commonwealth of the hospital] kisses and embraces each other in equity".[22]

In some respects, the text of *Información en derecho* is very close to *Utopia*. Quiroga envisaged, "a city of six thousand families, each comprising from ten to sixteen married family members, which is more than sixty thousand residents",[23] and the numbers on the Island of Utopia were exactly the same: "care is taken that each household (there are six thousand of them in each city, leaving aside the rural districts) has no fewer than ten or more than sixteen adults".[24] A little further on, Quiroga says that the administrative organization that they were seeking to set up would have one *jurado* [a sworn representative] for every thirty families, and one *regidor* [a councillor or alderman] for every four *jurados*; the corresponding structure described by Hythloday has one *Syphogrant* for every thirty families, and one *Tranibor* for every ten *Syphogrants*. These references indicate Quiroga's textual dependence on More, and when it comes to the *Reglas y Ordenanzas* (Rules and Regulations) of the Santa Fe Hospitals, it is even more obvious.

Vasco de Quiroga believed that *Utopia* itself actually answered the questions that would be raised in opposition to his project. Silvio Zavala claims that Quiroga himself translated at least the first book into Spanish,[25] although he may also have translated the second one too. He sent his translation, together with his *Información en derecho*, to his friend Luco, calling him "exceedingly wise and diplomatic" and able to resolve any possible queries concerning his calculations.[26] Unfortunately, this translation was lost and only a few words referring to it have survived.

At the end of the *Información en derecho*, there is a summary of More's intentions in writing *Utopia*: "to respond to and satisfy all opposing arguments and tacit objections to his commonwealth that this most prudent man felt could be brought against him, which are the same as those that have been and may in the future be brought against my own version".[27] In other words, it was not necessary for him to explain himself further, since the Englishman had already done this with the inspiration of the Holy Spirit.

Something was still missing, however. Quiroga concluded his comments on *Utopia* in his *Información en derecho* by quoting the letter-cum-preface of Guillaume Budé, "the honour and glory in these times of the School of France". He accepted every word that the Frenchman wrote and read the *libelus aureus* as if it were "a seminar on correct and profitable customs, from which it was up to each person to extract and adapt traditions for his own respective society".[28] He tried to do exactly that in his *Reglas y Ordenanzas* for the two *pueblos-hospitales*.

Utopia is rewritten

The *Reglas y Ordenanzas para el Gobierno de los Hospitales de Santa Fe de México y Michoacán, Dispuestas por su Fundador el Rmo. y Venerable Sr. Don Vasco de Quiroga, Primer Obispo de Michoacán* (Rules and Regulations for the Government of the Hospitals of Santa Fe in Mexico and Michoacán, drawn up by its founder, the Right Reverend Venerable Don Vasco de Quiroga, First Bishop of Michoacán) were probably written between 1554, when Quiroga returned to America from Spain, and 1565, the year of his death. It is a provisional text, with gaps and repetitions, which was discovered by Juan Joseph Moreno in the eighteenth century and published together with his biography of Vasco de Quiroga. Moreno states that Quiroga drew the rules up in "the final phase of his life, which is why they remained in draft form, without him being able to make a clean copy or ensure that they were observed, as the pious author of them wished".[29]

If Moreno is right, and it seems to me that he is, the *Reglas y Ordenanzas* do not describe the way the Hospitals actually functioned, but rather what their creator had in mind. There are certainly some glimpses of everyday life, but the text is basically *Utopia* adapted to the indigenous societies of New Spain.

Like More, Quiroga valued manual work and was concerned about the problem of idleness. He named "the trades of Weavers, and others attached or belonging to this trade, and Stonemasons, Carpenters, Masons, Blacksmiths and others like them that are both useful and necessary for the Commonwealth of the Hospital",[30] which are practically the same ones cited in More's *Utopia*. Everything was oriented to serving the common good.

In the *pueblos* of Santa Fe, as on the Island of Utopia, everyone would know how to work the land, since it would be an integral part of their childhood education. The working hours, six hours a day, were also the same, although in the Hospitals, the work could be concentrated into "two or three days of work, from sunrise to sunset, per week", depending on requirements.[31] The *Ordenanzas* included an exhortation both to work and to obey the *regidores*, a sign of the real world that was absent from the imaginary island. Disobedience has no place in an ideal society, where everything is perfectly ordered.

Those who had carried out all their duties and had time to spare would have to devote themselves to "any suitable work related to the trades or the needs of the said Hospital".[32] Everything would have to be carried out during the six hours, since there would be no extra time for work. Any time left over would have to be devoted to building maintenance; anything, in fact, to stop workers from lapsing into idleness or to prevent possible disruptions to the work, such as lack of materials.

Some families would be sent into the countryside for two years so that they could engage in agricultural production. After this period, they could return to the town and others would take their place, until everyone had fulfilled this assignment. The cycle would then begin again, making sure that there would always be someone with more experience in the fields capable of directing the rest. Those who enjoyed rustic life would be allowed to "remain there longer, with the express authorization of the *Rector* [Governor] and *regidores*, and there was no other way to do this". All this was exactly as in *Utopia*; Quiroga added only that the families in the countryside would "have to be properly trained and supplied with the tools and implements necessary for their labour".[33]

Overseeing the families in the countryside, there would be a "General Supervisor, who would oversee and visit them, and notify the *Rector*, the *Principal* and the said *Regidores* of any matter that concerned them that needed to be remedied, supplied, or reformed".[34] In *Utopia*, one of the phylarchs carried out this function, supervising thirty families. Since fewer people in the *pueblos* of Santa Fe lived in the countryside and the distances were also shorter, it would be possible for the supervisor to live in the urban area, go out to visit the families and return home at night to sleep.

Those who dwelt close to the main centre of the Hospital would be granted leave "to go one day to relax and enjoy [themselves] with the rustic families in the countryside".[35] This period of recreation was not the same as a holiday or period of idleness, since those who went to the countryside would work there like everyone else. This detail demonstrates that Quiroga anticipated the possibility of the inhabitants becoming weary of urban life, something that did not exist in *Utopia*. This may have been the influence of Antonio de Guevara (1480–1545), an author that Quiroga knew through *Menosprecio de corte y alabanza de la aldea* (Scorn of Court Life and Praise of Village Life), published in 1539, and other works in which he praised the simplicity of the countryside and contrasted it with the difficulties of urban life, or it may have been that the need for relaxation arose from the fact that the Indians of that time, who were accustomed to living in the fields and mountains, did not adapt well to urban life. This could only be done "with the permission of the *Rector*, *Principal* and *Regidores*, and in no other way" so that communal life would not break down.[36] They would be granted leave for a specified period and while they were away, they would receive food and live on equal terms with their companions in the countryside. In short, apart from the two years' stint common to everybody, there were two ways of moving to the country: to go for a specific period in order to recover from some unpleasant aspect of urban life, or permanently, because they adapted better to country life.

Both men and women would work. More criticized societies in which women and nobles did not do any work, because this placed too much of a burden on everyone else; it was necessary to share the workload so that it did not become too onerous. He proposed entrusting women with lighter tasks. Quiroga spoke of "women's occupations that suited them and that they had adopted, and that were necessary for their own benefit and good and for that of the commonwealth of the Hospital, [occupations] such as working with wool, linen, silk, and cotton, and anything necessary, pertaining to and useful for the trade of weaving".[37]

Neither of them regarded dividing the work according to gender as a problem, nor did they see it as oppressing women. On the contrary, doing so suggested equality in every sense of the word. Besides, the Purhépechas were used to the division of labour by gender, and the women even "helped their menfolk with sowing and harvesting, as well as weaving blankets, *paños de chocolate* [embroidered serviettes used for wiping chocolate from the mouth], and other things that were welcomed by society in general".[38] Equality was not the same as uniformity and it was accepted therefore that the work for men and women would not be exactly the same. Apart from that, each inhabitant was free to choose the trade he or she liked best out of all those that were useful to the hospital. Collective harmony was at the heart of everything and it was not necessary for all inhabitants to carry out the same functions.

Quiroga also imitated More's idea of storing and distributing the production of foodstuffs "rationally, comfortably and honestly, so that when each according to his type, need, manner and condition required it for himself and his family, nobody should want for anything in the Hospital".[39] This was one of the most important features that can be traced back to the communal living of the early Christian communities in Jerusalem, or Plato's Republic, and brings to mind Karl Marx's aspiration in his *Critique of the Gotha Programme*: "From each according to his ability, to each according to his needs!"[40] It is obviously not a question of a foreshadowing, but of analogous ideas circulating in history.

The Judeo-Christian tradition transmitted the well-known story of Joseph, narrated in Genesis 37–50. Among other lessons, it taught the principle of storing up provisions for future periods of scarcity, a precaution that was often ignored in history, especially in American colonial societies. Sérgio Buarque de Holanda has referred to the inertia and lack of foresight of the settlers as typical characteristics of colonial

Brazil;[41] even Spain's possessions in America suffered from food shortages in later years and it was necessary to use the gold obtained from the mines they exploited to remedy the problem.

This gave something of an idea of the sumptuary mentality that Bartolomé Bennassar talked about.[42] Quiroga foresaw this problem and was concerned to avoid it, even in the sixteenth century. He proposed that the Hospitals should follow the practice mentioned in *Utopia* of storing food supplies to avoid possible hardship in the years to come and instructed the inhabitants of the hospitals as follows: "each year, you will sow twice the amount that you need for a single year, which you will keep until such time as it is unlikely to be needed for the current year";[43] only after taking such precautions could the surplus be sold.

The Indians in the hospitals would guarantee the production of food using "certain indicators and likelihoods" and "certain assumptions about nature".[44] In *Utopia*, a primitive type of statistics was used to make precise calculations of consumption, while the Hospitals placed their trust in the knowledge of the indigenous peoples. Knowledge that we would consider today as traditional was a guarantee that the food supply of the whole community was assured, and indeed, what the Europeans learned about nature in the Americas depended fundamentally on the social mediation of the Indians.[45]

Farming would take place first on common land in the countryside, where some of the inhabitants would work for two years at a time and then be replaced by others, although families would also cultivate vegetable gardens and grow fruit trees close to their homes. While the question of land ownership was already settled in *Utopia*, Quiroga felt it necessary to stress that this formed the basis of his enterprise, because there were many disputes about land between natives and settlers. Land could not be transferred or sold at all because "otherwise, this good work would be lost".[46]

That principle turned out to be a shrewd one, because it protected the two hospitals for more than three centuries, until the second half of the nineteenth century, when they finally disappeared. The liberal reforms, by putting an end to the collective ownership of land, accomplished what the Spanish settlers, the Church and the Crown had been unable to do. As a result of this, "conflicts started to break out in the *pueblos* when the sale of properties to strangers began", so marking the beginning of the end of the *pueblos*, which finally occurred in the 1870s.[47]

Long before their final demise, however, still in the sixteenth century, the watchword was abundance. With everyone actively involved and nobody left idle, and the work being evenly distributed and not excessive, they would have lacked for nothing. It was neither poverty nor luxury, but frugality. The dignity of their existence lay in them having what they needed to live well, not working themselves to death, nor wishing for a plethora of superfluous things. With cooperation and frugality, there would be abundance; with competition and luxury, shortage.

Quiroga listed what would be raised and cultivated in the *pueblos*: "many birds of every kind, both from Castile and from the local areas, and turkeys, and other kinds, both useful and colourful, and livestock, such as ewes, rams, goats, cows, pigs, and domesticated animals, depending on the quality of the land, and oxen".[48] Whereas in *Utopia* only chickens, horses and oxen were mentioned, in Quiroga's writings European breeds were mixed with American ones, and indeed the exchange of animals was important in many ways and transformed the lives of both the Europeans and the Indians for ever.

He also referred to plant life. There would be "fruit trees from Castile and from this land, as well as all kinds of good garden produce, and useful, health-giving seeds, flax, hemp, wheat, maize and barley, and liquorice".[49] The prejudice against hemp had not yet been established at this time. Although only maize was mentioned among the indigenous crops, this does not mean that the inhabitants would not take advantage of local products, but simply indicates Quiroga's limited knowledge of America's native flora and fauna. The Indians would undoubtedly have used their traditional foods, as well as those from Europe that they found to their taste.

The principle of benefit or advantage determined what would be cultivated, and nothing that was useful would be discarded; the Europeans, however, had not yet studied the natural world of America sufficiently, so that they would not use many of the things that it could offer. What the Indians knew about the environment would have had to be put into a language that could be understood by educated Europeans like Quiroga before the knowledge could be used. If this task has still not been completed in the present-day, what chance was there of it happening in the sixteenth century when the process was just beginning and the arrogance of the colonizers held back progress?

Balancing production and storage would make it possible to care for those who were sick, by giving them the best of what was produced. The frailest would receive special attention and care, while carriers of contagious diseases would be isolated from the rest. This too relied on More's text, along with the growing practice of holding Mass in a chapel so that the sick could hear it.[50] While it is well known that there was religious toleration in *Utopia* and various religions existed together, for Quiroga, this was just not possible; his utopia, as was mentioned earlier, was completely Christian.

The tolerance that Raphael Hythloday referred to indicated that it was only possible to apply moderate persuasion to convert others to one's religion. In one case, a recent Utopian convert became "so carried away that he not only placed our faith above all others, but noisily condemned the rest as profane and their followers as impious, sacrilegious and destined to eternal fire".[51] As a result of this, he was imprisoned for causing a commotion. The Utopians believed that the truth would triumph in the end and that there was no need to force the pace.

Quiroga, on the other hand, had been a bishop since 1536, and his aim — and the principal mission of his utopia — was to evangelize the Indians; it was, furthermore, the only reason that justified the Spanish presence in the New World. It should also be added that he never used force to convert the natives and his missionary activity remained strictly based on persuasion and love. As he saw it, the Christian example of "love thy neighbour" came before preaching. Following the example of St Paul, and Erasmus (1466–1536), he used metaphors of war in his writings, but ruled out the use of violence for the purposes of evangelization. Although it has occasionally been claimed that he did use force, a careful reading of his texts allows us to verify that this was not the case.[52]

Evangelizing the indigenous populations would not solely be the work of foreign missionaries. The converted Indians would take the new religion to those they knew after receiving religious and political instruction. Anybody who wanted to leave the *pueblos* on friendly terms would need the permission of the *rector* and *regidores*, after the manner of *Utopia*. In addition, Quiroga instructed them, "you should teach, or be able to teach [what you have learnt], and use it to benefit our neighbours wherever you go, and may you always find someone who will welcome you because of it, and do you honour and profit".[53]

For the Christian education of the Indians, Quiroga left doctrine in printed form that had been "approved by his Holiness the Pope".[54] The author of that Christian doctrine was Gutierre González Doncel (1468?–1527) and his objective was "to promote a tangible improvement in the level of instruction and spiritual formation",[55] an idea that he shared with Erasmus (although Erasmus had already been condemned at the time that Quiroga printed the doctrine). Discreetly and at his own expense, the Bishop of Michoacán printed doctrine that was consistent with his religious views and at the same time ratified by the Church. His conciliatory attitude avoided problems.

Although movement between cities on the Island of Utopia was allowed after obtaining permission from the Tranibors and Syphogrants, there was no notion among the Utopians of setting up missions to teach their way of life to other peoples. This would only occur when they set out to colonize other regions in the event of over-population. Quiroga's *pueblos*, on the other hand, and the College of San Nicolás, which he also founded in Michoacán, were centres from which to spread the Christian faith and "good government" from Europe. He originally thought of ordaining Indian priests, although the idea met with resistance within the Church and was later officially prohibited.

Nonetheless, as with the early Christian Church described in the *Acts of the Apostles*, informal missions did not depend on the ordination of indigenous priests, since the Indians who went out from the *pueblos* of Santa Fe would spread the Christian message, as Quiroga wished. It is possible that this conception of missionary activity contributed to the rise of popular Christianity in Michoacán, where Vasco de Quiroga is remembered as a hero and founding father, the *tata Vasco* of the Purhépechas.

For religious reasons, the *Ordenanzas* do not follow *Utopia* on the question of marriage. One only has to recall the pre-nuptial ritual proposed by More: "the woman — be she virgin or widow — is shown to the suitor by some reliable and trustworthy matron while stark naked; and in the same way, some appropriate man shows the suitor naked to the prospective bride".[56] The reason for this was to make sure that there would be no scope for fraud or deception concerning the body, and no regrets afterwards. Even with these precautions, the marriage was not guaranteed. Divorce was permitted with the agreement of both consorts, but only with the consent of the Senate. It was not a simple matter and did not occur without good reason. As for adultery, it was considered abhorrent and was severely punished.

The monastic propriety of Europeans in holy orders would obviously not allow anything similar. Quiroga's *Ordenanzas* prescribed something simpler: the parents would marry off their sons when they were fourteen years of age and their daughters from twelve. The couple would preferably be formed from members of the families in the hospitals, although marrying someone who lived outside was not forbidden. The main thing was to do "everything always according to the order of the Holy Mother Church of Rome, and not clandestinely, although if possible, with the consent of the natural Father and Mother and family".[57]

The way households were governed, on the other hand, did follow the pattern set out in *Utopia*. The oldest man would preside, the women would obey their husbands, and the younger ones their elders. With such a hierarchy, "manservants and maid-servants, and other kinds of servant, who are generally very expensive and annoying to their masters" could be dispensed with.[58] Families would function without the need for help from outside, and everything would be organized and resolved through the wise rule of the elders, in the manner of the biblical tradition that led to the establishment of the priesthood.

Heads of families were charged with controlling the "excesses and disagreements" of all those under their authority, and if they did not do so, the *rector* and *regidores* would take over this task. In cases where "such excesses and mistakes occurred as a result of the incompetence of the *paterfamilias* or his failure to communicate orders properly, they should choose others who are skilful, as best suits the family".[59] Their authority was not absolute, but depended on the proper discharge of their obligations. In *Utopia*, public punishment set an example to prevent others from committing the same offence. This was not the case in the *Ordenanzas*, although fathers who failed to control their children would be accused of "shame and dishonour in correction" by their superiors.[60]

The good example of leaders, rather than punishment, should guide the actions of others, and this was the method chosen by Vasco de Quiroga to convert the Indians, even the fearsome Chichimecas, who were well known for their ferocity, as his aide, Cristóbal de Cabrera, recounted later.[61] A good example based on love would be more effective than punishment. Likewise, the authorities in the hospitals would serve as models and the rest would follow them.

Since life in the *pueblos-hospitales* was based on agricultural production, the heads of the peasant families would supervise the work. As well as supervising, it was important for them "to get involved in the work from time to time, chiefly at the beginning, so that the others would be shamed into doing likewise and not be lazy or have an excuse to stop doing what they should be doing".[62] A good example would be infectious.

After more than two decades of existence, the *pueblos* of Santa Fe proved to be successful, although without meeting initial expectations. The low number of inhabitants was in stark contrast to their original dreams. In *Utopia*, the cities had nearly one hundred thousand inhabitants, like the major cities of Europe.[63] In the Hospitals of Vasco de Quiroga, the population barely reached a few thousand, so that the original idea was only partly realized, and on a much smaller scale.

It should be stressed that the isolation of Utopia in the fictional account was not possible in the real historical and geographical setting of the Hospitals of Mexico and Michoacán. What Quiroga actually wanted was to protect the Indians from the undesirable influence of the Spanish settlers with their unbridled greed and bad company, which was not the same situation as on the Island of Utopia. Nor did he want isolation, in the sense of lack of communication, since his plan, as previously mentioned, included missionary activity.

For the same reason, and unlike *Utopia*, the hospitals did not renounce the use of money. The inhabitants did not need money because the plan of communal work and the distribution of production eliminated the need for commercial exchange within the *pueblos*. Money would be needed, however, for relations with the outside world; it would be important if they needed to buy food from elsewhere, for example. For such situations, Quiroga recommended having "a strongbox, or large coffer, securely fastened with three keys, one held by the *Rector*, another by the *Principal*, and another by the oldest *Regidor*" who were instructed always to keep them in a safe place, where the money belonging to the *pueblos* would be kept.[64] Likewise, the deeds to the lands purchased by Quiroga or obtained through donations would be placed in the coffer for safekeeping. This would prove to be sound advice, since many lawsuits were brought against the hospitals.[65]

Clothing was another characteristic that distinguished the inhabitants of the Santa Fe *pueblos* from outsiders. Everybody's clothes would be identical, as in a monastery, and following the model of *Utopia*. It was recommended to wear garments "of cotton and wool, white, clean and decent, plain, without costly embroidery and not too fancy".[66] This was in line with what More had written, but did the Indians readily accept it?

The Indians in New Spain were well known for their creative talents. The craft objects that Hernán Cortés sent to Charles V evoked great admiration for their beauty, magnificence and wealth of detail.[67] After the Conquest, many Indians succeeded in using their skills in order to learn the techniques of the Spaniards, which was fundamental to their survival in the new world that they were forced to enter.

The Indians, particularly the women, were highly skilled weavers. Their everyday garments were certainly simple, but there were some that were very rich in detail and it was for this reason that Quiroga advised against wearing over elaborate clothing; it was the price of equality. A social hierarchy that was reflected in what people wore had to disappear completely. We do not know exactly what the Indians thought about the matter, although the fact that Quiroga should bring it to their attention is enough to lead us to believe that the old habit of wearing clothes with bright colours and other costly embroidery continued. Since the process of changing a culture is neither easy, nor rapid even when it is successfully achieved, it is possible that uniformity in clothing caused them some distress and aroused their resistance. There were moral and spiritual reasons for this proposal. The stated objective was to ensure similarity among the residents, "so that envy and the pride of wanting to be better dressed, with some standing out more than others, which generally sows envy among men who are vain and imprudent, not to mention dissent and discord, may cease".[68]

Likewise, body and face painting was condemned. The Indians used both for festivities, religious rituals and war, although Quiroga did not want them to retain the practice in the Hospitals and warned them "neither to smear yourselves with red dye, nor paint, nor dirty your faces, hands and arms in any way, as you used to do".[69] This requirement formed part of the struggle to change specific aspects of the indigenous cultures.

Since evangelization was at the heart of Quiroga's utopian vision, it would not be possible to allow outward expressions that were linked to the old religions to continue. Painting their faces was an obstacle, which was why he referred to it as "dirtying the face"; the faith of the Indians' ancestors was seen as a vice, a stain that had to be cleansed. An exception was made, however, when painting the body was for medicinal purposes, although there is no explanation as to how the paints were made or what their ingredients were. Quiroga went so far as to say that clothing ought to express "whatever was in the soul".[70] If the inside was clean, the body would reflect it. The colour white symbolized spiritual purity, and simple clothing expressed the same quality that was expected of all good Christians.

Married women would comply with propriety by following the biblical custom of covering the head "and most of the body, on top of the other garments that they normally wear and without adornments or colourful embroidery and which should not be very expensive, or very fancy, especially when going to church".[71] This would differentiate them from the unmarried women, who were exempted from this obligation.

Their garments also had a practical function, such as protection from the heat and cold. They should be cheap, hard-wearing and appropriate to everyday life in the hospitals. One fatherly recommendation that stands out is the one that advises the Indians to wrap themselves up well in winter, to prevent the "pain in the side that it causes, and goes to the chest, and kills".[72] This is the *tata Vasco* that the Indians remember.

Everyone would have just two sets of clothes, one for ordinary working days and the other for "appearing in public in the square and at church on feast days".[73] Nothing else. Life would essentially be monochromatic after the style of a monastery, so that their minds would be fixed on important matters.

The authorities in the hospitals would be elected. Quiroga himself said in *Información en derecho* that this was the form used by the Indians so that the best man would govern. At that time, Quiroga was critical of this indigenous practice, but he later adjusted the norms of his hospitals to reflect their custom, perhaps because it also coincided with what was set out in *Utopia.*

Heads of families would elect one or two hospital *principales*, "from among the four representatives of the four groups, or *cuadrillas*, into which all the poor in the Hospital were divided, with each *cuadrilla* [choosing] its own, or all of them together". Voting would be secret, and the mandate would be for a period of three or six years, followed by more elections. It went further than the democracy envisaged by More, since the office of Prince in Utopia was for life. The *regidores* would be chosen via the same process, although for a term of one year "and in such a way that all competent married men take their turn".[74] The positions of *principales* and *regidores* would be held by Indians.

Both More and Indian custom stipulated that the man who was most suited to the functions of government should be chosen, while Quiroga, as might be expected, confined himself to stressing the religious component of this choice. The election would be held after the "Mass of the Holy Spirit, and having properly sworn that they will choose, without passion or preference, the man who, in their view, is the most competent, useful and sufficient for the common good of the Commonwealth of the Hospital".[75]

The *rector* would be above all of these and his duty would be to authorize free elections. The incumbent of this office, however, would not be determined by election and while Quiroga was alive, he himself indicated who he thought would be the best *rector.* For posterity, he left other instructions: "after we are no longer here, he [the *rector*] should be appointed every three years by the man in charge of the said hospitals".[76] This "man in charge" was the *Rector* of the College of San Nicolás, and he, therefore, would be the person who would appoint the *rector* of the hospitals after Quiroga's death.

Like the *principal*, the *rector* could remain in office for a longer period if he demonstrated "competence, honesty and prudence", although his mandate would always be renewed "every three years, as has been stated, and with express permission given *in scriptis*, signed by us and the man in charge and the rector and *lector* . . . of the said college of San Nicolás, with the opinion of the dean and chapter".[77] It was necessary to safeguard against tyranny.

In *Utopia*, the main authorities belonged to the class of *letrados* [those with legal training], to which Quiroga also belonged. Nevertheless, he did not make this a requirement for his Hospitals, but recommended instead that the *Principal* "be gentle, patient, and not more abrasive or harsh than is necessary [...] and does not allow himself to be treated with contempt, but should rather seek to be loved and honoured by all, as is reasonable, more through goodwill and love, than fear and strictness".[78] His description recalls the qualities required of bishops and deacons in the New Testament,[79] where St. Paul advised his disciple Timothy using similar words: "The Lord's bond-servant must not be quarrelsome but be kind to all, able to teach, patient when wronged."[80] This set him apart from Machiavelli who thought that it was "far safer to be feared than loved".[81] On the other hand, he was close to Erasmus: "let him love, who would be loved, so that he may attach his subjects to him as God has won the peoples of the world to Himself by His goodness".[82] Authority is obtained through setting a good example.

The *Principal* and *regidores* would meet every three days to take decisions, with two heads of families taking turns to attend. Quiroga added, "no other local council apart from this one should be held by anyone in the Hospital, under serious penalty of being expelled and so deprived of it".[83] The sanctions were different but both *Utopia* and the *Ordenanzas* included rules for upholding the institutions and controlling the authorities. Quiroga included another warning in the same vein. All decisions had to be "in accordance with these Regulations; anything else that they deemed appropriate, as long as it did not conflict with them [the Regulations], always required the consent of the *Rector*".[84] To guarantee the continuity of the communities, no authority was above the Hospital rules.

When the authorities made decisions, it was important to exercise prudence. For resolutions on "doubtful or difficult matters, they should first discuss the matter three days beforehand and decide what could be done, unless it be a matter of little importance or one that would not admit of such a delay".[85] After coming to a resolution on the matter, they would communicate their decision to the *rector*.

The *rector*, by ratifying decisions, appeared to be an essential figure in providing political stability. Since he was not subject to election, it seems that he could mediate more calmly in conflicts, and also exert some religious influence on secular policy, an important feature of the mixed *policía* that Quiroga aspired to implement.

In *Utopia*, there were three types of assembly: the Prince's Council, the Senate, and assemblies of the whole body of the people. More does not describe any of them in great detail. Similarly, because it was a provisional text, the draft of the *Ordenanzas* does not allow us to draw any conclusions as to whether there would be one or two types of assembly. What *does* emerge is that the text drew heavily on *Utopia* and the uncertainty, therefore, may be related to the fact that More's text is also rather vague on this point.

In communities of people, it is normal for conflicts to arise. Quiroga expressed harmony and friendship as an ideal and insisted that there should be few or no lawsuits; even so, he had to draft a rule in case they did arise. *Utopia* emphasized the absence of lawyers because: "In this way there is less scope for misunderstanding, and it is far easier to arrive at the truth".[86] In the *Ordenanzas*, the concerns are more moral in nature and biblical in inspiration.

Given this ideal, it would be the responsibility of the *rector* and the *regidores* to get to the root of grievances "in a straightforward and amicable fashion", guided by true testimonies. The first objective was to prevent the case being taken to external judges, "where you pay a fee and afterwards they throw you in jail". The second objective was spiritual growth, because it was better to lose the argument and have peace than to win and hate one's neighbour: "In this Hospital, you are all to be brothers in Christ, bound by peace and charity, as you were ordered to be and as is greatly to be commended".[87]

The conciliatory attitude was the same as the Sermon on the Mount,[88] seeking to avoid greater evils, such as imprisonment. St Paul too wrote to the Corinthians to correct their quarrelsome attitude, asking: "Why not rather be wronged? Why not rather be defrauded?"[89] When utopian morality deviated from Christianity, Quiroga frequently resorted to the Scriptures or ecclesiastical authorities to fill the gap. Even when he basically agreed with what More was saying, the bishop of Michoacán would add something to bring his text more clearly into line with Christian teachings. So he emphasized the Utopian ban on mocking the afflicted, by offering praise to God that they were not "like one of them, and this with the great compassion that you should have for them and also because, by doing so, you obey our Lord God's command [to love] our neighbour".[90] Utopian reason and Christian spirituality are bound together in the *Ordenanzas*.

Quiroga also advised attending mass, "because it is a holy pursuit, in which much is gained in everything".[91] The hospitals observed a series of religious festivals, the main one being the Exaltation of the Cross. These ceremonies were educational and filled the gap left when the Indians gave up their old indigenous festivals, which they must certainly have missed. There is, however, little sense in comparing religion in *Utopia* with the Christianity that Vasco de Quiroga professed; they were two different conceptions and the Utopian description was a non-historical abstraction.

So far, I have talked about how Vasco de Quiroga interpreted Thomas More's *Utopia*. I shall now try to show that while the text of *Información en derecho* and especially the *Reglas y Ordenanzas*, both draw heavily on More's *Utopia*, they do not constitute copies, but rather appropriations, re-readings of More's work, in which Quiroga dispensed with whatever did not serve his purpose, changed some points and included new ones. Hence, we can speak of a rewrite, by means of which Quiroga adapted More's text to the realities of the historical situation, which is more than an interpretation.

Irreconcilable differences

Before concluding, the most important difference between the two texts should be discussed: the subject of the penalty for those who disobeyed the rules. On this point, there was no possibility of agreement, since *Utopia* accepted slavery and the death penalty, two practices that Quiroga fought against during the four decades that he lived in New Spain. Hence the matter deserves to be treated at greater length.

In *Utopia*, it states, "serious crimes are punished with servitude, since it is considered that this is no less daunting to the guilty and far more profitable to the community than hurrying to be rid of them by immediate execution".[92] This was a criticism of the situation in England, where theft, for example, was punished by death, as More's book explains earlier. The idea was to moderate the punishment and allow the criminal to be rehabilitated, which could not happen if the person was dead. Those who rebelled against slavery, however, were not tolerated: "But if they rebel and resist such treatment, then — like wild beasts that can't be controlled by either bars or chains — they are put down".[93] On this point, *Utopia* takes a brutal and uncompromising utilitarian view: the life of a criminal was only protected if he was useful to the collective as a whole; More had no problem with killing rebels.

There was one possibility for releasing prisoners: "Once they have been tamed by long punishment, if they show by their contrite manner that they regret their crime more than the punishment it brought on them, then by authority of the governor or by a public vote their servitude can be mitigated or remitted altogether".[94] Deciding how to interpret this somewhat subjective criterion rested with the authorities; in other words, in this case, More was contradicting the principle of rulers being subject to the law, since they were given the power to determine what should be done, without having to adhere to any rule. The idea was that the just ruler was not going to commit an injustice.

On the Island of Utopia, prisoners of war were not regarded as slaves, and slavery was not passed on from one generation to the next. Only those found guilty of committing serious offences — whether native-born or foreign (the latter much more frequently) — were turned into slaves. For such as these — bought for derisory prices or given away — there was no remission; they were bound in chains and condemned to perpetual labour. The Utopians treated compatriots who were condemned to slavery with greater harshness because they thought that they deserved an exemplary punishment; they reasoned that they "couldn't be kept from crime in spite of an admirable education that directed them towards virtue".[95] There were no doubts about the quality of the education that they had received in Utopia, so that the responsibility fell wholly on the individual, and the sanction was necessary to maintain the social equilibrium.

There was a type of voluntary slave, the *mediastinus*. These were the poor from neighbouring countries who volunteered to serve the Utopians. They were treated much better than the others, almost as if they were citizens: "Should one of them want to leave, which doesn't occur often, they don't hold him against his will or send him away empty-handed".[96] This hardly seems like slavery, although that was what More called it.

None of these forms of slavery was acceptable to Vasco de Quiroga. Cruel historical reality prevented him from compromising on the matter. The Indians were being made slaves solely through the greed of the Spanish settlers, against the evangelizing mission that he wanted. Slavery frequently led to their deaths in the mines. Hence, neither slavery nor death entered into Quiroga's plans.

The most common forms of slavery in New Spain arose as a result of war or ransom. The first consisted of making slaves of prisoners in a "just war"; the second appeared when someone bought an Indian who was already a slave and was required to pay the price of his "ransom". The main objective of *Información en derecho* was to demonstrate that none of these types complied with Spanish legal requirements and were therefore illegitimate.

For Quiroga, the lax attitude towards the practice of slavery in Spanish America derived from lack of experience, poor understanding, or manipulation of information in order to favour particular interests, such as the exploitation of the mines using indigenous labour. His text, therefore, is presented as a kind of battle for the truth, so that the Council of the Indies would take the decision to abolish slavery in all its forms. Quiroga said that the slavery of the Indians meant "the total destruction of all the land";[97] in a physical sense, the Indians would die, the Spaniards would not be able to guarantee their livelihoods and the whole land would be destroyed. In short it would be a complete disaster, in a spiritual sense also. Everyone would lose: the Indians because they would remain ignorant of the Christian faith and the Spaniards because they had abandoned it.

Slavery among the Indians was associated with the old tyrannies, especially the rule of Moctezuma, the supreme chief of the Mexicas. The Spanish monarch should not be equated with such tyrants, and this would certainly happen if slavery were maintained. If the situation of the Indians stayed as it was before the arrival of the Spaniards, the conquest could not be justified. In addition, tyranny would challenge the authority of the Prince.

As for prisoners of war, Quiroga argued that there was no such thing as a just war and hence that this form of slavery was not in accordance with the legal system of Spain. He said that the Indians "do not invade us, are not a nuisance, nor do they resist the preaching of the Holy Gospel",[98] which were the reasons anticipated in doctrine for considering a war to be just. Nor could the Indians' natural self-defence against the attacks of the Spaniards be used to justify war. In the final analysis, slaves of war were all illegal.

As for the "ransomed" slaves, the line of reasoning was the following: true slavery did not exist among the Indians, only something that might be called "hiring labour in perpetuity". This custom was quite common and those who were hired did not lose their freedom or their family, and they did not become the property of the lord, so that when one of these was "ransomed", his status was changed illegally.

These were the general lines of Quiroga's argument. Slavery and war against the Indians were the worst things that could be done in America and nothing could justify them. Quiroga fought against the enslavement of the Indians and set out his project as an alternative to the way in which the Spaniards were treating them. From his point of view, colonization should be based not on violence, but on love. His *pueblos-hospitales* were the best way to preserve the land, evangelize the Indians and guarantee the sustenance of the Spaniards. Opposition to slavery was the cornerstone of the hospitals' creed; slavery, therefore, was unacceptable.

The maximum punishment envisaged in the *Ordenanzas* was expulsion. The *Ordenanzas* spoke of behaviours that would carry this sanction: "being unruly, or outrageous, or a bad Christian, or getting drunk, being too lazy, or refusing to obey these Regulations or doing anything against them, and being incorrigible in this, or going against the benefit and common good of the said Hospital".[99] These were potentially contagious behaviours that could compromise the work as a whole. Whoever did not want to follow the rules of this utopia was invited to leave and live his life elsewhere, where the world was much worse, and slavery and death probably awaited him.

The meaning of utopia

Vasco de Quiroga interpreted *Utopia* on the basis of his experience in the New World. It is a reading based on his actual experience of the daily life of the Indians in New Spain, particularly in Michoacán. More's *Utopia*, set behind a literary artifice, was detached from history in order to better criticize it. The bishop of Michoacán set out to make the journey in reverse and bring utopia back as a project with a date and place for its realization. The existence of the two communities of Santa Fe bears witness to this.

The writings of Vasco de Quiroga are sharply critical of the colonial system, especially the violence perpetrated by the Spaniards against the Indians, and slavery. It was necessary to return to the central values of the Christian faith — "love thy neighbour", a simple life, and humility — although the settlers were too bent on becoming rich at all costs, which left only the Indians.

Quiroga did not think that radical critiques, such as the one made by Las Casas, would have any effect. The Spanish settlers would not return to the peninsula, the Crown would not give up its possessions and the Church would not renounce its authority over the Indians. What options remained then? He thought it necessary to create an alternative within the colonial system, one under the authority of the Church and Crown, but avoiding abuses and enabling the lives of the Indians to be reorganized with some dignity; and, indeed, historical testimonies show that life in the *pueblos* of Santa Fe that were governed by the norms of Vasco de Quiroga was, in fact, incomparably better than outside them. This probably explains why *tata Vasco* has remained a hero in the memory of the indigenous peoples.

Información en derecho and the *Reglas y Ordenanzas* are at one and the same time a critique of the colonial system and an attempt to harmonize interests. It was not possible to extend the solution that they proposed more generally because of the short-term goals of all the actors. The settlers wanted only to accumulate wealth quickly. The Crown needed to defray the costs of colonial administration and to increase its revenues. The Church wanted rapid results in terms of evangelization. Quiroga's work, on the other hand, took a long-term view and was not adapted to the frenetic rhythm of the dawn of modernity.

Quiroga's utopia can certainly be criticized. It was a top-down, European project, created by an intellectual for the people, by the metropolis for the colony. It did not break with the colonial system and, to some extent, contributed to colonization, even though that was not Quiroga's intention. Criticism, though, does not need to be simplistic or to ignore history. It should be recognized that Quiroga proposed a workable alternative to colonization, one that was actually implemented, even if only in part. The proposal had some colonial features, obviously, but there was also a genuine desire to dignify the lives of the Indians. The fact that the communities lasted for more than three centuries and the testimony of the Indians themselves show that the objective was achieved to some extent.

Vasco de Quiroga's interpretation and rewriting of Thomas More's *Utopia* deserves, nevertheless, to figure more prominently in the universal history of utopias. More's *Utopia* inaugurated the utopian genre and its success over the following centuries. It was, however, no small matter that barely twenty years after its publication, a Spanish lawyer should interpret it as a project inspired by God to correct the course of history. While it was not possible for Quiroga to fully realize his dream, he did manage to achieve something. And in hard times, it is always as well to remember that something is better than nothing.

Notes

1 Fintan B. Warren, *Vasco de Quiroga and his Pueblo-Hospitals of Santa Fe* (Washington DC: Academy of American Franciscan History, 1963).

2 Bernardino Verástique, *Michoacán and Eden: Vasco de Quiroga and the Evangelization of Western Mexico* (Austin: University of Texas Press, 2000); Fernando Gómez-Herrero,

Good Places and Non-Places in Colonial Mexico: The Figure of Vasco de Quiroga (1470–1565) (Lanham: University Press of America, 2001).

3 Silvio Arturo Zavala, *Recuerdo de Vasco de Quiroga* (Mexico: Porrúa, 2007), p. 66.

4 Vasco de Quiroga, "Carta al Consejo", in *La utopía en América* (Madrid: Dastin, 2002), p. 62. "ordenados en toda buena orden de policía y con santas y buenas y católicas ordenanzas"; Alejandra B. Osorio, *Inventing Lima: Baroque Modernity in Peru's South Sea Metropolis* (New York: Palgrave MacMillan, 2008), p. 220, glosses the sixteenth-century concept of *policía* as follows: "*Policía*. The good government and civilized life made possible by the laws and ordinances of a well-ordered community or *república*; civic order; civility; urbanity, refinement, and manners".

5 *Ibid.*, p. 63: "tabla rasa y cera muy blanda".

6 Vasco de Quiroga, "Información en derecho", in *La utopía en América* (Madrid: Dastin, 2002), p. 93: "dócil e hecho de cera para todo bien".

7 Sonia Corcuera de Mancera, *El fraile, el indio y el pulque: Evangelización y embriaguez en la Nueva España (1523–1548)* (Mexico: Fondo de Cultura Económica, 1991).

8 Pedro Márquez Joaquín, ed., *¿Tarascos o Purépecha?: Voces sobre antiguas y nuevas discusiones en torno al gentilicio michoacano* (Morelia: UMSNH, Instituto de investigaciones históricas-Universidad Intercultural Indígena de Michoacán-El Colegio de Michoacán-Grupo Kw'aniskuyarhani de Estudiosos del Pueblo Purépecha, 2007).

9 Quiroga, "Carta al Consejo", p. 64: "pues todo es para ellos mismos y para sus hijos y descendientes y deudos y para pro y bien común de todos donde se han de recoger los huérfanos y pobres de las tales comarcas y ser doctrinados y enseñados en las cosas de nuestra santa fe".

10 Quiroga, "Información en derecho", p. 75: "la confusión e infierno de las minas".

11 *Ibid.*, p. 195: "contarla y afirmarla por cosa vista y hecha y experimentada, y porque, si esto una vez no se experimentase, parece que no se podría creer".

12 *Ibid.*, p. 196: "quien lo tiene experimentado ninguna duda pone en ello".

13 Frank Lestringant, "O impacto das descobertas geográficas na concepção política e social da utopia", *Morus — Utopia e Renascimento* 3 (2006): 155–73.

14 Quiroga, "Información en derecho", p. 218: "el arte y manera de las gentes simplicísimas de este Nuevo Mundo"; "conformes y semejantes a aquéllas de aquella gente de oro de aquella primera edad dorada".

15 *Ibid.*: "varón ilustre y de genio más que humano"; "inspirado del Espíritu Santo".

16 *Ibid.*: "ordenanzas y muy buen estado de la república"; "para hacerlos bastantes para no se consumir ni acabar, y para introducir la fe y policía mixta que solamente les falta".

17 *Ibid.*: "gran griego y gran experto y de mucha autoridad".

18 *Ibid.*, p. 219: "para tal gente, tal arte y estado de república convenía y era menester, y que en sola ella y no en otra se podía conservar por las razones todas que dichas son".

19 Agnes Heller, *Renaissance Man* (New York: Routledge, 2016).

20 Marcel Bataillon, "Vasco de Quiroga et Bartolomé de Las Casas", *Revista de Historia de América* 33 (1952): 83–95.

21 Quiroga, "Información en derecho", p. 223: "policía humana en tanta perfección no se podría conservar, si todos no fuesen buenos, lo que parece imposible".

22 *Ibid.*, p. 224: "las raíces de toda discordia y desasosiego y de toda lujuria y codicia y ociosidad y pérdida de tiempo mal gastado"; "se introduce la paz y justicia, y en ella [la república] se besan y abrazan con la equidad".

23 *Ibid.*: "una ciudad de seis mil familias, y cada familia de diez hasta diez y seis casados familiares de ella, que son sobre sesenta mil vecinos".

24 Thomas More, *Utopia* (London: Penguin, 2012), p. 68.

25 Zavala, *Recuerdo*, pp. 161–165.

26 Quiroga, "Información en derecho", p. 234: "harto sabio y sutil".

27 *Ibid.*: "para responder y satisfacer a todos los contrarios y tácitas objeciones que sintió este

varón prudentísimo que se le podrían oponer en su república, que son las mismas que se le han opuesto y podrán oponer a la de mi parecer".

28 *Ibid.*: "honra y gloria en estos tiempos de la escuela de Francia"; "un seminario de correctas y provechosas costumbres, de donde cada uno ha de sacar y acomodar tradiciones para su respectiva sociedad".

29 Rafael Aguayo Spencer, *Don Vasco de Quiroga: Taumaturgo de la organización social, seguido de un apéndice documental* (Mexico: Oasis, 1970), pp. 243–244: "el último periodo de su vida y por esta razón quedaron en borrador, sin haberse podido sacar una copia en limpio, ni haberse puesto en observancia, como lo deseaba su piadoso autor".

30 Vasco de Quiroga, "Reglas y Ordenanzas para el Gobierno de los Hospitales de Santa Fe de México y Michoacán", in *La utopía en América* (Madrid: Dastin, 2002), p. 253: "los oficios de Tejedores, y otros todos a este oficio anexos, y pertenecientes, y Canteros, Carpinteros, Albañiles, Herreros y otros semejantes útiles y necesarios a la República del Hospital".

31 *Ibid.*: "dos, o tres días de trabajo de sol a sol en la semana".

32 *Ibid.*, p. 263: "obras que convengan para los oficios, y necesidades del dicho Hospital".

33 *Ibid.*, p. 261: "estar allí, más tiempo, que con licencia expresa del Rector y Regidores, y no de otra manera alguna lo pueda hacer"; "han de estar bien instructas, y proveídas de herramientas, e instrumentos necesarios para la labor".

34 *Ibid.*: "Veedor general de ellas, y que las vea, y visite, y avise al Rector, y Principal, y Regidores dichos los que hubiere de remediar, proveer, y reformar en ellas".

35 *Ibid.*, p. 266: "ir algún día a recrear, y os desenfadar por las familias del campo rústicas".

36 *Ibid.*: "con licencia del Rector, y Principal, y Regidores, y no de otra manera".

37 *Ibid.*, p. 259: "los oficios mujeriles dados a ellas, y adoptados, y necesarios al pro, y bien suyo y de la república del Hospital, como son obras de lana, y lino, y seda, y algodón, y para todo lo necesario, accesorio, y útil al oficio de los telares".

38 María Guadalupe Chávez Carbajal, "Visión y condición de la mujer en Nueva España: el caso de Michoacán", *Historia y Espacio* 19 (2002): 1–10, esp. p. 2: "participaban junto a sus hombres en las siembras y cosechas, además de tejer mantas, paños de chocolate y otras cosas más que eran bien recibidas por la sociedad en general".

39 Quiroga, "Reglas y Ordenanzas", p. 255: "congrua, cómoda y honestamente, según que cada uno, según su calidad, y necesidad, manera, y condición, lo haya menester para sí, y para su familia, de manera que ninguno padezca en el Hospital necesidad".

40 Karl Marx, *Critique of the Gotha Programme* (Beijing: Foreign Languages Press, 1972), p. 17.

41 Sérgio Buarque de Holanda, *Raízes do Brasil* (São Paulo: Companhia das Letras, 1995).

42 Bartolomé Bennassar, *La España del Siglo de Oro* (Barcelona: Crítica, 2004).

43 Quiroga, "Reglas y Ordenanzas", p. 263: "sembréis en cada un año doblado de lo que hayáis menester, lo cual guardaréis hasta que no pueda faltar verosímilmente el año presente".

44 *Ibid.*, pp. 263–264: "ciertos indicios, y verosimilitud"; "ciertas conjeturas naturales".

45 Mauricio Nieto Olarte, "Ciencia, imperio, modernidad y eurocentrismo: el mundo atlántico del siglo XVI y la comprensión del Nuevo Mundo", *Historia Crítica*, Special Issue (2009): 12–32.

46 Quiroga, "Reglas y Ordenanzas", p. 256: "si de otra manera fuese, se perdería esta buena obra".

47 Paz Serrano Gassent, *Vasco de Quiroga: utopía y derecho en la conquista de América* (Madrid: Fondo de Cultura Económica-Universidad Nacional de Educación a Distancia, 2001), p. 229: "empezaron a aparecer conflictos en los pueblos, al iniciarse la venta de las propiedades a extraños".

48 Quiroga, "Reglas y Ordenanzas", p. 262: "muchas aves de todo género, así de Castilla, como de la tierra, y Pavos, y de otros géneros provechosos, y vistosos, y ganados, como son

Ovejas, Carneros, Cabras, Vacas, Puercos, y animales serviles, según de la calidad de la tierra, y Bueyes".

49 *Ibid.*: "árboles fructíferos de Castilla, y de la tierra, como de todo género de hortaliza buena, y de todas las semillas saludables, y provechosas, lino, cáñamo, trigo, maíz, y cebada, u orozuz".

50 *Ibid.*, p. 271.

51 More, *Utopia*, pp. 108–109.

52 Ernest J. Burrus, "Cristóbal Cabrera on the Missionary Methods of Vasco de Quiroga", *Manuscripta* 5, 1 (1961): 17–27; Ross Dealy, *Vasco de Quiroga's Thought on War: Its Erasmian and Utopian Roots* (PhD diss., Indiana University, 1975).

53 Quiroga, "Reglas y Ordenanzas", p. 258: "que enseñéis, o podáis enseñar [lo que habían aprendido], y aprovechar con ello a nuestros prójimos do quiera que fuéredes, y halléis siempre quien por ello os acoja, y os hagan honra y provecho".

54 *Ibid.*, p. 257: "aprobada por su Santidad del Papa".

55 María Amparo López Arandia, "De Castilla a Michoacán. La obra de Gutierre González en la educación del siglo XVI", *Revista de Antropología Experimental* 10 (2010): 131: "una mejora en el nivel de instrucción y formación espiritual en un ámbito tangible".

56 More, *Utopia*, p. 92.

57 Quiroga, "Reglas y Ordenanzas", p. 258: "todo siempre según orden de la Sta. Madre Iglesia de Roma, y no clandestinamente, sino si posible es con la voluntad de los Padres, y Madres naturales y de su familia".

58 *Ibid.*, p. 259: "criados, y criadas, y otros servidores, que suelen ser costosos y muy enojosos a sus amos".

59 *Ibid.*, p. 260: "los excesos y desconciertos"; "aconteciesen los tales excesos y descuidos por la inhabilidad, y mal recaudo de los dichos Padres de familia, elíjanse otros, que sean hábiles que más convenga de la misma familia".

60 *Ibid.*: "vergüenza y deshonra en la corrección".

61 Burrus, "Cristóbal Cabrera", p. 19.

62 Quiroga, "Reglas y Ordenanzas", p. 260: "poner algunas veces las manos en la obra, mayormente a los principios, porque los demás hayan vergüenza, y hagan lo mismo, y no tengan pereza, ni excusa para dejar de hacer lo que deban".

63 Aires A. Nascimento, note 48 in Thomas More, *Utopia ou a melhor forma de governo*, 2nd ed. (Lisbon: Fundação Calouste Gulbenkian, 2009), p. 310.

64 Quiroga, "Reglas y Ordenanzas", p. 264: "una caja, o cofre grande barreteado de tres llaves, una que tenga el Rector, otra el Principal, y otra el más antiguo Regidor".

65 Warren, *Vasco de Quiroga*, chapters 5 and 7.

66 Quiroga, "Reglas y Ordenanzas", p. 265: "de algodón, y lana, blancos, limpios, y honestos, sin pinturas, sin otras labores costosas, y demasiadamente curiosas".

67 José Luis de Rojas, "Inventiva y picaresca indígena en el trabajo en la Nueva España en el siglo XVI", *Relaciones. Estudios de Historia y Sociedad* 11, 44 (1990): 7–31.

68 Quiroga, "Reglas y Ordenanzas", p. 265: "y así cese la envidia, y soberbia de querer andar vestidos, y aventajados los unos más, y mejor que los otros, de que suele hacer envidia entre hombres vanos, y poco prudentes, y disención, y discordia".

69 *Ibid.*, p. 270: "ni os imbixéis, ni pintéis, ni os ensuciéis los rostros, manos, ni brazos en manera alguna, como lo solíades hacer".

70 *Ibid.*: "que haya dentro en el alma".

71 *Ibid.*, pp. 265–6: "y lo más del cuerpo, sobre las otras vestiduras, que suelen traer y sin pinturas, ni labores de colores, que no sean muy costosas, ni muy curiosas, mayormente cuando vais a la iglesia".

72 *Ibid.*, p. 265: "dolor de costado, que de ello se causa, y da en los pechos, y mata".

73 *Ibid.*: "en público en la plaza, y en la iglesia los días festivos".

74 *Ibid.*, pp. 266–67: "de cuatro que de sí mismos todos los pobres del Hospital, divididos en

cuatro partes, o cuadrillas, de cada cuadrilla, el suyo, o todos juntos"; "y de manera que ande la rueda por todos los casados hábiles".

75 *Ibid.*, p. 266: "la Misa del Espíritu Santo, y habiendo jurado en forma, que elegirán a su entender el más hábil, útil y suficiente al pro, y bien común de la República del Hospital, sin pasión ni afición".

76 Vasco de Quiroga, "Testamento de Don Vasco de Quiroga", in *La utopía en América* (Madrid: Daston, 2002), p. 286: "después de nuestros días se ponga de tres en tres años por el patrón de los dichos hospitales".

77 *Ibid.*, p. 287: "suficiencia, honestidad y prudencia"; "siempre de tres en tres años, como es dicho, y con expresa licencia dada *in scriptis* firmada de nos y del dicho patrón rector . . . de dicho colegio de San Nicolás, con parecer del deán y cabildo".

78 Quiroga, "Reglas y Ordenanzas", p. 267: "sea manso, sufrido, y no más áspero, ni riguroso de aquello que convenga [...] y no consienta ser menospreciado de nadie antes procure ser amado, y honrado de todos como sea razón, más por voluntad, y amor, que por temor, ni rigor".

79 1 Timothy 3:1–13; Titus 1: 6–9.

80 2 Timothy 2: 24. New American Standard Bible (NASB).

81 Niccolò Machiavelli, *The Prince* (New York: P. F. Collier & Son, 1909–14; Bartleby.com, 2001), p. 32.

82 Desiderius Erasmus, *The Education of a Christian Prince* (New York: Octagon Books, 1963), p. 206.

83 Quiroga, "Reglas y Ordenanzas", p. 267: "Y otro Ayuntamiento fuera de éste no se haga por nadie en el Hospital, so penas graves de ser lanzados, y privados de él".

84 *Ibid.*, p. 268: "conforme a estas Ordenanzas, y lo demás que les pareciere que convenga no yendo contra ellas, con acuerdo del Rector siempre".

85 *Ibid.*: "cosas dudosas, o dificultosas, lo platiquen primero tres días antes, que se determinen en lo que se hubiere de hacer, salvo si fuere cosa de poca importancia, o que no sufra tanta dilación".

86 More, *Utopia*, p. 95.

87 Quiroga, "Reglas y Ordenanzas", p. 269: "llana y amigablemente"; "donde paguéis derechos, y después os echen en la cárcel"; "Habéis de ser en este Hospital todos hermanos en Jesucristo con vínculo de paz, y caridad, como se os encarga, y encomienda mucho".

88 Matthew 5: 25–26.

89 1 Corinthians 6, 7. NASB.

90 Quiroga, "Reglas y Ordenanzas", p. 270: "como uno de ellos, y esto con mucha compasión, que de ellos hayáis, y porque también en esto cumpláis lo que Dios nuestro Señor nos manda de nuestros prójimo".

91 *Ibid.*, p. 272: "pues es santa ocupación, en que se gana mucho en todo".

92 More, *Utopia*, p. 94.

93 *Ibid.*

94 *Ibid.*

95 *Ibid.*, p. 91.

96 *Ibid.*

97 Quiroga, "Información en derecho", p. 73: "la total perdición de toda la tierra".

98 *Ibid.*, p. 83: "no nos infectan, ni molestan, ni resisten a la predicación del Santo Evangelio".

99 Quiroga, "Reglas y Ordenanzas", p. 272: "ser revoltoso, o escandaloso, o mal cristiano, o se emborrachar, o demasiado perezoso, o que no quisiere guardar estas Ordenanzas, o fuere, o viniere contra ellas, y fuere en ello incorregible, o fuere, o viniere contra el pro y bien común de este dicho Hospital".

4 Where is Columbus's Helmsman Taking Us?: *The City of the Sun* of Tommaso Campanella as a Utopia Critical of the Iberian Empires

Carlos E. O. Berriel

In 1602, when Tommaso Campanella wrote *The City of the Sun,* there were three wholly Catholic countries: Spain, Portugal and Italy. These same countries also constituted the richest states in Europe. The Italian states were advanced centres of Mediterranean trade at the time, while the Iberian peninsula was united under a single crown and controlled the New World, which included America, India, Africa and other colonial territories. Three hundred years later, these three countries were among the poorest in Europe and their societies lay in ruins. In large part, they survived by exporting their peasants to the emerging Latin American economies that had once been their possessions. They all remained impeccably Catholic. They all adopted fascism shortly afterwards as a form of state.

None of this could have been foreseen by the thinkers and philosophers of the Counter-Reformation. Nevertheless, Campanella already perceived some of the essential elements of the situation, which makes him a shrewd observer and analyst of his time and its problems. In order to understand the complex structure of the utopia that he created, *The City of the Sun*, it is necessary to relate it to the events of that historical period, by which we mean not only the religious crisis and the reorganization of the Catholic Church during the Council of Trent, but also the creation of the Iberian colonial system, the consolidation of the Spanish monarchy under absolutism, the wars of religion in France, the Thirty Years' War, the transition from a merchant economy to a manufacturing one and, finally, the scientific revolution.

Campanella's utopia arose in the context of the Counter-Reformation. It is both a criticism of the Counter-Reformation and hostile to what the Iberian monarchies stood for in their alliance with the Roman Catholic Church. In this respect, therefore, the City of the Sun would be the antithesis of the model that Spain and Portugal imposed on their respective empires; it represents a proposal that charts a different course, one which, had it been followed, would have led Latin America towards a political, economic and cultural model more in accord with modern science and the developments being promoted by the bourgeoisie in the Protestant countries of Northern Europe.

Campanella: A utopian, a rebel and a prophet

The author of the utopian work that best expressed the complex problems of the Reformation and Counter-Reformation and the turbulent atmosphere that surrounded them was Tommaso Campanella (1568–1639). He set out his ideal republic in his best-known work, *The City of the Sun* (1602).[1] This republic, in his view, offered a political model that was able to save the Church from the devastating blows of the Protestant Reformation and lead it once and for all to a prime position among the kingdoms of this world. His idea of salvation for the Church, however, did not coincide with the hegemonic view, strongly influenced by the Jesuits, that was conceived at the Council of Trent. His is a theocratic polis based on the natural philosophy of Bernardino Telesio, the same philosophy that was sending Giordano Bruno to the stake, and which would soon bring Galileo Galilei before the court of the Inquisition.

Campanella, like Thomas More, was a scholar and an enemy of scholasticism and both were political martyrs. Campanella came from the poverty of the Calabrian countryside, where he was born in 1568, the son of an illiterate shoemaker. Calabria was a land of medieval communists, later rocked by heresies and religious movements, and was also the birthplace of the medieval millenarist Joachim of Fiore. In order to obtain a good education, he entered the Order of Preachers, or Dominicans, since he was an ardent admirer of the doctrine of the great saints of the Order, St. Thomas Aquinas and St. Albert the Great. In the libraries, he read everything, from medicine to astrology, books of magic and prophecy, works of superstition and science. At the age of twenty, he read *De rerum natura iuxta propria principia* by Bernardino Telesio (1509–88), a work in which Telesio explained physics as based entirely on nature and was the complete opposite of the Aristotelianism that was prevalent in the schools.[2] The influence of Telesio would be enduring, providing Campanella with a mental framework that would be crucial for writing *The City of the Sun*. His utopian polis was, as we shall see, a vast, complex and bold structure, which can only fully be understood by reference to the circumstances of his time. Campanella, like Thomas More, was an astute reader of the contemporary world, which he portrayed in the *City of the Sun*.

Time and circumstance: the revolution in the sciences

The most original aspect of Tommaso Campanella's thought was rooted in his aspiration to reconcile the new Renaissance natural philosophy with a proposal for the radical reform of the sciences and society.[3] The image of nature as the bringer of harmony, truth and justice — and understood as the expression of the *ars divina* — became a model that served as the inspiration for re-establishing the encyclopaedia of knowledge and, in particular, for reflecting upon life associated with man. Violence, injustice and the human errors that unsettle civil societies all stemmed from the fact that men had strayed from the model of nature, in which it was necessary once more to find fresh inspiration in order to reform communal life effectively.

It was not a question of evils — violence, injustice and errors — seen in terms of ethical abstractions, but specific aspects of earthly life. One of the most terrifying questions at that time was the hostility of the Catholic Church towards scientific discoveries. After the Council of Trent, the Church — which in previous times had been objective and neutral with respect to both these advances and the philosophy that gave rise to them — identified the scientific revolution as one of the main sources of Protestantism, because it was a strand of thought that released science from the custody of religion, and regarded man — based on the humanism of Pico della Mirandola and Erasmus, among others — as an analogue of God and fully capable of initiating a satisfactory relationship with the creator of the world and his work, without the need for intermediaries like the Church.

Campanella's work was well expressed and directed towards the single aim of re-establishing the power of the Church; this was the complete opposite of what his many enemies claimed. He considered that the Tridentine reform was not only insufficient to lead the world back to the Catholic Church and its control, but also inadequate for restoring Rome as the appropriate entity for accomplishing that task.

In writing *The City of the Sun*, Campanella relied on the social imaginary in its utopian variant in order to undertake a thoroughgoing criticism of the Tridentine Church, its decisions and commitments, and at the same time, to advocate a complete overhaul of the theoretical and practical assumptions of the Counter-Reformation. In essence, Campanella wanted to reconcile faith and reason by redirecting science towards the inner workings of the Church; this action was indispensable for saving it from imminent irrelevance with respect to those states that were already Protestant, supporting the scientific revolution with its corresponding economic repercussions, and which would soon become the principal centres of power and colonial metropolises. More specifically, Campanella wanted the expression of faith conceived by the Counter Reformation to yield to the way of thinking of the scientific revolution in order to conquer the throne of the world. This would involve a radical change in Church policy, which saw the essence of its new identity in the structural alliance with the Iberian metropolises. Ultimately, what the philosopher wanted was to save the Church from itself and to rectify the path that it had taken.

Campanella's universalist ideal was born and manifested itself immediately in mature form. In his first political text, he already addressed the questions that would occupy his mind for forty-five years: the state of primordial innocence in nature, the loss of that innocence — which brought all the evils into the world — and the hope for a cosmic regeneration that would lead humanity back to the golden age, the moment when, with all antagonism overcome, a single universal, priestly, Christian monarchy would be instituted, which would rule the world in an age of peace and supreme blessedness.

These objectives of Campanella's are intrinsic to the Catholic conception of universalism. The sin of Adam continued to be the origin of the destruction of the Garden of Eden and the coming of Christ brought redemption. That redemption, however, was merely potential, and was not immediately effective in social and political affairs without the efficacious collaboration of Man. According to Campanella, human action for conscious salvation was still at an early stage because Christ's words were neither faithfully heard nor followed, so delaying the return of peace and happiness on earth.

To hasten the prophesied coming of this golden event announced in the scriptures, Campanella became involved in eliminating the obstacles represented by international political agreements and in correcting the Church's drift off course, a Church that was slow to renew and purify itself of excesses in order to match up to the task of universal government that he himself was preaching.

Even before *The City of the Sun*, Campanella had already drawn up a bold and ambitious plan. With the Lutheran Reformation and its consequences for European politics, the separation between temporal and spiritual powers intensified, a division that Campanella rebelled against. In his view, the Christian world ought to have a single government, exercised by an authority who was a political and a religious leader at one and the same time; logic dictated that it should be the pope. In practice, Campanella advocated that a European sovereign sufficiently powerful to make the conquest of that universal monarchy a viable proposition should mobilize economic, political and military resources and place them at the feet of the throne of the pontiff of Rome. Campanella initially proposed that Philip II, king of Spain, should lead this mission *(Discorsi ai prìncipi d'Italia*, 1595), and forty years later, that it should be the king of France, Louis XIV *(Monarchie delle nazioni*, 1635) and his minister, Cardinal Richelieu.

The circumstances of a work: The Calabrian revolt and Spanish domination

For Maria Moneti Codignola, with *The City of the Sun*, Campanella joined in the debate of an issue that was already settled and his contribution was to reinforce certain *topoi* and characteristic notions, such as the Calabrian heretical-utopian tradition and the dramatic panorama of Italy, especially southern Italy, between the end of the sixteenth century and the first half of the following one.[4] This panorama was characterized by Spanish domination, which would immediately adopt those colonial features for which it unfortunately became notorious, namely, economic and cultural decline due to the loss of political freedom, the strong presence of the Church and its fight against heretics and reformers, spearheaded by the Holy Inquisition. For Moneti Codignola, reading Campanella means picking one's way through the language of someone who is forced to live a life of constant pretence and dissimulation out of the sheer need for self-preservation and so expresses himself in cyphers and codes.

Historical elements make it necessary to establish a connection between the creation of this utopia and the crisis caused by Spanish domination of the kingdom of Naples, and more specifically, the movement known as the Calabrian revolt. The "real" world of Campanella was, first and foremost, Spanish domination of the country where he was born. Campanella was a subject of Naples and also, therefore, of the Spanish Empire, because the monarchs of Aragon had occupied the throne of Naples. Between 1503 and 1713, Naples was the second most important city in the Spanish monarchy. As well as harbouring the biggest fleet in Europe, it was the city with the largest number of inhabitants. This feature helped to make the city the major cultural hub of the empire and an important centre of political and economic power.

In addition to being a philosopher, Campanella was a supporter of direct political action, and the 1599 revolt in Calabria, just after the death of Philip II, provided him with an exceptional opportunity. Campanella threw himself body and soul into this visionary enterprise. Camouflaged among the multitude of dreamers, rebels and bandits, he had a plan to liberate Calabria, which, under the yoke of the feudal nobility, high-handed clergy and oppressive Spanish rule, lived in a state of dire poverty and discontent. This situation persuaded him that it was possible to organize a large scale movement of popular revolt by taking advantage of the internal divisions in the

Kingdom of Naples and seeking assistance from the Turks, who wielded great power in the Mediterranean. His denunciation of the oppression and decadence of local public life took on a utopian, prophetic tone. He preached a republican polis in Calabria, one that he intended to establish by force. It would be a communitarian, theocratic polis, without private property or classes, and he himself would be its legislator and leader, organizing the region according to the principles of a City of God.

Campanella openly incited rebellion against the Church, which had dismissed his ecumenical proposals. Based on his studies of astrology, he announced future transformations of the world, prophesied, on the basis of omens, the imminent return of political and religious unity, and forecast significant wonders for the year 1600, the moment when the Christian republic promised by St. Catherine and St. Bridget would arrive. That republic was the one he set out in *The City of the Sun.*

Two traitors betrayed him to the Spanish government and the revolt failed. Together with hundreds of rebels, Campanella was taken prisoner and, as the recognized leader of the revolt, was accused of treason and heresy. Brutally tortured, Campanella spent twenty-seven years in the Spanish prison in Naples, where he wrote his utopia. The objective of this work was to explain his political project and the theory that informed it and — as we have already mentioned — perhaps, in this way, secure his release.

The Solarian programme

Campanella wrote *The City of the Sun* while in prison in Naples. The work was written primarily with the intention of demonstrating that the union of faith and reason, of the Church and the new science, would lead to Caesaropapism. The symbiotic relationship between Rome and the Iberian metropolises that the Jesuits had fostered, on the other hand, had turned reason and faith into irreconcilable polar opposites. The result was the rejection of the modern world by the Church, cloistered in its particularism and contrary to the universality of scientific reason.

The name of the City of the Sun derives from the practice of sun worship that was observed there. Initially it was written in the Italian vernacular *(Città del Sole*, 1602), but there are numerous later versions in Latin (*Civitas solis*). The novelesque plot owes a good deal to More's *Utopia*, the literary framework, and even the significant number of utopias published during the ninety years that separate these two central works of the genre. It has the most obvious and commonly found external characteristic of utopias — insularity — since utopias are always isolated from history, even when they are not actually islands; Campanella's city of the sun, for example, is to be found "rising from a broad plain".[5] This insularity is not merely a geographical fiction, but a mental attitude in which the island has the classic function of an ideal representation. The objective of using the island as a metaphor is to preserve the utopian community from outside corruption. In this way, it is possible to represent a closed world, a microcosm ruled by specific laws that enable a society uncontaminated by evolution to exist, a contrast to the "real world", in which both the author and the readers of the utopia are situated.

According to Germana Ernst, at the same time, *The City of the Sun* was drawn up as a programme for the failed insurrection and its philosophical idealization. The appeal to nature, understood as the intrinsic expression of divine art, and the criticism of existing society, unhappy and unfair precisely because it had strayed from that model or because it did not imitate it correctly, are the keys to a simpler, more persuasive reading of Campanella's utopia. In his utopia, Campanella sketches the philosophical city he wished to found. The correct imitation of the natural model would initially involve readjusting the relationship between social roles and individual attitudes and intentions. When this relationship is upset, the roles in the social comedy are assigned without rhyme or reason and what predominates is a separation between being and appearance. Thus, false kings like Nero prevail, whereas wise men like Socrates, a king *in veritate* and by nature, are persecuted and put to death. The aim of the Solarians is to reject the dominant madness in order to re-establish a proper connection between society and nature. This would help to avoid the distortions and ills stemming from the prevalence of chaos over reason and appearance over truth.

In the City of the Sun, autonomous natural science and faith are not incompatible but complementary, contradicting the ideals of the Council of Trent. Given that nature is the living statue of God, studying the laws and phenomena of nature is to draw nearer to God, since He is present in them. Scientific research is an activity similar to prayer, which is why Hoh, the great ruler, is both the chief priest and the principal scientist.

The *City of the Sun* is an unusual utopia, which differs from other common expressions of utopia by presenting itself, not as a satire, but as clear scientific prediction, an inevitable, fateful event destined to happen in the real world once the stars were favourably aligned. Hence, it brings together, with remarkable energy, prophetism and natural philosophy.

In its original conception, the City of the Sun occupied the whole of the earth; only later was it reduced to a single city. The city reproduces Copernicus's rational model of the solar system; it comprises seven concentric circuits with the names of the seven planets, at the centre of which there is a round temple, which acts as the heart and brain of this philosophical polis.

The work takes the form of a poetical dialogue between a Grandmaster of the Knights Hospitallers and a Genoese Sea-captain who had been Christopher Columbus's helmsman and who, upon disembarking at the island of Taprobane, discovered the City of the Sun in the middle of a large plain.

The political regime of the City of the Sun is reminiscent of a theocratic system. Above them all is the *Metaphysic*[6] or the Sun, named Hoh, "their spiritual and their temporal chief".[7] He is both priest and prince, whose pre-eminence shines out because of his immense wisdom. The power exercised by the ruler emanates from God and belongs, on Earth, to Hoh. This priesthood invested with political power owes its prestige to metaphysical wisdom and historical and technical experience, and is consecrated by means of the sacramental act of ordination and divine investiture.

In the exercise of power, Hoh receives the support of the Supreme Council, a triumvirate of princes comprising:

~ *Pon*, or *Power*, with jurisdiction over all matters pertaining to war and peace. Under his orders, women and children are continually prepared for war, in spite of the fact that Solarians — like More's Utopians — are peace loving and fight only to defend themselves or to keep the balance of power in the region.

~ *Sin*, or *Wisdom*, coordinates the production of knowledge and education, which is universal. The seven walls of the city are decorated with frescoes of geographical maps, mathematical proofs and theorems, great people and figures, animals, plants, and so on, turning the city itself into a kind of book. This radical educational method generates wise men such as Hoh the Metaphysic, an all-knowing creature, at once a priest, scientist and philosopher, a mediator between the divine will and human intelligence. Particular importance is attached to the sphere of education, which is regarded as all-powerful and the fundamental method for acting on human material; its intention is to eradicate the anarchic, individualistic, primitive nature present in the citizens. Its *paideia* seeks the introjection of order and rules to eliminate dissidence and opposition, with the fundamental aim of adapting the structure of the mind in accordance with the social structure. The boys are taught from their earliest years, as they play and walk along the city walls. It is not a question of guiding them according to their personal inclinations, but of ensuring that they gain an understanding of all areas of knowledge. In this way, the boys learn all the trades, choosing a definitive occupation only when their general education is finished. For the Solarians, the man who masters all branches of knowledge is suited to the science of government, for he is the one who best understands the knowledge that is essential to social order.

~ *Mor* or *Love* is in charge of procreation and matrimony, two strictly controlled activities that are not left to the free will of individuals, who obediently hand their bodies over to the State for the essential function of reproduction. The men are married at the age of twenty-one and the women at nineteen. Matrimony is medically prescribed and regulated, according to circumstances, by eugenic criteria typical of stock breeding in order to improve the race and produce the type of Solarians that the State wants. The doctors choose the couples in accordance with physical characteristics determined by the State. Copulation takes place at a time determined by astrologers in order to favour characteristics also imposed by power. The future spouses, while they wait separately, recollect themselves in prayer. The woman must contemplate statues of distinguished men, whose images will have a positive influence on their progeny. Sterile women remain at the service of the community and may attend to the needs of men who are unable to remain chaste. The boys, as in Sparta, belong to the State. It is not the citizens who choose the city, but the city that chooses its citizens.

In principle, nobody has absolute power in the *City of the Sun*. The people's Assembly has the authority to remove any magistrate from office, except the four principal ones. Removal from office is decided in a consensual manner, because the magistrates are so wise and virtuous that they resign their posts of their own accord if they find more suitable and worthier citizens to carry out their functions.

The egalitarian nature of their laws brings about social uniformity; dissidence, objections or demands are unknown among the Solarians, and there are no active minorities or political parties. The citizen is regarded as part of the whole, a whole in which he is a mere particle. His will blends instinctively into the State's, which is rational and just, and where the individual is always subject to the requirements of order and general equilibrium.

State control over all aspects of life is revealed in the geometrical structure of the city. The passion for symmetry reflects a love of order taken to mystic extremes. Its inner workings run like clockwork.

This polis is organized along the lines of a convent, with perpetual worship before the altar of the temple, whose interior is decorated like a representation of the Copernican universe. The divine will is the inspiration for everything. The city's hierarchy goes from God to man, passing first through the priests, who "serve as mediators between God and man"[8] due to the greatness of Hoh the Metaphysic. The religious abstraction of deism prevails, supported by rational inquiry. The main pillars of Solarian religion are: a generalized worship of the Creator and a belief in divine providence, the immortality of the soul, reward and eternal damnation. They worship no physical creature and there are no manifestations of idolatry. The Sun is venerated only as the visible image of God. Public and secret confession is practised, although the objective of this is more ethical and social than religious.

In the *City of the Sun* there is only a residual form of commerce. The singular energy of Campanella's proto-economic ideas is one of the profoundest results of his links with Telesio's philosophy, which, as we saw, is reflected in his concept of religion. For Agnes Heller,

> The utopias of More and Campanella should not be treated simply as ideal demands for ethical norms, springing purely from moral indignation. They wished (...) to satisfy at one and the same time the demands of both optimalization and humanization. This optimalization (...) explored the possibilities of making maximum use of the means of production (...) or of raising productivity.[9]

Under these conditions, there was no asceticism or sharing of poverty in these utopias, so that the concept of the satisfaction of basic human needs would be a static one. Consequently, Campanella and More "desired and premised a way of life in keeping with the needs of their time, but truly humanistic, and one that could be enjoyed at the existing standard of the age"[10]. The combination of the ethical and economic perspectives gives rise to a special style, which is precisely utopia:

> Passionate indignation and deep moral pathos are combined with an expert, objective, scientific, and detailed description of the model proposed. (...) Objectively, the paradoxical thing about both works is that (...) their critical sections have universal validity, while their objective, 'scientific' sections turn out to be naïve.[11]

As Campanella is more specific with respect to the participation of the Solarians in science and its technology, he would come across as more exaggeratedly "naïve". Nevertheless, as Heller says,

> But the brilliance of both works shines through the naivety. It is particularly apparent in the fact that when the economic structure is described, (...) the emphasis falls at least as much on production as on distribution and consumption. More and Campanella instinctively knew that 'inequities' in distribution are a function of how production is organized. Here, moreover, the seemingly more naïve Campanella was actually the more perceptive. For while More thought that the elimination of private property would be sufficient to set a new productive mechanism in motion, that was still not enough for Campanella. In More, production is carried on in the family. Sons follow their father's trade, the family makes its own clothing, and so forth; only distribution, and in part, consumption, are in common. Campanella, however, envisages production too as social, in large-scale workshops and work brigades. The fact that

> he could conceive of such a thing only at the cost of abolishing the family (among other things) bears witness not only to the Platonic tradition, but also to the influence of certain contemporary, especially Anabaptist, examples.[12]

There is no idleness or exclusion in the City of the Sun, no corrupt clergy, and the strong do not abuse the weak; in fact, there are no strong or weak. There is no poverty, hunger or shortages, no disorder, civil disobedience or revolts, no crime and no violence directed at the institutions. Order reigns supreme and there is respect for the sacred, which is subjected to empirical checks.

There is no place for fantasy or exception in collective life, only the rule of rationality implicit in the city. There is a fear of "natural" development, subject as it is to the unsettling actions of history and the everyday events of real life. For this reason, the City of the Sun has no past and is not the result of a historical evolution. The Solarian utopia just *is*; it exists in an immutable present that knows neither past nor future, since, being perfect, it will never need to change. The visitor to this utopia, the Sea-captain who had known Columbus, only discovers it once it is completed.

The life of the Solarians is ordered entirely by reason, which means that they live according to the dictates of Campanella's metaphysics. Their religion falls only slightly short of being Christian; the Solarians, "who follow only the law of nature are so near to Christianity".[13] The difference lies in the absence of the revelation, in other words, the coming of Christ, and since they have not heard of his coming, the Solarians do not believe in the Trinity, but in a single God, represented by the life-giving Sun. For Campanella, Christianity "adds nothing but the sacraments to the law of nature"[14] and is therefore "the true law and . . . once its abuses have been corrected, it will become mistress of the world".[15]

The theoretical assumptions of utopia

The City of the Sun is, by common consent, one of the most significant works of political thought of its time, one that combines a theory of society and of the state, a philosophy of nature, a theology and a set of ethics. And it is precisely because it possesses these formal requirements that it can be classed as a utopia; otherwise it would be a political fiction.

Utopia is one of the most complex literary genres. Its scope extends beyond literature to reach philosophy — in its various branches — history, political thought, religion, town planning, psychology, aesthetic movements, and so on. There are few literary genres whose birth certificate is clearer than that of the utopian genre. With a date (1516) and an author (Thomas More), from its cradle *Utopia* lent its name, defining characteristics and procedures to this genre. Throughout its history, utopia has acted as a constant interlocutor with a variety of societies and their corresponding political theories, very often unequivocally taking the form of a political theory and a blueprint for society. Since the publication of Thomas More's work in 1516, *utopia* has been used to describe any society supposedly perfect in every respect, and indeed his work became the prototype of a literary genre that has continued to evolve and to split into numerous sub-genres.

The birth of utopia was roughly contemporary with the publication of Niccolò

Machiavelli's *The Prince* (1513) and Baldassare Castiglione's *Book of the Courtier* (1528), two works that jointly marked the culmination of quattrocento humanism and perhaps also its limit: the idea, constructed through social praxis, that man could take control of his own destiny. Humanism saw individual existence and collective life as historical — human — and, therefore, pliable material that could be moulded with reference to a teleology that had always existed, but was then attaining a fleeting emancipation. According to Quentin Skinner, "Some scholars have treated *Utopia* simply as a contribution to a more general 'programme' of humanist reform, a programme which More is said to have worked out in close agreement with Erasmus, Vives, Elyot and their various followers."[16]

The key notion that drove the genesis of utopia was the underlying belief in the perfectibility of society; on the other hand, what was intrinsic to the Christian conception was the perfectibility of the human being. Utopia pointed out that society was incomplete but that that imperfection could be resolved. The literary expression of the complete remission of social problems is a utopia in itself: a text that constructs the perfect polis with words, presumes itself to be complete and considers that a perfect society is possible, once the dictates of Reason have been applied. Utopia as allegory formalizes the contradictions of the time in which it was written, and projects a notion of the eternal that is the product of those circumstances. The Platonic ferment is self-evident: utopia is the image of social perfection inherent in a specific historical moment.

It is reasonable to believe that the characteristic set of problems associated with the utopian genre originated with Plato's *Politeia* (*The Republic*) and also legitimate to consider any utopia as always being a rewriting of that Republic. We may, in short, say that the circumstances of Plato's work were determined by the decline of the model of life in the polis of Attica, and consequently, the old political community being overtaken by history. In the Athenian model of the community, individual existence depended on his relationship with the aristocratic collective, and with the polis, politics, and the public interest. In this context, individual and community formed a single indissoluble unit, to the extent that, without the community, the individual immediately lost his very means of survival, and not only citizenship. Belonging to the community preceded possession of the means of production and subsistence. The citizen possessed economic resources because he was a member of the community, not the other way round (as is the case in bourgeois society). The individual had an acceptable standard of living because he was a citizen; he did not become a citizen because he possessed material goods. This situation ran counter to economic law, as would be the case in the literary utopias. Economics was subordinate to politics, and to the interests of the polis. In other words, having money neither conferred citizenship, nor did it mean having political power. This made the Greek aristocrat a link in the community chain, one that could not be broken. Consequently, individual and community were inseparable. The crisis and collapse of the Attic community in the latter half of the fifth century BCE came about as the result of the increasing dominance of relations based on money; gradually and inevitably, money became the centre of economic life and so the means of access to control of the State. The result was that the community ethos gave way to the metic, the proto-bourgeois selfish individual opposed to the collective, the mortal enemy of Plato and future utopian systems.[17]

The most common explanation of the origin of this literary genre that was so close to history, philosophy and politics essentially follows the idea that utopia was generated by the bourgeois process of rationalizing life so characteristic of the Renaissance. In the atmosphere of philosophical humanism encapsulated in the pre-cogito of Pico della Mirandola, the idea was toyed with of man as the author of his own existence, and the individual as analogue of the architect of the world.

In his *Discorso sopra la dignità umana* (1486), Giovanni Pico della Mirandola introduces the concept of man as microcosm of the universe. Starting from the myth of creation, Pico imagines that God, after creating the world, decided to complete his work by giving life to Man. However, in spite of the fact that all possible spaces were filled, his creative will was unabated. He decided therefore that Man would not have a definite nature or a specific place to live in, so that he himself would be completely free to choose and find the position where he would be most satisfied. So, in contrast to a natural world governed by fixed immutable laws decreed by God, there was the human being with the power to choose his essence, so that he could sink to the level of a beast or soar until his spirit merged with the divine. According to Pico, Man does not have a nature of his own, but creates his essence through action. Hence, human evolution is open to the possibilities of growth and improvement, of transforming himself and the world with no other limits than the attainment of perfection and eternal bliss. Essentially, the dignity of man of which Pico speaks does not consist of *being* but of *becoming*, which does not happen with natural things. Pico's ideas contain enormous possibilities and it is no coincidence that his *Discorso* was always considered to be the manifesto of Humanism; without it, there is no bourgeois ethic. Without this philosophical heritage, utopian ideas are simply not possible. It is no coincidence either that More translated his work into English.[18]

The historian, Luigi Firpo, considers that, in order to define the characteristics of Renaissance utopias, specifically the Italian series, it is necessary to understand the spiritual climate at the beginning of the cinquecento.[19] This was a period when humanist rationalism prevailed, characterized by a longing for human autonomy, the idea of the supremacy, almost the omnipotence, of intelligence, as well as an attitude to life that was heroic, courageous and vital. Ultimately, after centuries living under the discipline imposed by the rigid absolute rule of Catholic dogmatic culture, the humanist discovered sufficient reason intrinsic to the physical world to guarantee him harmony: nature. As the new science would confirm, nature becomes aware of itself through the human essence and takes the name of *reason*, the new model and measure of action. For More, *virtù* (virtue) was to live *secondo natura* (according to nature); the man who followed the model of nature was the one who obeyed reason, which was nature aware of itself, in its own essence.

This intellectual process of the scientific revolution transformed the medieval conception of nature as a stage created by God for man to that of an actor alongside man. The question was: what would the purpose of nature be? Self-recognition *in* and *by* man. The senses — communicating vessels between man and nature — became a criterion of knowledge. Modern Galilean science depended on this new principle: using the mind, which was both the bearer of reason and also in nature, as the starting point for forging the world. Following the same reasoning, connecting to nature in a rational way meant breaking with the traditional form of collective living. On the political plane, the journey towards nature to extract the complete set of rules for collective living directly involved the concept of equality, which led in turn to the concept of legality.

Science and religion: Is harmony possible in the City of the Sun?

The "natural sciences", as they were called at the beginning of the sixteenth century, were stated to be a new style. Several authors highlighted the fact. In a letter sent to Galileo in 1632, Campanella claimed that he could see "a new science" heralded by "new worlds, new stars, new systems, new nations".[20] The Aristotelian system, the basic framework for many European intellectuals since the thirteenth century, was frequently called into question. In the natural sciences, as well as in the arts, a concern for the difference, or even contrast between appearance and reality emerged, which justifies the idea of the "mechanization of the world picture" used to refer to this period. The experimental method, a term associated with the concept of *experience*, marked a clear difference with regard to the precepts of faith. The new philosophy was taking a route that was leading inexorably to collision with the Catholic Church. In the twentieth century, historians like Herbert Butterfield used the phrase *scientific revolution* to summarize all these advances in natural philosophy.[21]

The scientific revolution then underway started from the empirical axiom that scientific truth was reached via the human senses, the seat of experience. Thus, the criterion of truth was shifting from revelation to man's physical senses. The senses, formerly a source of error and perdition, were now becoming the basis of reliable knowledge of the physical world. In consequence, science became independent and broke away from the field of theology, which undermined the authority of the Church in practice and made her redundant in this area. This brings us to Telesio.

The philosophy of nature that underpins *The City of the Sun* is the result of the influence of Bernardino Telesio's work, *De rerum natura iuxta propria principia*, which Campanella read in his youth and which left a permanent impression on his thinking and work. Telesio did for the philosophy of nature what Machiavelli had done for politics: transform it into an area of reflection and discussion that was independent of other spheres of thought, such as morality or religion. His basic aim was to reveal the objective reality of nature, since, if they were observed correctly, the things themselves displayed their intrinsic nature and characteristics. This explanatory principle had universal validity; in other words, nature is unique, at all times and in all places. And what was most important, man possessed the capacity to learn the secrets of nature because he himself formed part of it. The senses were an effective means of gaining access to knowledge and man, like nature, had sense perception. Consequently, what nature revealed of itself coincided with what the senses bore witness to. Sense perception, therefore, was nature revealing itself through one of its constituent parts, man. Galileo would adopt this principle as his thesis and take it to its logical conclusions.

In Telesio, Campanella discovered a way to reach the truth via the direct observation of nature. As he would write later, the whole of his work seemed to him to be both coherent and liberating, and he was completely fascinated and convinced by it. In 1590, Campanella followed in Telesio's footsteps by writing *Philosophia sensibus demonstrata*, a heterodox text in which it was blatantly obvious that his philosophy departed from the prevailing rules of the Church.[22]

Telesio developed a rational method for apprehending tangible reality, which postulated that there were only natural forces in nature and that these should be explained only in accordance with their intrinsic principles, without recourse to metaphysics. Men had the capacity to know nature as an entity and so had power and dominion over everything that was part of nature. Hence the natural world was an extensive estate at man's disposal, an emporium to be appropriated, in a material as well as in an ideal sense. Nature could therefore be appropriated in a profane way. This philosophical principle was established in the early days of manufacturing, which was clearly an opportune moment for such ideas to emerge and had crystal-clear historical implications for the logic of manufacturing: the complete modification of social relations by means of a new modern society, new social classes, a new form of state and a new outlook on the world.

If Campanella's thesis were to prevail, the Church and the empires allied to it — Spain and Portugal, as well as the Italian states — could welcome the new science with open arms, could consider nature an entity accessible to the profane mind, which would notwithstanding still be able to find God. The sovereign would be a scientist as well as a high priest. The investigation of nature by means of rational science and prayer could be the same activity. This philosophy would therefore bring the Roman Catholic faith and the scientific revolution together, since it acknowledged the existence of the sacred in the world, even as it laid down the philosophical foundations of modern science. A close link between reason and faith was therefore possible, which meant that the Church could ally itself with the system of modern manufacturing that was already being established in Northern Europe. In the end, this would be more effective for safeguarding the interests of the Church than what the three Catholic states were implementing in the world of the Counter Reformation.

The City of the Sun as the antithesis of the Spanish Empire

The Spanish Monarchy, which was being kept afloat at that time by the gold and silver from the New World, had become an irrational, ecclesiastic, aristocratic system in economic terms. Work was considered dishonourable. The Church and the nobility owned most of the wealth, which was tied up in sumptuous buildings or government bonds or squandered on luxuries, and so diverting it away from productive economic activities. The result was that Spain, in its imperial role, was in no condition to satisfy the demand for material goods and equipment from her American dominions, so making her dependent on those countries that were open to manufacturing and the sciences, which were precisely the Protestant countries. So, those that were growing rich at the expense of the Spanish Empire were the same ones that were practising what the Inquisition had condemned, and even prohibited in its dominions.

To reproduce itself, this empire had no need to take advantage of the latest advances of the scientific revolution that was already underway, but instead relied on the predominantly Jesuit leadership of the *Propaganda Fide*, the inquisitorial Sacred Congregation for the Propagation of the Faith, which set out to prevent the new science from being absorbed by categorically prohibiting its conceptual and economic structures from taking root in the Iberian countries and Italy. Indeed, the Spanish Inquisition behaved in all respects as a veritable economic agent. By sending scientists to be burned at the stake and banning the scientific revolution, the Inquisition forced countries under its control to follow the economic and political path prescribed by the colonial system and its underlying practices: the colonial pact, slavery and exploitation of the indigenous population (via the system of *encomiendas*), production of primary products, monoculture farming on large estates, and so on. The persecution of science, in short, was an integral part of the framework of the colonization of America, an expression of Iberian power.

This was where the crux of the matter lay. Which political path would best enable the Church to achieve its aim and become the Mistress and Saviour of the World? Campanella did not need to invent a scenario that he could then oppose; it was enough to observe what was happening around him. The inquisitorial policy and its impact on the world that was still Catholic provided all the material he needed to construct *The City of the Sun*. Campanella's utopian polis is an ideal republic in which everything, or almost everything, that is done is the opposite of the prevailing policies of the Inquisition. In short, the City of the Sun is the antithesis of the Iberian countries and Italy and expresses what the world might look like without the Counter-Reformation alliance of the Church and the Iberian empire: a universally Christian, rational world united under the Throne of St Peter.

Since Campanella's utopia is the union of scientific reason and faith, it is accordingly a complex system that brings together scientific rationality and prophetic irrationality, a modern mentality and religious traditionalism. Campanella is not a supporter of the Counter-Reformation but a bizarrely rational reformer. And, in his game of smoke and mirrors, which is both a criticism and defence of the Church, his City of the Sun is, perhaps contrary to expectations, the most baroque of seventeenth-century utopias. At first, as we know, scientific reason was not opposed to faith, but enjoyed a certain autonomy of its own. It was at the Council of Trent that the Catholic Church opted to transform simple autonomy into opposition, with serious consequences, the most obvious being the violent persecution of scientists and the bloody episodes that everyone is familiar with; the whole process, after all, was associated with the Protestant Reformation.

The fact is that the Reformation freed a substantial part of Europe from the authority of the Church, thereby reducing the Catholic circle unambiguously to Portugal, Spain and Italy. It is equally true that these countries, the three greatest European powers at the time of the birth of the Reformation, became the poorest and weakest on the continent within a few centuries. Tridentine Catholicism — Italy, Spain and the subordinate kingdom of Portugal — refused to adopt the bourgeois methods that used the new scientific reason to lead their economies to manufacturing and the unfettered development of trade. Their own actions locked the triad of Catholic countries into the circle created by the Inquisition, which became the architect of an economic system based on colonial exploitation. Inquisitorial irrationalism also became a political and economic system. It is possible to argue that this downward spiral was not yet apparent at the beginning of the seventeenth century; nonetheless, Campanella's *The City of the Sun* does seem to contradict this idea.

To sum up, after the Reformation, Europe was basically split into two blocs. On the one hand, there were the Catholic nations, the major imperial metropolises of the time, which would stagnate in their regime of colonial pacts; on the other, there were the Protestant nations, as yet without significant colonial empires but which, over time, would stimulate their economies by following the path of manufacturing and become more modern nations, open to future industrialization.

The Tridentine Church, in effect, gave expression to the papacy's pact with the Iberian monarchies, proceeding to become both dependent on them and part of them. In other words, the Church merged with the Spanish and Portuguese monarchies.

Rome's destiny remained locked into the process of merchant capitalism, which was both its economic manifestation and, at the same time, its limit; and this defined the way societies in the Spanish and Portuguese colonies in America were organized. This alternative entailed the exclusion of manufacturing — associated with bourgeois rationality — as the hegemonic option for the Catholic economic circuit. The result was that the Jesuit missionary Church, opposed to science and actively involved in inquisitorial trials, was ultimately charged with the task of deciding the fate of the two Iberian monarchies and the shape that they would take in the future. The destiny of the Church was to join forces with the colonial pact and that particular brand of Iberian merchant capitalism that would lead to the incorporation of new territories occupied by the Jesuit missions. In the end, the Inquisition was not an organ subordinated to the Church; on the contrary, the Church was subject to the power of the Inquisition.

The Counter-Reformation confirmed the radical split between Catholic orthodoxy and the new attitudes of religious consciousness and secular culture. The generation of Giordano Bruno (1568–1600), Galileo Galilei (1564–1642) and Tommaso Campanella (1568–1639) was present as the restoration of the principle of authority was being enforced and saw, from the start, how all opportunities for developing scientific knowledge were vanishing and, in consequence, how the opportunities for social development associated with that knowledge and economic practice were also vanishing. In stark contrast to this would be a Church open to the new sciences, actively encouraging modern methods of research, well disposed towards books that had not yet been read and eager for new knowledge. In short, this would be a Church whose destiny would be bound up with the manufacturing bourgeoisie and that would sit alongside it on the throne of the World. What might have happened but did not is precisely the meaning of *The City of the Sun*.

The City of the Sun is an essential document for understanding Campanella's deepest, truest intentions, a text written to synthesize and define his own religious, philosophical and social ideas, and to set out his proposed reforms for the Church and the political institutions. This is the only way to explain the origin of this utopian dialogue read as a literary document. *The City of the Sun* is not — at least, not exclusively — a book written to attract the goodwill or protection of judges and powerful men. Written in the squalor of an inquisitorial prison, this work was taking stock, among other things, of the episode of the Calabrian revolt, a re-affirmation of the thesis of the advent of the "new Golden Age" heralded by the imminent "great conjunctions" and an alternative political model for the Catholic Church and the monarchies of Spain and Portugal. Along with theocratic universalism, which he would assert once more in 1637 in *De regno Dei*, Campanella retained the social idealism of the City of the Sun as the highest value of his political speculation throughout his life.

In *The City of the Sun*, therefore, faith addresses God the creator, as revealed in the spectacle of nature and perceived by means of reason. *The City of the Sun* appears as the philosophical discovery that demonstrates that the truth of the Gospel is in harmony with nature.

Campanella presents us with a republic that is founded, not by God, but by philosophy and human reason, in order to demonstrate that the truth of the Gospel and rational philosophy do not contradict each other. The natural religion of the Solarians, discovered through philosophy and based on reason, derives its power from a rule that allows them to compare historical religions, choose the one that is true, and reduce it to its core essence, eliminating the abuses that are harmful to society. In this respect, he considers that natural religion, which is innate, is always true, whereas constructed religions are imperfect and may even be false. Natural religion is a characteristic of all living beings, and since God created them, they will all return to Him. In this way, a rational polis demonstrates its superior capacity for constituting a power that would secure for the Church control of the world, which the policy of alliance with the Iberian metropolises formulated at the Council of Trent was already showing signs of not being able to achieve. We may conclude, then, that *The City of the Sun* would be the Iberian colonial system turned on its head, in other words, its antithesis.

Notes

1 Tommaso Campanella, *La Città del Sole: Dialogo Poetico* (*The City of the Sun: A Poetical Dialogue*), translated into English by Daniel J. Donno from the Italian text, edited by Luigi Firpo (London: University of California Press, 1981).
2 Bernardino Telesio, *De rerum natura iuxta propria principia, libri IX* (Naples: Horatium Salvianum, 1586).
3 Germana V. Ernst, *Tommaso Campanella. Il libro e il corpo della natura* (Rome-Bari: Laterza, 2002).
4 Maria Moneti Codignola, "Campanella, a cidade historiada", *Morus — Utopia e Renascimento* 5 (2008): 86–106.
5 Campanella, *La Città del Sole: Dialogo Poetico*, p. 27.
6 *Ibid.*, p. 31.
7 *Ibid.*, pp. 31–32.
8 *Ibid.*, p. 103.
9 Agnes Heller, *Renaissance Man*, translated by Richard E. Allen (London: Routledge & Kegan Paul, 1978), p. 359.
10 *Ibid.*, p. 359.
11 *Ibid.*
12 *Ibid.*, pp. 359–360.
13 *Ibid.*, p. 121.
14 *Ibid.*
15 *Ibid.*
16 Quentin Skinner, *The Foundations of Modern Political Thought*: Volume One: *The Renaissance* (Cambridge: Cambridge University Press, 1978), pp. 255–256.
17 José Chasin, "O Futuro Ausente", in *Ensaios Ad Hominem*, vol. 3, 1 (São Paolo: Estudos e Edições Ad Hominem, 2000).
18 Giovanni Pico della Mirandola, *Discorso sulla Dignità dell'Uomo*, intro. by Eugenio Garin, edited by Giuseppe Tognon (Brescia: La Scuola, 1987).
19 Luigi Firpo, *Per una definizione de Utopia*, in *Utopie per gli anni ottanta: Studi interdisciplinari sui temi, la storia, i projeti*, edited by Giuseppa Saccaro Del Buffa and Arthur O. Lewis (Rome: Gangemi, 1986).
20 Tommaso Campanella, *Apologia di Galileo. Tutte le lettere a Galileo Galilei e altri documenti*, intro and ed. by Gino Ditadi (Este, Padua: Isonomia, 1992), p.128.
21 Sir Herbert Butterfield (1900–79) was a British historian, remembered basically for his book *Whig Interpretation of History* (London: Bell, 1931). Other historians, such as the American, Steven Shapin, *The Scientific Revolution* (Chicago: The University of Chicago Press, 1996), by restoring the nineteenth century dispute over the "Revolt of the Medievalists", disagree with several aspects of this theory.
22 Tommaso Campanella, *Philosophia sensibus demonstrata*: *Testo latino a fronte*, Vivarium series (Naples: La Scuola di Pitagora, January 2013).

5 Utopian Imagination Across the Atlantic: Chile in the 1820s

Carlos Ferrera

In the transition from the eighteenth to the nineteenth century, the Atlantic world was shaken by the liberal revolutions. Historiography that tended to look beyond the nation as its object of analysis, whether the focus was the Atlantic or transnational, has presented this geographical space as a framework for the multi-directional movement of goods, people and ideas, and driven by trade, emigration and exile. Pratt Guterl has advocated the study of specific topics as the best way to achieve a transnational approach and, in this respect, utopia may play a primary role, since, among those ideas, we find many associated with utopian language; this is not in the least surprising, given the importance of formulations of this kind in liberal and republican proposals of that period. The hypothesis of this study is precisely that utopia permeated the political thought of the time and filled it with images, a significant point in a culture characterized by *sensismo* [sensualism], the doctrine that proclaimed the power of those images, their capacity to affect the imagination and to use feelings to drive knowledge and action.

Utopian approaches circulated on both sides of the Atlantic in more or less explicit form in countless contemporary literary works, as well as in specific policies. They were well received in Latin America, which was trying to keep abreast of what was happening in other parts of the world during the period of independence. This occurred at a time when expectations of progress and civilization were spreading in the wake of the sweeping transformations brought about by the political crisis in the Spanish monarchy and a foundational moment in politics seemed to be emerging. In that climate, the American continent, in turn, was seen in Europe as a new space, open to all possibilities.[1]

Exile and other forms of displacement played a major role in that process. In this connection, we need to consider both the gravitational pull of London as a place on which exiled Latin Americans and Europeans converged, to which should be added the drive and vibrancy of the political and economic scene in England at the time. The common thread running through this essay will be José Joaquín Mora, who represented the Spanish liberalism that first arose in Cadiz in 1812, and who lived in exile in London and later travelled around South America, invited by various rulers to collaborate in establishing a new society. Without, of course, being the only one, since many more embarked on that transatlantic voyage, he was, nevertheless, typical because of his intense public activity and tireless travelling from place to place. Although Mora worked in the United Provinces of the River Plate and in Santa Cruz's Bolivia, he was mainly active in Chile, a place regarded from the earliest moments of its independence as a territory that lent itself to an ideal society.

An age of utopias

The importance of utopia in the eighteenth and early nineteenth centuries seems to be beyond all doubt. According to Christine Rees, it burgeoned in response to the anxieties deriving from the intense transformations experienced during that period. The proposal of a "new city" implied both a radical critique of the previous society (defined by despotism and the predominance of religious order) and a break with it.

In general, ideal models remained anchored in the usual assumptions made of a Golden Age; nevertheless, they differed from those models because they did not simply boil down to natural abundance, but rather their existence came to depend equally on a proper social organization, based on reason. Although the traditional models were kept, utopian language was renewed with new approaches in which egalitarian experiments coexisted with others based on private property, urban models together with rural ones, and so on. Likewise, the aspiration to attain the ideal was occasionally tackled in sub areas, such as perpetual peace, happiness, national regeneration or the creation of new men by means of education. There was, simultaneously, a transformation of this discourse because of its tendency to reveal itself as something possible and pragmatic, the fruit of reform and of anticipation of the future, and because of its refusal to be branded as utopian or unrealistic. The result of this was that many accounts changed direction. The customary travellers in that narrative genre started to arrive in societies where utopia was not so much a finished reality as had been the norm laid down by More, but was something potentially within their grasp. Antoine Hatzenberger has highlighted the dynamic nature of the utopias of the time, their adaptability to the new scientific doctrines emerging in that period, for example, the Hollow Earth theory that gave rise to a flourishing genre of utopias in subterranean worlds. Such works even influenced contemporary scientific and political language, which borrowed the ideas of utopia, such as the progressive regenerative myth of a society created out of nothing.[2]

As European travel to other continents intensified, utopias tended to be located in territories that were just beginning to be explored. From the seventeenth century, such enclaves came to represent a state of nature where utopia could be realized, as was demonstrated, for example, in the works of De Foigny, who situated his rationalist utopia in *La Terre Australe Connue* (1676) in the Antipodes. Likewise, Defoe constructed utopias on islands in the South Pacific in *Robinson Crusoe* (1719) and *A General History of the Robberies and Murders of the Most Notorious Pyrates* (1724), while the characters in Robert Paltock's *The Life and Adventures of Peter Wilkins* (1750) or Simon Berington's *The Memoirs of Signior Gaudentio di Lucca* (1737) found themselves overseas with worlds already set up under a utopian philosophy. In the same way, Jacques-Henri Bernardin de Saint Pierre recalled, in the preamble to *L'Arcadie*, that after sailing the seven seas he ended up establishing an ideal society in the Amazon, where men lived in tolerance and in accordance with the laws of nature.

The role of spaces overseas was reinforced as a result of the success, in the eighteenth century, of philosophical Geography, which defined territories according to moral criteria. Locations, often little known or never visited, were described as places of barbarism, but also as utopias. As Wolff has emphasized in the case of Eastern Europe, their circumstances fuelled projects of social experimentation — including those of Voltaire and Rousseau — and of the reorganization of territory and civilization assigned to enlightened monarchy. Something similar occurred in America, where the earliest samples of geographical analysis had certainly had a negative bias, represented in Buffon's formulations and his defence of the natural inferiority of America with respect to Europe. The development of Natural History, on the other hand, which Mary Louise Pratt situated from the time of La Condamine's expedition to the Royal Audience of Quito in 1735, made it possible for knowledge to be systematized in accordance with Linnaeus's systems of classification and for them to spread from the coasts to the interior of the continents. That approach, however, was accompanied by a Romantic vision based on a sentimental, aesthetic perception. America became a sublime utopian setting and hence external to the everyday world. It was filled with static, luxuriant, breathtaking lands ready to be turned into a place for ideal experiments and promotion of the capitalist economy. That vision was strengthened in the early decades of the nineteenth century, as demonstrated by Humboldt's work, which was well received by the Latin American elites and a distinguished group of European geographers, invited first by the colonial authorities — as had been the case of Humboldt himself — and then by the leaders of the new independent republics.[3]

Chile was not left out of this scenario. In *Viaje a Chile en la época de la independencia*, written in 1817, Samuel Haigh began the account of his travels by recalling how he had dreamt of walking along streets paved with gold when they had offered him the chance to go there. Of course, he found nothing of the kind, but he did find what he defined as "natural potential", capable, using a good irrigation system, of turning the country into the "granary of America". For his part, Alexander Caldcleugh in his similarly named *Viaje a Chile*, of 1819, spoke of a "privilegio geográfico" (privileged geography), the provider of a happiness, truncated only by the continual wars suffered during the process of achieving independence; its natural blessings were similarly transferred to its population, whose upper classes were endowed with attributes that were foreign to corrupt traditions, and the lower orders to vice. One final example, whose experiences in the country left a greater mark, was Claude Gay, who arrived in Chile in 1828 as a teacher and was hired by the Ovalle Government to undertake a geographical study of the country. After travelling around the country for twelve years, he published the *Historia física y política de Chile* (Physical and Political History of Chile), accompanied by two atlases. His work is interesting from our point of view because it includes elements of the utopian genre. His narrative was characterized by a tendency to *costumbrismo* with very rich descriptions, similar to contemporary portraits of ideal societies. At the same time, the adventures of the narrator throughout his journey harked back to the accounts by utopian travellers, since they were liberally sprinkled with hardships and hazards; this was also true of his feelings of nostalgia, solitude, love of freedom or awe at whatever he had seen. Apart from that, Gay combined the system of classification of Linnaeus with the sublime synthesized vision of Humboldt for the utilitarian purpose of attaining universal scientific knowledge and placing some of the virgin, fertile territories under cultivation.[4]

The foundational moment of independence

The independence process bore a strong political foundational stamp; here we adopt the sense attributed by Rancière to historic moments when the permanent constitutive elements of society are being rejected and a language of emancipation and openness to possibility is being formulated. Although many who supported independence resorted to cultural expedients of the past, they experienced the process of the collapse of the Spanish monarchy as a new moment, a return to a state of nature and the prelude to a possible ideal society.

Rafael Rojas, for his part, has recognized the presence of a republican utopia in that period, a synthesized version of the sentimental language of Italian Enlightenment of Beccaria and Muratori, the neo-Thomism of Francisco Vitoria and Suárez and the traditional *ius gentium* of Grotius and Pufendorf. The republic's political project aspired to the formation of a homogeneous community of a Christian stamp, founded on virtue and with a regenerative mission, which was where its utopian component would be. The geographical context chosen was initially restricted to domestic territories, although, throughout the process, there was a desire to transcend national entities, as is indicated by the attempts at continental unity noted at the Congresses of Panama (1826) and Tacubaya (1828).[5]

In the case of Chile, utopianism was present from the years prior to independence, and was associated in large measure with its geography. The so-called conspiracy of the Three Antonios in 1781 had already tried, unsuccessfully, to replace the monarchy with a republic where slavery and hierarchies would be abolished and the redistribution of the land would be undertaken. In 1804, Juan Egaña's *Oración inaugural para la apertura de los estudios de la Real Universidad de San Felipe* (Inaugural speech to open the academic year at the Royal University of San Felipe) highlighted the utopian potential of a country so well endowed by nature:

> The idea of a city formed for wisdom and with only the wise for inhabitants, if it were to be accomplished, nowhere in the world could it be better located than in Chile. Because of its situation, 300 leagues away from that great world where war, intrigue and urgent necessities occupy the lives of more than a third of its inhabitants, and free, because of its political destiny, from that tumult of passions that excite ambition, ideas of superiority and balance of power, we retain an existence that is orderly and without differences, where birth and death regularly form a circle, and where we always remain at the same point in the hierarchy.[6]

According to Sagredo Baeza, it stemmed from the idea that its geographical uniqueness and isolation would guarantee freedom and the arrival of the "republican Eden", as the eighth verse of the national anthem put it

> By sea and land the henchmen
> Of the vile despot threaten,
> But the whole of nature
> Awaits them to fight:
> The Pacific to the South and West
> To the East the Andes and the Sun
> To the North an immense desert
> And the Centre freedom and unity.[7]

Or as Camilo Henríquez himself put it, in 1813:

> This is a visible fact of geography and makes Chile's situation tangible to us. This vast region, being able to survive on its own, and having, in the bowels of the earth and on its surface, not only what is necessary for life but even for the recreation of the senses; being able from its ports to engage in useful trade with all nations, to produce sturdy men to cultivate its fertile fields [...] finding itself enclosed as if within a wall, and separated from other peoples by a chain of extremely high mountains topped with everlasting snows, by an extensive desert, and the Pacific Ocean, would it not be an absurdity contrary to the destiny and order inspired by nature to go in search of a despotic government, a venal, corrupt ministry, harmful obscure laws, or the biased decisions of ambitious aristocrats, on the other side of the world?[8]

Nonetheless, this vision was not always restricted to Chile but, as we pointed out earlier, embraced the continent as a whole. In *La Camila*, the play written by Henríquez, the leader of the independence movement, some people from Quito who were fleeing the Spanish repression of 1809, found a utopia in the privileged environment of the Amazon — in the same setting chosen by Saint Pierre in *Arcadie* — where it was possible to live according to natural law under a philanthropic government. Just as in Argentina, the discourse of independence rested on the creation of a community of patriots, which could be developed through the idealization of the territory. Those blessings of geography were enriched with a temporal dimension, because pre-Hispanic populations, such as the Incas and Araucanians, were included at its heart. It was an appeal to an idealized past that heralded a future of perfection with millenarist echoes, motivated and legitimated by the sacrifices made in the heroic struggle for independence.[9]

The transatlantic dimension of utopia

In such a climate, the movement of people, including exiles, adventurers or those driven by economic motives, had a major role to play. In this context, London came to prominence by becoming a meeting point for Europeans and Latin Americans. The linguistic and ideological affinities of Spanish liberals taking refuge in the metropolis favoured contact with diplomatic agents from Latin America who were asking for their translations and publications; at the same time, the new disorganized republics represented an opportunity for European military men, scientists and artists. The Latin Americans, for their part, who had sought refuge during the last years of Spanish rule, went in search of British political and financial support in the early years of independence. This objective was attained in part by means of the proliferation of publications that described the political situation and the economic and investment potential in the new states of Latin America. A case in point was the anonymous *Colombia: Being a Geographical, Statistical, Agricultural, Commercial, and Political Account of That Country*, published in London in 1822, which, according to its sub-title, was "Adapted for the General Reader, the Merchant, and the Colonist". When it came to attracting European colonists, its pages were full of the superiority of this new republic over the United States, because of its proximity to Europe, its climate, what it produced and "its distinguished geographical situation"; likewise, the advantages were also of "a moral nature" because its constitution guaranteed civil, political and religious liberties, and an educational policy designed to raise the cultural level of the common people.

Not all publications originating among groups of exiles followed a similar course. The more conservative *El Emigrado Observador*, without relinquishing utopian imagery, appealed to English capitalists to invest in Spain, "el verdadero jardín de las Hespérides" (the true Garden of the Hesperides) and not South America because of its instability. Nevertheless, the general trend was in the opposite direction; the *Repertorio Americano*, for example, talked of a new era, the potential of the Orinoco river, or the need to introduce Lancasterian schools and hydraulic systems in order to spread civilization.[10]

In this context, the work of the London publisher, Ackermann, was indispensable, because of his contacts with members of the exiled communities and representatives of the new Latin American states. Through periodicals and catechisms published on general or specifically Latin American topics, he promoted an educational programme for that continent. His works, aimed at broad sectors of society, also pursued the dissemination of knowledge that tended to promote British trade in the area. Utopian imagery was not unusual among his proposals either, and could already be seen on the December 1811 cover of his monthly publication, *The Repository of Arts*. In its centre was a bust of Athena who observed, as if in a panopticon, an ordered environment with looms, the attributes of culture and the arts and cornucopias.

Eighteenth-century catechisms, an example of the dialogue literary genre with strong echoes of religion, grew in importance during the revolutionary process because of their simple didacticism. They were also important for the topic that concerns us here. On the one hand, they were different from those aimed at the British public, which sought to educate the lower classes in order to prevent social unrest, whereas in Latin America they were positioned to create citizenries; on the other hand, their literary structure was similar to the utopian one, since they set up a dialogue with scarcely any debate, where the evidence of a more perfect world was unveiled. Whether the subject was astronomy or political economy, knowledge, expounded in a clear structured manner, showed reality as ordered and deeply harmonious. At the same time, the texts showed South America with blank territories, ready to be exploited according to the geographical principles pointed out earlier. Equally, they showed a world in which racial relations were structured hierarchically between whites and natives, with the latter on a lower rung because of their inability to convey feelings. This racial focus was not, of course, alien to the utopian genre. In this respect, a contemporary example was set in the vicinity of Chile. In *El descubrimiento austral por un hombre-volador o el Dédalo francés* (The Southern Discovery by a Flying Man or the French Daedalus), various rational, egalitarian societies appeared, ruled by the French. Their rulers, in a show of social engineering, mixed the native races together and allowed the French elite to have concubines; however, they made sure that the resulting "offspring" could not mix with the French again in order to avoid the racial degeneration of the latter.[11]

Ackermann's collaboration with the Spanish liberals in exile centred in turn on the figures of José María Blanco White and José Joaquín Mora, to whom were added other authors such as José de Urcullu, Joaquín Lorenzo Villanueva and Pablo Mendíbil. Blanco White, who stood out because of his political and cultural Anglophilia, handled utopian categories in the quarterly publication, *Variedades o El Mensajero de Londres*, in which England is described in idealized images as he sails into port, for the language of freedom "resonaba en sus oídos" (rang in his ears). Blanco confessed his pleasure at seeing himself in possession of the impossible "porque había penado toda mi vida" (because I had suffered hardship all my life); the inn he stayed at in the first village he came to had seemed like a palace to him; the houses had no bars on the windows, which gave an idea of the general security, there were no deserted places, everywhere was cultivated and roads were plentiful. These ideas were simultaneously combined with a rejection of "the panaceas of charlatans" — which was an allusion to events experienced during the French revolutionary process — promising to "regenerar un estado por medio de algunas máximas" (regenerate a state by means of a few maxims); he also recalled the need to put order before anarchy, for anarchy was the danger facing Latin America, and the reason why he made constant appeals not to surrender to excess and, in short, to complement the political changes — a utopia of order — with a programme of education and virtue.

In his *Consejos importantes sobre la intolerancia, dirigidos a los hispanoamericanos* (Important advice on intolerance, addressed to the Hispano-Americans), he lamented the idea of women losing "their shyness", young men feeling "more republican" through ignoring serious and sensible men, and frankness turning into "a Bacchanal". Notwithstanding, in the picture he presented of Latin America he praised the native-born for having risen several notches on the intellectual and civil scale during the wars of independence; he valued the constitutional effort deployed in the new republics to show a "creencia en la perfectibilidad que en Europa se había abandonado" (a belief in perfectibility, which had been abandoned in Europe) and felt he could not blame them for the succession of such constitutions and systems and the heightened passions because of the "estado del que se partía" (situation they had started from).[12]

The end of the collaboration between Blanco and Ackermann was due, according to Tully, to Blanco's growing interest in more aesthetic matters. Durán, however, has insisted on the political dimension of that literary approach, as evidenced in his works of criticism. Blanco White had adopted a gloomy view of Spanish literature, which he considered had been the product of a history of intolerance and despotism since the sixteenth century. According to Durán, the break came about because Ackermann was looking for literature of a lighter kind, suitable for a mostly female readership. In any case, he was not the only one in that circle to mix literature and politics or to consider literature as the barometer of the development of a country. The Duque de Rivas once again took up the idea of a mythical Spain, or "Hesperia", in his ode to "El Desterrado" (The Exile), published in *Ocios de Españoles Emigrados* (Leisure Pursuits of Spanish Émigrés). The country in the poem was moribund, even though it had been an "emporio de riquezas y de placeres" (emporium of riches and delights); its inhabitants lacked vital energy, which led the poet to call up the heroic spectres of the past. However, the voices of brave living Spaniards could soon be heard above the others and, with them, hopes of regeneration were reborn, in which the future assumed the form of an ideal model, one that was "libre, triunfante y glorioso" (free, triumphant and glorious).[13]

After severing relations with Blanco White, Ackermann approached José Joaquín Mora with a venture that in no way entailed a shift away from politics; this would, in any case, have been a difficult thing to do at that point in the nineteenth century when the various manifestations of literature and science were inextricably bound up with politics, and therefore utopias.

Having received an enlightened education, Mora went on to teach Bentham and Condillac at the Colegio de San Miguel in Granada during the reign of Charles IV.

After the return of absolutism in 1814, he was involved in a heated polemic with Böhl de Faber, whose irrational Romantic metaphysics he censured. In the course of the Liberal Triennium (1820–1823), he became much more radical and, when the Triennium ended, had to go into exile, following a course that reflected the situation of the Atlantic world, mentioned earlier. He lived in London and wrote for Ackermann's publications. He crossed the Atlantic to Buenos Aires, summoned by Rivadavia; from there, he moved on to Chile, where he stayed rather longer, working with General Pinto's government, and ended up in Bolivia as an adviser to Santa Cruz. In every case, he was the driving force behind numerous reforms with a strong utopian resonance, curtailed by the early political demise of his mentors. [14]

Earlier, in *La Crónica Científica y Literaria*, he had welcomed the Congress of Vienna for opening up a space to the Enlightenment and the technical, industrial and artistic discoveries in Europe that enabled the ruins of war to be repaired promptly. There is no doubt that in his ambivalent discourse, he seemed to reject utopias, as we saw in the case of Blanco White. In a commentary on George Dyer's *On the theory and practise of benevolence*, Mora preferred the term "benevolence", used by the author, to the one frequently used by others, "humanity"; although both implied the same thing, humanity referred to "las grandes teorías, a conquistas gigantescas y a los abusos de fuerza" (grand theories, giant conquests and the abuse of force), whereas "benevolence" referred to "la virtud doméstica" (domestic virtue). After that, he found that the key to re-establishing social harmony lay in the education of the poor by the rich, although, shortly afterwards, he was praising Hofwil, a clearly Utopian Swiss educational establishment. A mixture of rural colony and educational experiment, it brought together both poor and comfortably-off young people who studied according to the Pestalozzi Method, which involved working the land, using modern methods such as underground irrigation, under the watchful eye of the headmaster from a tower. In that panopticon, the "espíritu de orden, moderación y regularidad" (spirit of order, moderation and regularity) reigned and the young men behaved with "la moral más pura" (the purest morality). Together with education, technology was taking on a utopian dimension as it enabled the shaping of a new man, characterized by his ability to excel and imaginative power. From that point on, the euphoria they felt because of their accomplishments reached sublime heights at the prospect of the sheer scale of unstoppable progress.[15]

In London, he collaborated with Ackermann on the publication of the *Museo Universal de Ciencias y Artes*, which was also published in Colombia, Buenos Aires, Chile, Guatemala and Peru. Its structure was typical of enlightened publications, with many articles containing useful information pertaining to agriculture and chemistry, expressing great confidence in machines, and plenty of literary and geographical texts. The image of the American continent that it conveyed reiterated the idea of a future full of promise. A note in the 1825 issue included an allegory of the "Triunfo de la Independencia Americana" (Triumph of American Independence). In its description, the genius of independence appeared as a young woman, "lozana, bella y con un gorro frigio", (glowing with health, beautiful and wearing a Phrygian cap) sitting in a "carro triunfal tirado por seis caballos en representación de las seis repúblicas independientes" (triumphal chariot drawn by six horses representing the six independent republics). As always in compositions of this kind, the image was surrounded by perfect allegories of Prudence, Hope, Temperance, Justice, the Sciences and the Arts, Eternity, Unity, Commerce and Plenty.[16]

In keeping with the context indicated earlier, Mora praised the discipline of geography because displaying the "public riches" of a place and the "moral state" of its inhabitants showed that it was ready for good government and the impulse of trade. At the same time, it showed a bountiful America in which each individual state had a larger expanse of territory than, and riches on a par with, the most powerful nations in the world. Eliminating the perverse colonial government would do the rest. As he demonstrated in a review of the progress made in the new countries, Colombia was showing signs of regeneration, West Indian women were displaying their serenity as good wives and mothers, and Haiti had become a civilized nation twenty-five years after independence. In the same issue, he summarized a letter from the philanthropist, John Thomas Barber Beaumont, who asserted that America would shortly be supplying the world with its products and perhaps "nobles y útiles ejemplos de organización social" (noble and useful examples of social organization). Naturally, the continent welcomed utopian experiments. In 1826, Mora wrote in the *Museo Universal* about the New Harmony community, founded by Robert Owen in the United States. After pointing out that the project had earned the approval of all "profound economists" and "all friends of humanity", he explained that Owen's ideas had been poorly received in Europe, which was "devastated by revolutions". This was why he had concentrated his efforts in the United States, where the experiment was developing successfully, as was apparent from the increase in its population, the multiplication of schools and the stimulus to productive agriculture. He went on to advise the republics to the south to copy those plans as a way of increasing their population. Finally, he dismissed those who considered such measures as pipe-dreams by citing the example of the opening of a canal between the Great Lakes and the Atlantic, which had been considered a fantasy ten years earlier but was now a reality.[17]

Likewise, in accordance with the colonizing tradition of the Enlightenment, he lauded the creation of colonies in the Argentinian territory of Calera de Barquín in Entre Ríos, to which colonists who were "healthy and of good conduct" would be sent.[18] The *Museo* included the "Bosquejo de un plan de colonización en la América del Sur" (Outline of a colonization plan in South America), published by the company run by an Englishman, Barber Beaumont. This project included plans of the chief towns of the colonies, each one intended for some six hundred inhabitants. These towns were of extreme geometrical regularity, a square mile in area, ringed by trees, vegetable plots and flower gardens that guaranteed recreation and healthy living. They were set out in great detail: the width of the streets, which were numbered, and of the houses, and even the sizes of the blocks of dwellings were all indicated. Most of the colonists were tenants whose passage to America had been paid for, although the possibility that they might, in time, become owners was not ruled out. These proposals, together with his praise for Owen's experiments, reflected utopianism imbued with the Enlightenment and a republican tradition in which the identification of liberalism with private property was not as strict as it became later. In any case, a property of modest size was desirable, in accordance with the republican tradition, so as not to spoil in any way the dream of a community of economically independent equals that prevailed in the revolutionary process. The utopian ideal did not depart from this model and was described in some enlightened utopias written at the time, such as Saint Pierre's *Arcadie*, which was constituted as a society of frugal peasants, or Jean Baptiste Say's *Olbie*, a utopia whose political economy was governed with frugality and economic equality. Mora did not depart from this model either; in his

Catecismo de Economía política (Catechism of Political Economy), which was published by Ackermann for the Latin American market, he argued in favour of the role of capital as a tool to drive mechanization and bring about a society of plenty, as well as high wages, which would guarantee high consumption and morality by promoting education and good manners in people.[19]

Mora viewed the political economy, which he thought exemplified rationality and the scientific spirit, as the key to social happiness, since it released the sources of production and trade, eliminated obstacles to consumption and maintained the "equilibrio entre la opulencia y la penuria" (balance between opulence and penury). In this way, he was following the eighteenth-century trail that led from the old utopias, based exclusively on agriculture, to those that introduced trade, as well as contemporary theories about the market as an instrument capable of accelerating the decentralization of society and displacing the political power that had traditionally intervened, in an arbitrary way, in the allocation of resources. The result of his applying it to the real world opened the door to a multitude of ideal images:

> if we were to see all the land in that country cultivated with the utmost perfection, all its inhabitants applying themselves to useful, productive work; the interests of the government identified with those of individuals; public authority employed in alleviating public ills and advancing the common good; order consolidated on the foundations of social happiness and the whole nation turned into one vast family whose concord would be born of its own interest, (would not the theoretical instrument deserve the name of science?)[20]

The new economy could not be understood without a utopia of individual sovereign beings, in which moral education was just another economic variable. The importance of this was confirmed in Lorenzo Villanueva's *Catecismo de moral* (Catechism of Morality), which equated society with a moral body, safeguarded by citizen virtue. This meant placing the common good first, subordinating leisure to work and, in short, attaining a "remedo de la bienaventuranza eterna" (imitation of eternal bliss) on Earth. To that virtue, José de Urcullu, for his part, added manners, which made relations in the "human family" run more smoothly. Likewise, *No me olvides* (Forget-me-not), published in 1829 in London and several Latin American capitals by another exile, Pablo Mendíbil, included a moral tale, entitled "La esclava de Booroom" (The Booroom Slave), which was a story of civilization and racial subjugation with a gender dimension. Its plot recounted the vicissitudes of a black princess who was captured by slave traders after she had gone for a walk in the jungle without the permission of her parents. She soon managed to escape and after many hardships she reached a natural paradise of plants and animals, inhabited, of course, by English people who boasted that they never made slaves of anybody and educated her in humility and respect for her superiors. After a while, the princess regained the "virtud de la obediencia" (virtue of obedience), was calling them (in the Spanish version) "amitos" (little masters), and living happily and surrounded by the advances of civilization, such as "peines y vestidos" (combs and dresses).[21]

The situation in republican Chile

Mora's arrival in Chile from Buenos Aires was not something that was particularly exceptional. José Victorino Lastarría, in his memoirs, highlighted the arrival of men of learning from abroad for the purpose of improving education in the country, and their fleeting role in attempts to inculcate independence of thought and regeneration. Miguel Amunátegui, who took a conservative standpoint, attributed the invitation extended to Europeans to the "utopian vision" in favour of changing things rapidly, which prevailed among the Chilean elites of the day.[22]

The 1820s were a time of struggle between conservative liberals, called 'pelucones' (bigwigs), and more radical ones, known as 'pipiolos' (upstarts), who, by 1828, were in power with General Pinto as President. Gabriel Salazar considered the struggles as expressing the conflict between conservative centralism and federalism, with the latter in favour of greater sovereignty for the local powers, the so-called "pueblos", and also of more direct political participation through assemblies rather than a representative regime. Other authors, on the other hand, have emphasized the similarities between the two groups. For Jocelyn Holt, the differences were circumstantial and complicated by the presence of other sectors with particular claims, such as the tobacco dealers — a conservative group that controlled part of the overseas trade through their tobacco monopoly and advocated strong government — the federals and followers of the former "supreme director" Bernardo O'Higgins, or of the late Carrera, who was O'Higgins' rival during the independence process. In this regard, Ana María Stuven has stressed the ideological similarities of the Chilean elite, influenced by the culture of the Spanish Enlightenment and the French Revolution. As successors to the colonial power, the members of the elite believed in a natural order of things. Even though they could differ about whether that order was the fruit of divine creation, in the conservative case, or of history and compromise, in the liberal case, both sides defended the maintenance of constitutional order and its reforms within a gradual process. We can also point out that the elites as a whole shared, at least initially, the existing utopian climate, which focused on three areas: the educational, the festive and the constitutional.[23]

Education fulfilled a prime role in the enlightened project and in eighteenth-century utopianism, as well as in early nineteenth-century liberalism, since it allegedly had the power to create citizens and law-abiding people, in short, the new man. Hence its progress was considered "la panacea de todos los males políticos y morales que aquejaban a Chile" (the panacea for all the political and moral ills that afflicted Chile). From the earliest stages of independence there was consensus about the need to develop education. Manuel de Salas had argued for a utilitarian education; Camilo Henríquez had set up the National Institute and promoted Lancasterian schools; and in 1810, Juan Egaña had proposed a plan to foster scientific and industrial education for the purpose of "suavizar las costumbres" (refining people's manners).[24]

This educational programme was linked to the catechisms mentioned earlier, and to the Lancasterian system of mutual instruction, for which the catechisms served as manuals. This system was based on a cooperative methodology, whereby pupils at more advanced levels were responsible for teaching their younger companions and which made it easy to spread education at a low cost. At the same time, it was undeniably in the utopian tradition, as demonstrated by the fact that it was the system employed in utopias like Henríquez's *Camila* or Cabet's *Voyage en Icarie*.

Alongside that model of general education, there were more selective experiments in the 'pipiolo' period, such as the *Plan de Estudios del Liceo de Chile* (syllabus of the Chile Lyceum) drawn up by Mora, which aimed to reform society by starting with the elites, and targeted the children of the well-to-do. With a bias that was typical of the Enlightenment, it included, for the first time, studies of the humanities, with special emphasis on literature — whose mission was "fecundar un campo tan vasto en un país tan favorecido por la naturaleza" (to fertilize such a vast field in a country so favoured by nature) — geography, eloquence and Bentham's utilitarianism. It also tackled the creation of a new language, a characteristic aspect of eighteenth-century utopianism, which had a strong presence in Latin America after independence. Mora attributed the lack of intermediate sounds in the Spanish language to the fact that the speech of the average inhabitant was unintelligible gibberish. He installed, in its place a simple clear language and replaced Nebrija's *Gramática de la Lengua Castellana* (Grammar of the Castilian Language) with another concocted by himself. Furthermore, he believed that all the things that were wrong with the language had a political repercussion, since they perverted ideas and opened the way to despotism. For that reason, the language of the peninsula had little to offer the liberated American continent, which led Mora to argue for a Latin American literature that adopted models of the British writers, Shakespeare, Hume and Milton, the French, Molière, Racine and Montesquieu, or the Spanish, Jovellanos and Quintana.[25]

Just as significant was the organization of the Lyceum as an institution. Student life was regulated along typically utopian lines, down to the last detail; these included references to the size of the dormitories, the distance between the beds, as well as other arguments of a hygienist nature, such as prescribing cleanliness, the existence of a pool for bathing and swimming, and light meals. Moreover, a disciplinary organization was added in accordance with the monastic principles of utopianism that were clearly apparent in Salomon's House in Francis Bacon's *New Atlantis*, which Mora so admired, with its community of wise men, guides to the ideal city. Inside the Lyceum, the rule of silence was respected, there was absolute equality, "uproar and shouting" were forbidden, as were conversations with those who had no connection to the school or the servants.[26]

In keeping with enlightened principles, education was to be extended to women with the aim of improving their capabilities as mothers and as companions to men. In his *Cartas sobre la educación del bello sexo* (Letters on the Education of the Fair Sex), Mora echoed the expectation aroused by America, where "todas las esperanzas de los filántropos sobre la mejora de la especie humana" (all the hopes of philanthropists concerning the improvement of the human species) were directed. The need for a moral revolution after the political one of independence could only be based on education. Girls should be inculcated with civic virtues: the value of effort, compassion, and moulded with an austere rather than capricious character. Their education should avoid all specialization and be acquired by means of "gently sentimental" literature. To this end, Mora recommended learning English instead of French because of the "gravity of its literature" which led to "reflection and a devotion to freedom". The ideal scenario for the educational project was to be found in the rural environment, where fraternal ties could be established more easily than in the city, as well as a methodical life based on work and the development of practical philosophy, not on the fanciful dreams that were habitual in French literature. In this way, he turned utopia into reality and connected with the enlightened ideal of the creation of utopian enclaves, such as the one represented in Lady Hamilton's *Munster village*, where a benefactress created an ideal society on one of her estates, with schooling for everybody, religious tolerance, manufacturing work and stage plays. These ideas came to fruition in the School for Young Ladies that Mora's wife set up in Santiago. Inspired by Lancasterian principles, it was similar to one she had previously opened in Buenos Aires. Eventually it had sixty pupils who received an education that centred on Christian Morality and Religion, sewing, the inevitable English and French, Spanish Grammar and descriptive Geography.[27]

Along with education, public festivals and ceremonies constituted another essential element in utopian programmes and were, as such, habitual in the landscape of the ideal societies that featured in the literature of the genre. Maurizio Ridolfi has stressed the role of ritual as an instrument in the sacralization of politics and its use of the symbolic dimension to construct identitarian social links. Bronisław Bazcko, for his part, saw the utopian side of festivals as a way of reflecting the dreams of a society, as well as its capacity for becoming a place of renewal and of exercise of the collective imaginary. Celebration also became an island with its own rules and rhythms, set within everyday society, and contributing, albeit episodically, to the shaping of the new man. Displays laden with ritual significance expressed this kind of aspiration during the Pinto government. According to the historian, Pinto Vallejo, the political programme of the 'pipiolos' rested on the creation of an abstract people, more imagined than real, which had closer links to the idealized image of the (Ancient) Romans than to the labourers and craftsmen of contemporary Chilean society. This model was no doubt consistent with the project of a new man that matched up to the utopia being invoked. In the brief period of the 'pipiolo' mandate, festivities commemorating Independence or the proclamation of the new Constitution followed hard on the heels of each other, all serving educational purposes and with the aim of reinforcing the political community. Accordingly, they included orderly parades of all representatives of society "con semblantes de alegría" (with happy faces), collective meals, stage plays, public games and dances, "alejados de toda inmoralidad" (with no suggestion of any immorality), austere vegetable decorations and pyramids "como en las repúblicas antiguas" (as in the ancient republics), bells, masses, public prayers led by members of parliament and actors, and night-time illuminations of symbolic value because they represented the triumph of "light over the darkness".[28]

In this area, theatre and drama played an important role. Enlightened tradition had given it an educational civilizing function that the Chilean elite had shared since Independence. *El Boletín del Monitor*, for example, celebrated the première of the anticlerical play by Nevares, *Aristodemo*, and recommended it be staged once a month. Its packed audiences would applaud "las razones contra la superchería" (the reasons against fraud). According to the periodical, its message made a deep impression on all strata of society: on "enlightened men" because they discovered the bishop of Rome and his "secuaces del clero fanático enriqueciéndose a costa de la credulidad" (henchmen, the fanatical clergy, enriching themselves at the expense of [people's] credulity); and lesser thinkers, who learned that those societies, where ignorance ruled, were looted by "hordas de sacrílegos" (sacrilegious hordes) in the name of God. The theatre was a way of "polishing manners and language", encouraging sociability, without which man was a "semi-wild" creature. For that reason, and as had been the case with the men of wisdom, the theatre had to be imported from Europe because of the cultural backwardness of the country. With such ideas floating in the air, it was no surprise that during those years the Spanish theatre companies of Teresa Samaniego and Emilia Hernández, Francisco Villalba or Francisco Rivas Prádenas should disembark there in quick succession. The arrival of Teresa Samaniego, who was a disciple of the famous Spanish actor Máiquez and a committed liberal in the Triennium of 1820–1823, was greeted as the dawn of civilization in Chile.[29]

Through the use of the emotions, theatre has frequently brought community links closer. Mora himself had already acknowledged the potential of performance accompanied by tasteful, harmonious music when he published *Doce canciones en español* (Twelve Songs in Spanish). The work included a hymn to Simón Bolívar, whose "solemn and religious" melody could easily be adapted, according to the author, to grand patriotic sentiments; and another to the sixteenth-century *comunero*, Juan Bravo, backed by the "lively warlike" music of Weber's opera, *Der Freischütz*, in which the word "Liberty" repeated in lively energetic notes "set the pulse racing".[30]

Tragedy played a prominent role in this educational project; Camilo Henríquez in his time had considered it fundamental on stage, and there were frequent appeals to put on new tragedies during the years that concern us here. In this respect, the leaders of the independence movement elevated Vittorino Alfieri, a dramatist from Piedmont, to mythic status, and the press of the day called for his works to be performed for their civic value. Piero Gobetti highlighted his utopian dimension because he presented solitary heroes, locked in the eternal struggle between freedom and tyranny, characters living in a timeless golden age governed by models of virtuous behaviour. In short, the sort of ideal people that the Chilean elite yearned for, according to Pinto Vallejo, ready to sacrifice life and self-interest for the community. Another case was *Virginia*, performed at commemorations of Chilean independence. Set at an unspecified time in Ancient Rome, it staged the story of a heroine who preferred to kill her children and commit suicide rather than live under tyranny. Her example stirred up the erstwhile submissive Roman people, who rose up against their ruler. In this way, the play showed that the ultimate alternative to tyranny was exemplary sacrifice in order to achieve a future of virtue.[31]

Civic theatre of this kind had a recognized place in many of the utopian projects of the era, such as Louis Sébastien Mercier's *L'an deux mille quatre-cent-quarante*, published in 1786, Jean-Baptiste Say's *Olbie* or *An Account of the First Settlement, Laws, Form of Government, and Police, of the Cessares, A People of South America*, a work of 1764 by James Burgh, in which the mission of the theatre was to prevent the corruption and "effeminization" of the people. In the Chilean case, it never concealed its political content, since the works were put on with proclamations delivered by members of parliament or actors. After a performance of the tragedy, *La condesa de Castilla* by Álvarez Cienfuegos, for example, the Spanish actress Emilia Hernández recited the following poem, foreshadowing an ideal future of freedom and progress:

> Free; be satisfied
> That living in peace and unity
> Chile will be a great nation
> Wise and opulent
> Forever free of tyrants
> And of superstition
> May heaven grant you to see,
> Freedom of conscience
> And the sciences will come to Chile
> As announced by Volter (sic)
> Then, oh what pleasure

The Arts will be reborn
Everyone will love God
Even though in different ways
Since, there being one God for all
All will enjoy one God
But let not impious fate
Wish that this fortunate land
Should find itself marked
By fanaticism like mine [Spain] . . . [32]

Theatrical language also affected interpretations of events at the time. The failed coup against Pinto's government in 1828, for example, served to re-launch Alfieri's republican heroes:

> The people's feeling that only over their dead bodies would independence be taken from them opposed [the attempted coup]. Pinto held out, surrounded by people of all classes. Disappointment of the rebels, for they had been led to believe that they would be received in triumph. Instead of those false promises, there were expressions of loyalty and patriotism at every turn. The apparatus of force did not impose silence on the unarmed crowd. In the midst of their ranks, voices cried out 'Long live the people! Long live Pinto!' The scene we have just described was a sublime spectacle. It should be enshrined in our annals as the most solemn testimony of the progress that the people have made on their road to freedom.[33]

Regrettably, the tragic tension appeared to be too much for the cultural level of the time. Years before, *El Censor de Buenos Aires* had pointed out the dangers of a poor acting. Both overacting and acting without feeling turned the sublime into the ridiculous, removed the sense of illusion and pressed the audience into participating directly by praising or censuring. In general, the enlightened elites in Chile (and in Argentina, too), regardless of their political affiliation, were wary of the lower orders and sought to limit their expansion. In the case of the theatre, an attempt was made to discipline the audience, just as the enlightened European theatre of the day — which was accepted as the paragon of utopias — had advocated. They called for clean auditoriums, theatre judges to censor works considered immoral and publicly condemn their authors, as Mercier did in his utopia, *L'an deux mille quatre-cent-quarante*. Likewise, they decried the kinds of performance considered base and common, even if they were much more successful, such as magic comedies. These spectacles were condemned because they were too far removed from rationality and because they occasionally included disturbing political content, such as defence of guerrilla leaders. Something similar happened with the *chinganas*, dances that, according to Mora, degraded the "orden físico y moral de la sociedad" (physical and moral order of society) by encouraging the promiscuity of men and women with gestures of wild abandon.[34]

Despite the good intentions, civic theatre did not prosper. Tragedy seemed to be a genre that was too elevated to be understood by the general public, so that intellectuals like Andrés Bello and Mora himself recommended staging comedies that were well done and had some kind of a moral content. In this regard, Mora staged *El marido ambicioso* (The Ambitious Husband), which followed in the wake of the neoclassical Spanish playwright, Moratín, and did not forgo passages with strong utopian echoes. The leading character, a determined liberal ready to sacrifice his married life to obtain an important political post, expressed his anticipation of a society in which his ideals would triumph:

> Reason is queen of the world
> Interests, in accord
> Only with reason hold sway
> From the south to the north
> May Your Excellency make easier
> With wise rulings
> The enlightenment of the masses:
> May you shape an education
> In line with the principles of the century;
> May the Gothic frameworks of our grandfathers fall
> Before the truth; and may the woods
> Become splendid cities where burgeon
> Virtues, gold and work /
> Like flowers in the meadow.
> Command them, Your Excellency, to study
> From the old to the young,
> Chemistry and Economics
> And the rights of man.
> Let unjust exemptions from birth cease;
> Let work be given to the wise man
> Not to the schemer, nor to the noble;
> Let there be severe punishment
> For those fierce tyrants
> Despots of thought
> And persecutors of opinions.
> Cleanse the secretariats
> of those stupid vermin.[35]

In short, it was a project for a new society that led his interlocutor in the scene to respond that if he expressed those theories, he would be considered "a lunatic".[36]

The final element of the project to create a new society rested on the drafting of a constitution. Projects of this type proliferated in the decade following independence, but were criticized by conservatives in subsequent decades for being out of touch with reality and for the chaos they caused. Against this, Renato Cristi and Pablo Ruiz Tagle have re-evaluated their role in the later consolidation of republican Chile, as well as their egalitarian content associated with the progress of civilization and ideas about promoting wealth through work. This process of permanent testing was a conscious thing, as General Pinto himself demonstrated when, in his speech introducing the 1828 text, he maintained his opinion of the "perfectibilidad de las constituciones" (perfectibility of Constitutions) until a republic like the Chilean one was attained.[37]

In that same year, in fact, a text was approved, the drafting of which was attributed to Mora. Egaña's earlier conservative version of 1823 had merged enlightened elements with utopian-sounding republican ones, such as the establishment of a moral power as being, "único capaz de llegar a un estado ideal" (the only one capable of arriving at an ideal state), and the promotion of civic festivities, plays and prizes for virtue. For its part, the 1828 Constitution — to which Mora had added a variety of elements inspired by the Cadiz Constitution of 1812 — incorporated equal rights (including the end of ecclesiastical privileges and primogenitures), freedom, property, the primacy of the Legislative body, the extension of the right to vote, freedom of the press, as well as trial by jury and religious tolerance. It also envisaged a federal component, which, for Salazar, represented the triumph of the provincial middle classes over the Santiago oligarchy.[38] This was not altogether an anomaly, because in the years following independence, federalism, throughout Latin America, reflected the autonomous tradition of the town councils of the colonial period, and was considered by many a source of freedom. The *Mercurio de Valparaíso*, for example, praised the advance of federalism on the continent as the only means of resolving conflict and spreading freedom.[39] At the same time, another London periodical aimed at the Latin American market, the *Repertorio Americano*, which was published by exiles and included contributions from Pablo Mendíbil and Andrés Bello, commented on the *Columbiad* by Joel Barlow, a radical republican from the United States. Throughout its pages, Hesper, the guardian genius of the western continent (America), after appearing to Columbus and showing him the horrors of the conquest, encourages him with the triumph of independence:

> A work so vast a second world required,
> By oceans bourn'd, from elder states retired;
> Where, uncontaminated, unconfined,
> Free contemplation might be able to expand the mind,
> To form, fix, prove the well adjusted plan
> And base and build the commonwealth of man.
> A new creation waits the western shore,
> . . .
> Here social man a second birth shall find,
> And a new range of reason lift his mind
> Feed his strong intellect with purer light,
> A nobler sense of duty and of right,
> The sense of liberty; . . .

Freedom spread across the world thanks to education and federal government and the last book of the work confirmed the triumph of a federative utopian vision on a world-wide scale with

> freedom of trade, advances in communications as a result of the opening of the Suez and Panama canals, the appearance of cities in America, the social pre-eminence of scientists, whose researches were moving in the direction of prolonging the life and happiness of man, and the men of letters and philosophers, instigators of the love of order and of moral actions. Finally, a conference of all the nations was called to achieve human harmony. The delegates met in a temple of reason, situated in Egypt, in whose porch stood the figure of earth's genius with Truth's mirror in his hands. On his pedestal the symbols of the most noble arts were engraved; at his feet, the symbols of war and superstition. The delegates entered the temple of reason and received from her the order to strengthen their links until they formed a federal empire. As a result, they avoided war and misfortunes forever, and peace and happiness reigned in the universe.[40]

The end of utopia

The 'pipiolo' liberals were not in power for long, being ousted by the conservatives after the Battle of Lircay in 1830. Earlier there had been the failed coup, mentioned above, and a tense political climate, described in terms that were once more indebted in large measure to utopian language. The *Mercurio de Valparaíso* associated the drafting of the Constitution with material and spiritual regeneration, with a strong classical tone:

> Hail, Chile the fortunate! After many vicissitudes, you raised, in the year 28, a monument as astonishing as the lofty Andes! Hail, you mansion of free men, nation of smiling Amalthea, sanctuary of candour and spirituality! You are born anew![41]

Their rivals also resorted to a similar type of language and announced a future full of promise after the fall of Pinto:

> . . . And a resounding voice at once inspires
> The longing for a happy future
> Then a gracious nymph appears,
> In shining chariot
> And with loud echo declares imperiously thus
> Chile will be for ever fortunate
> Unity will reign in it eternally,
> It will serve as a refuge for the innocent,
> And a happy mansion for the unfortunate;
> Upon its fertile soil will not enter
> Horrendous tyranny [...]
> Its soul [i.e.] industry, sciences,
> Trade, the arts,
> Will everywhere make
> More than a thousand advances
> The powers will populate
> Our ports with ships,
> Even as far as the deserts
> The ingenuity of manufacturing
> And useful agriculture
> Will create abundance [...]
> Public spiritedness will be revived
> In glorious scenes,
> That made Rome, Sparta, Athens
> And Greece eternal.[42]

Shortly afterwards, Mora was expelled from Chile and took refuge in Peru. Utopia had crumbled and with it his confidence in the country, which lost its idyllic qualities and became an "Arabia"; the Chilean new man likewise went from being a "Roman hero" to being branded a "coarse Boeotian".[43] It was perhaps because of this that Mora backed the strong government of Marshal Santa Cruz in his new country of residence. The trend was general. Mora had shown the importance of utopianism in early liberalism. He rejected, of course, great transformations, which explained why he rejected the French Revolution. He was more in favour of a gradual strategy, one that could change society in the medium term, but also in a fundamental way. By contrast, the end of the 1820s marked a break in Latin American liberalism. One of the most outstanding cases was Bolívar; he censured the constitutionalist utopianism of those who conceived "republics of air" and set out to attain a political perfection that was not connected to the social reality surrounding it. In the Chilean case, Andrés Bello, who had previously been close to Mora and was now the inspiration behind the new conservative government of Portales, represented that appeal to a historical rationality and the search for stability. These objectives were only possible if republican power, built on strong centralized institutions, was handed over to the oligarchy until the people were ready, forgetting the attempts to erase the past and start with a clean slate.[44]

Notes

1 Matthew Pratt Guterl, "The Future of Transnational History", *American Historical Review* 118, 1 (2013): 130–139. For the idea of Latin America as a consumer of European ideas that brought hope of change, see Miguel Ángel Centeno and Fernando López Alves, "The Other Mirror: Grand Theory through the Lens of Latin America", in *Imported Modernity in Post Colonial State Formation*, edited by Eugenia Roldán and Marcelo Caruso (Frankfurt am Main: Peter Lang, 2007), pp. 7–28.

2 Christine Rees, *Utopian Imagination and Eighteenth-Century Fiction* (London: Longman, 1996), p. 9. For changes in utopian models, see Bronisław Baczko, *Lumières de l'Utopie* (Paris: Payot, 1978), pp. 29, 37 and 410; Antoine Hatzenberger, ed., *Utopies des Lumières* (Lyon: Ens, 2010), p. 36.

3 The role of overseas in utopias, in Rees, *Utopian Imagination*, pp. 55 and 80ff; Larry Wolff, *Inventing Eastern Europe: The Map of Civilization on the Mind of Enlightenment* (Stanford: Stanford University Press, 1994), p. 395; for Buffon's view of American inferiority, see Antonello Gerbi, *La disputa del nuevo mundo. Historia de una polémica, 1750–1900* (Mexico: FCE, 1960); Mary Louise Pratt, *Ojos imperiales: literatura de viajes y transculturación* (Mexico: FCE, 2010), p. 75.

4 Samuel Haigh, Alexander Caldcleugh and Max Radiguet, *Viajeros en Chile, 1817–1847* (Santiago de Chile: Editorial del Pacífico, 1955), pp. 14 and 25, 148 and 160, respectively. Zenobio Saldivia, *La visión de la naturaleza en tres científicos del siglo XIX en Chile: Gay, Domeyko and Philippi* (Santiago de Chile: Universidad de Santiago de Chile, 2003), pp. 125 and 138.

5 The application of Rancière to the moment of independence, in Patrice Vermeren, "El desplazamiento de la filosofía, el no-lugar de la democracia y la lengua de la emancipación", in *República, liberalismo y democracia*, edited by Marcos García de la Huerta and Carlos Ruiz Schneider (Santiago de Chile: LOM 2011), pp. 161–175; Rafael Rojas, *Las repúblicas de aire. Utopía y desencanto en la revolución de Hispanoamérica* (Madrid: Taurus, 2009), pp. 38 ff and 244.

6 The quotation by Egaña, in Enrique Fernández, *El nacimiento de la cultura política de la nación en el Río de la Plata y Chile (1808–1818)* (Zaragoza: Institución Fernando el Católico, 2011), p. 84: "La idea de una ciudad formada para la sabiduría y habitada solamente de sabios, si fuese capaz de realizarse, en ningún punto de la tierra podría colocarse mejor que Chile. Distante por su situación a 300 leguas de aquel gran mundo donde la guerra, la intriga y las necesidades urgentes ocupan la vida de más de un tercio de sus habitantes y libre por su destino político de aquel tumulto de pasiones que excitan la ambición, las ideas de superioridad y el equilibrio de poder conservamos una existencia metódica y uniforme donde el nacimiento y la muerte regularmente forman un círculo que nos viene a dejar en el mismo punto de jerarquía."

7 Rafael Sagredo Baeza, "Chile: de finis terrae imperial a 'copia feliz del edén autoritario'", in *Crear la nación. Los nombres de los países de América Latina,* compiled by Juan Carlos Chiaramonte, Carlos Marichal and Aimer Granados (Buenos Aires: Sudamericana, 2008), pp. 41–67: "Por el mar y la tierra amenazan / Los secuaces del déspota vil / Pero toda la naturaleza / Los espera para combatir: / El Pacífico al Sud y Occidente / Al Oriente los Andes y el Sol / Por el Norte un inmenso desierto / Y el Centro libertad y unión."

8 Henríquez's text, in Enrique Fernández, *El nacimiento,* p. 127: "Esta es una verdad de geografía que se viene a los ojos y que nos hace palpable la situación de Chile. Pudiendo esta vasta región subsistir por sí misma, teniendo en las entrañas de la tierra y sobre su superficie no sólo lo necesario para vivir, sino aún para el recreo de los sentidos; pudiendo desde sus puertos ejercer un comercio útil con todas las naciones, produciendo hombres robustos para la cultura de sus fértiles campos[...] hallándose encerrada como dentro de un muro, y separada de los demás pueblos por una cadena de montes altísimos de eterna nieve, por un dilatado desierto y por el mar pacífico, ¿no era un absurdo contrario al destino y orden inspirado por la naturaleza, ir a buscar un gobierno arbitrario, un ministerio venal y corrompido dañosas y oscuras leyes, o las decisiones parciales de aristócratas ambiciosos, a la otra parte de los mares?"

9 Camilo Henríquez, *La Camila o la patriota de Sud-América* (Buenos Aires: Imp. de Benavente y cía, 1817). The idea of an idealized past heralding a perfect future, in Enrique Fernández, *El nacimiento,* p. 178.

10 For the role of exile in the Atlantic world, see Juan Luis Simal, *Emigrados. España y el exilio internacional, 1814–1834* (Madrid: Centro de Estudios Políticos y Constitucionales, 2012), pp. 37 and 327; Eugenia Roldán Vera, *The British Book Trade and Spanish American Independence: Education and Knowledge Transmission in Transcontinental Perspective* (Aldershot, Hampshire: Ashgate 2003), p. 47; *Colombia: Being a Geographical, Statistical, Agricultural, Commercial, and Political Account of That Country,* vol. 1 (London: Baldwin, Cradock and Joy, 1822), p. civ; *El Emigrado Observador* (February 1829): 41; *El Repertorio Americano* 1 (1829): 80.

11 For the role of catechisms in America, see Roldán, *The British Book Trade,* pp. 93, 142 and 164. For the link between structured knowledge and harmony, see José Joaquín Mora, *Museo Universal de Ciencias y Artes* vol.1 (1825): 271; Nicolas-Edme Restif de la Bretonne, *El descubrimiento austral por un hombre-volador o el Dédalo francés* (Santiago de Chile: Centro de Investigaciones de Historia Americana, 1962).

12 José María Blanco White, *Variedades o El Mensajero de Londres* 1, 5 (1824): 17 and p. 404. The advice is in vol. 2, 7 (1825): 95.

13 The praise for constitutionalism, in *Variedades o El Mensajero de Londres* 2, 6 (1825): 2; Carol Tully, "Ackermann, Mora, and the Transnational Context: Cultural Transfer in the Old World and the New", in *Londres y el liberalismo hispánico,* edited by Daniel Muñoz and Gregorio Alonso (Frankfurt am Main: Iberoamericana Vevuert, 2011), pp. 153–164; Fernando Durán López, ed., *José María Blanco White. Artículos de crítica e historia literaria* (Sevilla: Clásicos Andaluces, 2011), p. 62; Ángel Saavedra, Duque de Rivas, *El Desterrado* [manuscript], Biblioteca Nacional de España [National Library of Spain], online version, https://beta.worldcat.org/archivegrid/collection/data/430966189; for the relationship of the latter with exile, see Peter Cooke, "Nation, Myth, and History in *Ocios de Españoles Emigrados* (London, 1824–27)", in *Londres y el Liberalismo,* Muñoz and Alonso, pp. 95–109.

14 For Mora's life and career, see Luis Monguió, *Don José Joaquín Mora y el Perú del Ochocientos* (Madrid: Castalia, 1967).

15 *Crónica Científica y Literaria,* 4 April 1817, and 29 April 1817. Mora's technological utopia, in Pilar Asensio, "Mora en Londres: Aportaciones al Hispanoamericanismo", in *Londres y el liberalismo,* Muñoz and Alonso, pp. 111–123.

16 *Museo Universal de Ciencias y Artes* 1 (1825): 219.

17 *Museo Universal de Ciencias y Artes* vol. 1 (1825): 359, 375 and 315. Utopian experiments are in vol. 2 (1826): 61 and 62.

18 In 1825, the British company, River Plate Agricultural Association, settled fifty colonists in Calera with the objective of growing wheat, although the settlement was soon abandoned.

19 *Museo Universal de Ciencias y Artes* 1 (1825): 251. For limitations on property in early liberalism, see Pierre Rosanvallon, *The Society of Equals* (Cambridge, MA: Harvard University Press, 2013). José Joaquín Mora, *Catecismo de Economía Política* (London: R. Ackermann, 1825), pp. 22 and 25.

20 Mora, *Catecismo de Economía,* p. 90: "si viésemos en aquel país cultivado con la mayor perfección todo su terreno, dedicados todos sus habitantes a trabajos útiles y productivos; identificados los intereses del gobierno con los de los individuos; empleada la autoridad pública en aligerar los males públicos y en fomentar el bien; consolidados el orden en las bases de la ventura social y convertida la nación entera en una vasta familia cuya concordia naciera de su interés propio (¿no mereciera el nombre de ciencia el instrumento teórico . . . ?)". For the change to utopias based on trade, see Gregory Claeys, *Utopias of the British Enlightenment* (Cambridge: Cambridge University Press, 1994). For the role of the market as the basis of utopia, see Pierre Rosanvallon, *El capitalismo utópico* (Buenos Aires: Nueva Visión 2006), p. 13.

21 Joaquín Lorenzo Villanueva, *Catecismo de Moral* (Paris: Garnier Hermano, 1896), first published in 1826. José de Urcullu, *Lecciones de Moral, Virtud y Urbanidad* (Madrid: Librería de Antonio Calleja, 1845), p. 53 (a work of 1825). *No me olvides. Colección de producciones en prosa y verso, orijinales (sic), imitadas y traduzidas (sic): para MCCCCXXVIII por D. Pablo Mendíbil* (London: Ackermann, 1828), p. 63.

22 José Victorino Lastarria, *Recuerdos literarios. Datos para la historia literaria de la América Española y del progreso intelectual de Chile* (Santiago de Chile: Librería de M. Servat, 1885), pp. 12 ff. Miguel Amunátegui, *Don José Joaquín de Mora: apuntes biográficos* (Santiago de Chile: Imprenta Nacional, 1888), p. 101; in a similar vein, Hernán Godoy Urzúa, *Apuntes sobre la cultura de Chile* (Valparaiso: Universidad Católica, 1982), p. 80.

23 Gabriel Salazar, *Construcción de Estado en Chile (1760–1860): democracia de "los pueblos", militarismo ciudadano, golpismo oligárquico* (Santiago de Chile: Sudamericana, 2005); Alfredo Jocelyn-Holt, *La Independencia de Chile: tradición, modernización y mito* (Madrid: Mapfre, 1992); Ana María Stuven, *La seducción de un orden. Las elites y la construcción de Chile en las polémicas culturales y políticas del siglo XIX* (Santiago de Chile: Universidad Católica, 2000).

24 The new man, in *Museo Universal de Ciencias y Artes* 1 (1825): 216. The panacea of ills in *El Mercurio chileno,* 1 April 1828.

25 José Joaquín Mora, *Oración inaugural del curso de oratoria del Liceo de Chile* (Santiago de Chile: Imp. de R. Rengifo, 1830).

26 For the rules of El Liceo, see *El Mercurio de Valparaíso,* 17 December 1828.

27 José Joaquín Mora, *Cartas sobre la educación del bello sexo por una señora americana* (London: Carlos Wood, 1824), pp. 5, 12 and 42; Lady Mary Hamilton, *Munster Village* (London: Robson & Co, 1778) in Digital Library Project, http://digital.library.upenn.edu/women/hamilton-mary/munster/munster.html (accessed 20 September 2015). For the College of Young Ladies, see Amunátegui, *Mora,* p. 134.

28 Maurizio Ridolfi, "Fiestas y conmemoraciones", in *Historia cultural de la política contemporánea,* edited by Jordi Canal and Javier Moreno (Madrid: Centro de Estudios Políticos y Constitucionales, 2009), pp. 59–96; Bazcko, *Lumières de l'utopie,* pp. 236 and 244; Julio Pinto Vallejo and Verónica Valdivia, *¿Chilenos todos? La construcción social de la nación (1810–1840)* (Santiago de Chile: LOM, 2009), p. 161. Examples of civic festivities in support of independence and the Constitution are in *El Mercurio de Valparaíso,* 22 September 1827 and 24 September 1828.

29 For the civilizing value of the theatre, see *El Argos de Chile*, 1818; *El Boletín del Monitor*, 5 November 1827; *El Verdadero Liberal*, 9 March 1827.

30 For emotional utopia in the theatre, see Jill Dolan, "Utopia in performance", *Theatre Research International* 31, 2 (2006): 163–173. Enthusiasm for the arrival of Samaniego in Miguel Amunátegui, *Primeras representaciones dramáticas en Chile* (Santiago de Chile: Imprenta Nacional, 1888), p. 80. The twelve songs are in *Museo Universal de Ciencias y Artes* 1 (1825): 82.

31 Henríquez's defence of tragedy is in Amunátegui, *Primeras representaciones*, p. 102. Giuseppe Bellini, *Storia della Relazioni Letterarie tra l'Italia e l'America di lingua spagnola* (Milan: Cisalpino-La Goliardica, 1977), pp. 95 ff. The request for new plays by Alfieri in *El Argos de Chile*, 3 September 1818. Piero Gobetti, *La filosofía política di Vittorio Alfieri* (Turin: Piero Gobetti, 1923), p. 125. José Victorino Alfieri, *Virginia* (Madrid: Imp. Repullés & Vargas, 1813), p. 11.

32 El *Mercurio de Valparaíso*, 22 September 1827: "Libres; estad satisfechos
Que viviendo en paz y unión / Chile será una nación / Grande, sabia y opulenta / De tiranos siempre exenta / Y de la superstición / El cielo os conceda ver, / La libertad de las conciencias / Y a Chile vendrán las ciencias / Como lo anunció Volter (sic) / Entonces, ó (sic) que placer / Las artes renacerán / Todos a Dios amarán / Aunque de diversos modos / Pues siendo un Dios para todos / Todos de un Dios gozarán / Mas no quieras suerte impía / Que esta tierra afortunada / Por el fanatismo hollada / Se encuentre como la mía . . . "

33 *El Mercurio de Valparaíso*, 30 July 1828: "Se opuso el sentimiento del pueblo de que sólo muertos se les quitaría la independencia. Pinto se mantuvo rodeado de gentes de todas las clases. Decepción de los rebeldes, se les había hecho creer que serían recibidos en triunfo. En lugar de esas falaces promesas se veía por todas partes expresiones de lealtad y patriotismo. El aparato de la fuerza no impuso silencio a la muchedumbre inerme. En medio de las filas se gritaba ¡viva el pueblo!, ¡viva Pinto! La escena que acabamos de referir formó un espectáculo sublime. Debe consagrarse en nuestros anales como testimonio más solemne de los progresos que el pueblo ha hecho en su carrera a la libertad."

34 *El Censor de Buenos Aires*, 5 September 1815. The misgivings of the Chilean elites, in Pinto Vallejo and Valdivia, *¿Chilenos todos?*; those of the Argentinian elites, in Gabriel Di Meglio, *¡Viva el bajo pueblo! La plebe urbana de Buenos Aires y la política entre la Revolución de Mayo y el rosismo* (Buenos Aires: Prometeo, 2007), p. 115. Carmen Maturana, "La Comedia de Magia y los efectos visuales de la era pre-cinematográfica en el siglo XIX en Chile", *AISTHESIS* 45 (2009): 82–102. Criticism of the *chinganas* in *Mercurio de Valparaíso*, 24 July 1829.

35 José Joaquín Mora, *El marido ambicioso* (Santiago de Chile: Imp. de R. Rengifo, 1828), p. 28: "Reina es la razón del mundo / Los intereses, acordes / Con la razón solo imperan / Desde el mediodía al norte / Facilite Vuecencia / Con sabias disposiciones / La ilustración de las masas: / Una educación conforme / Con los principios del siglo; / Las góticas armazones / De nuestros abuelos caigan / Ante la verdad; los bosques / Se conviertan en ciudades / Espléndidas, donde broten / Virtudes, oro y trabajos / Como en el prado las flores. / Mande Vuecencia que estudien / Desde el anciano hasta el joven, / Química y Economía / Y los derechos del hombre. / Que del nacimiento cesen / Las injustas exenciones (sic); / Que el empleo se dé al sabio / No al intrigante ni al noble; / Que con rigor se castiguen / Esos tiranos feroces / Déspotas del pensamiento, / Y verdugos de opiniones. / Limpiad las secretarías / De esas sabandijas torpes."

36 Mora, *El marido ambicioso*, p. 31.

37 Renato Cristi and Pablo Ruiz-Tagle, *La República en Chile. Teoría y práctica del Constitucionalismo Republicano* (Santiago de Chile: LOM, 2006); Rojas, *Repúblicas*, p. 110.

38 Salazar, *Construcción de Estado en Chile*, p. 327.

39 *El Mercurio de Valparaíso*, 12 September 1827.

40 *Repertorio Americano* 2 (1829): 20: "libertad de comercio, avances en las comunicaciones

por la apertura de canales en Suez y Panamá, la aparición de ciudades en América, la preeminencia social de los científicos, cuyas investigaciones se encaminaban a la prolongación de la vida y la felicidad del hombre, y de los literatos y filósofos, impulsores del amor al orden y a las acciones morales. Finalmente, se convocaba un congreso de todas las naciones para lograr la armonía humana. Los delegados se reunían en un templo de la razón, situado en Egipto, en cuyo pórtico figuraba el genio de la tierra con el espejo de la verdad en la mano. En su pedestal aparecían engastados en oro los símbolos de las artes más nobles; a sus plantas, los símbolos de la guerra y la superstición. Los delegados entraban en el templo de la razón y recibían de ella la orden de estrechar sus vínculos hasta formar un imperio federal. Como resultado, huían para siempre la guerra y las desgracias, y la paz y la felicidad reinaban en el universo."

41 *Mercurio de Valparaíso*, 31 December 1828: "Salve ¡o Chile venturoso! que después de una larga serie de vicisitudes alzasteis el año 28 un monumento tan asombroso como los elevados Andes! ¡Salve mansión de hombres libres, patria de la risueña Amalthea, asilo del candor y la espiritualidad! ¡Renacisteis!"

42 *El Crepúsculo*, 3 October 1829: " . . . Y una sonora voz al punto inspira / La ansia de un porvenir venturoso / Mas se aparece en resplandeciente carro / Una ninfa graciosa, / Y alzando el eco dice así imperiosa / Chile será por siempre afortunado / La unión reinará en él eternamente, / El servirá de asilo al inocente, / Y de mansión feliz al desgraciado; / En su suelo feraz no tendrá entrada / La horrenda tiranía [...] / La alma industria, las ciencias, / El comercio, las artes, / Harán en todas partes / Progresos más de mil / Poblarán las potencias / De buques nuestros puertos, / Y aun hasta los desiertos / El ingenio fabril / La útil agricultura / Creará la abundancia [...] / Revivirá el civismo / En gloriosas escenas, / Que a Roma, Esparta, Atenas / Y a Grecia eternizó."

43 Mora's contemptuous descriptions of Chile and the Chileans in Amunátegui, *Mora*, pp. 274 and 316, respectively.

44 The republics of air, in Rojas, *Repúblicas*, p. 244. Iván Jaksic, *Andrés Bello: La pasión por el orden* (Santiago de Chile: Editorial Universitaria, 2001). Miguel Ángel Centeno. "Republics of the Possible: State Building in Latin America and Spain", in *State and Nation Making in Latin America and Spain*, edited by Miguel Ángel Centeno and Agustín Ferraro (Cambridge: Cambridge University Press 2013), pp. 3–24.

6 Cabet's Utopia, from Minorca to Argentina: Bartolomé Victory y Suárez

Horacio Tarcus

After the Battle of Caseros in 1852, Argentina experienced a historic moment of such intense and far-reaching transformations — driven by the dominant class —, that it was recently referred to as a "veritable process of social engineering".[1] During the course of this transformational period, three major objectives were achieved. In the first place, Argentina completed its integration into the new world capitalist order, so establishing the foundations of bourgeois order in that corner of the world. This was mirrored not only in the extremely rapid urbanization of Buenos Aires and other provincial capitals, the modernization of transport and means of communication, the commercial exploitation of all the factors of production in anticipation of a national market, but also in the transformation of regional bourgeois groups born under colonial or post-independent rule into a modern agricultural bourgeoisie that strove to control the levers of political power and set itself up as a dominant class of national scope.

Secondly, the country finally managed to form a unified system of political representation during this period. According to Marta Bonaudo:

> The constitution that was approved in 1853 asserted the criterion of the sovereignty of the people and situated the figure of the citizen as the basis of all legitimacy. However, starting with the specific practices of power that emerged and were developed during those thirty years, the elites systematically violated fundamental aspects of the thinking that was at the heart of that legitimacy, but which did not prevent them from consolidating a narrative of legality that shored up the edifice of the nation state.[2]

Such factious forms of political practice were not, however, an obstacle to the creation of a public sphere, which was envisaged as a space for discussion and deliberation in which private individuals met in public and appealed to reason and argument, with a view to mutual understanding.[3] Creoles and immigrants, bourgeois sectors and those associated with the workplace all contributed to what came to be called the Argentinian "associative explosion" that followed Caseros, and which encompassed clubs, cultural associations for the people, mutual, cooperative, literary, scientific and masonic associations and so on.[4] No less crucial for the development of the public sphere was the expansion of an extensive network of newspapers and periodicals, the space of public debate par excellence. This press network was complemented by increasing numbers of pamphlets, magazines and books being published, fuelled by a growing network of bookshops and printing houses, which, shortly afterwards, would become the publishers that emerged at the end of the nineteenth century.[5] All this meaningful expansion and diversification of the reading public was, in turn, favoured by a flourishing school system, which had been established round about 1884 when the National Education Act (law 1420) was passed, but enjoyed its most rapid expansion at this time. During this expansionary cycle, a field of intellectual endeavour gradually emerged, as a glittering array of writers, journalists, educators and political essayists helped shape and organize the different spaces within it.

The third objective reached was the organization of the State. After a decade of coexistence between two quasi-states at loggerheads with each other, a nationwide state apparatus began to be put in place, as the constitutional rules required, under the presidencies of Bartolomé Mitre and Domingo F. Sarmiento. Its sovereign power was strengthened in the sense that it took over and centralized in Buenos Aires prerogatives and functions that had hitherto been in the power of the provinces, namely, those concerned with fiscal and monetary affairs and those of a military, judicial, administrative and educational nature.[6]

In the context of this far-reaching process, the concerns and interests of the Creole intellectual elites shifted from the problems associated with revolution and "regeneration" to those of social order, from romanticism to liberalism, from socialism to positivism. Chiaramonte has described this process in terms of a shift in the mentality of the elites:

> In Argentina, after the so-called Generation of 1837, the mission of intellectuals ceased to be thought of in terms of revolution. The writings of Esteban Echeverría and some of his contemporaries are the last in which nineteenth-century intellectuals conceive of their task as a revolutionary one. After that time — and the subsequent works of various members of that generation also seem to bear witness to the changes — Argentinian intellectuals seem resigned to the fact that certain aspects of society could not be modified. Across Ibero-America as a whole, the prospects of brilliant economic expansion that were offered by the international division of labour tended to diminish the emphasis on social inequalities, so that the central problem was social order itself, which became the key slogan of end-of-century governments, in Argentina, Brazil and Mexico too. This was not a fashionable phenomenon, not just the result of positivist influence — to the extent that positivism accepted the existing social order and accentuated order so as to make progress — but constituted a true shift in mentality among the Latin American political elites, induced by the prospects of establishing a connection with the world economy and by the very changes in the social structure of those countries.[7]

For Tulio Halperin, this change in the ideology of the Generation of 1837 had operated "bajo el doble estímulo del fracaso de las tentativas de liberar a Argentina y de la frustración de la experiencia revolucionaria francesa de 1848" (under the twofold stimulus of the failure of attempts to liberate Argentina and the frustration of the French revolutionary experience of 1848).[8] Nonetheless, romantic socialism did not disappear from the River Plate after this major shift in the way of thinking of the intellectual elites or with the failure of the 1848 revolutions. It would be precisely those on River Plate soil who were exiles of the counter-revolution, which followed the burgeoning of social and political activity during those years, who would encourage a second phase of social romanticism. These were the men who had been expelled as a result of their political, journalistic or intellectual activity, especially from France or Spain, who arrived in Latin America, not as simple immigrants eager to seek their fortune, but with the hope of realizing their dream of an emancipated social order on a new continent, without the encumbrances of the Old World. They are the ones who, in tribute to the splendid study by Edward H. Carr, we refer to here as the "Romantic Exiles".

Marcelo Segall pointed out that "cada jornada revolucionaria vencida en Europa está señalada por la llegada de exiliados a la América" (every day of revolution in Europe ending in defeat is marked by the arrival of exiles in South America). The first European outlaws were the liberal soldiers defeated by the Holy Alliance, who fought alongside the Creole liberators. The second wave of refugees was made up of the opponents of the July Monarchy, especially teachers and journalists committed to the ideas of romantic socialism. The third wave of political immigration coincided with the 1848 Revolutions. Those who fought in Paris in the Days of the June Uprising were the first to arrive; next came the revolutionaries from Germany, followed by the Mazzinianos (from Italy). Most headed for Chile, others for Uruguay and Brazil. Later, a great number chose to take refuge in Argentina.[9]

Arriving as part of this third wave of exiles were the Chilean, Francisco Bilbao (who arrived in Buenos Aires via France), the Minorcan Bartolomé Victory y Suárez, the Frenchman, Alejo Peyret and the Spaniard, Serafín Álvarez, the main bearers of the books, pamphlets, periodicals, in short, the ideas, of Fourier, Saint-Simon, Leroux, Owen, Cabet, Lamennais, Proudhon and Fernando Garrido; they were the ones who circulated these ideas and texts not only in the intellectual spaces of the elites, but also in the Argentinian workers' movement, which was in the early stages of organization and expression. The new socialist ideas, promoted by the "romantic exiles", would of necessity circulate via the new liberal press media, in the nascent literary circles, the education system and the reformist periodicals, whose thought the Creole elite hegemonized. It was no coincidence, then, that we find Peyret and Bilbao collaborating with the project of President Urquiza, the mutualist Victory y Suárez managing the Rural Society no less, and the Utopian, Serafín Álvarez becoming a trial judge.

Nonetheless, even if the exiles occupied a relatively minor place in the process of "national organization" — as progressive educators, people who drove the new journalism forward, encouraged mutualism and cooperativism, or promoted colonization policies — their socio-political conceptions, midway between "utopian socialism", radical 1848-style democracy and modern socialism, often brought them into conflict with pragmatic liberalism and the incipient conservatism of the dominant Argentinian elite. This elite, for its part, firmly anchored in *realpolitik*, would resist the renewal of the *romantic socialist* ideas that it had largely helped introduce in 1837, to the extent that the whole programme that aspired to link together intense civil associationism with a radical democratic system and a federalist state, would be repeatedly discredited by the elites as "utopic". Those who cultivated these ideas were branded as "dreamers" and "visionaries". The romantics could not but pick up the gauntlet. Francisco Bilbao assumed the risk involved: "Ni el rubor de pasar en política por utopistas puede ser una dificultad a los propagadores de la idea" (Not even the embarrassment of being taken for dreamers in politics can be an obstacle to those who propagate the idea).[10] The romantics, for their part, questioned the "realism" of the liberal generation in power, their "worship of facts" and contempt for the law and their paternalistic attitude towards the citizenry. If the liberals argue that "the citizenry was not mature", the romantics will maintain that the citizenry is being educated in servitude and obedience. The romantics' verdict was, at times, categorical: "Los evangelistas de la víspera son los apóstatas del día siguiente" (Yesterday's evangelists are tomorrow's apostates).[11] Those who sought to renew those ideas, the "romantic exiles" of the River Plate of the 1850s and 1860s would have to challenge their authority and powerful resistance.

These romantic exiles of the River Plate were not utopian in the strict sense of the term, even if metaphorically speaking their audacious ideas ahead of their time have often warranted the term. They did, however, correspond to a certain notion of utopia in the sense of their bold, visionary anticipation of the social and political realities of the morrow. If utopia means, as Quevedo translated it, "there is no such place", the River Plate romantics were utopian in the sense that they did indeed postulate "ideas out of place".[12] In the second half of the nineteenth century, at a time when the countries of Latin America were starting to establish highly centralized presidentialist political systems, were not the efforts of people like Alejo Peyret — seeking to make Proudhon's *Principle of Federation* available to the press of the Argentine Confederation — absolutely "out of place"? Was not Francisco Bilbao's insistence on offering Argentina, his country of asylum, a democratic Constitution that would peacefully integrate immigrants, Creoles and Indians, simply "absurd"? Was not Victory y Suárez being utopian when, in 1863, he postulated that the social rights of workers should be built into the Constitution?

If the romantic socialists were labelled utopian, they accepted it only grudgingly, since they committed themselves in practical terms to the creation of modern Argentina. On the one hand, they supported the liberal nature of the Constitution, the accelerated modernization of civil society, secular educational policies, the inflow of capital, technology and immigrants, and the setting up of a state that would guarantee economic progress and public civil liberties. On the other, however, they would not accept Juan Bautista Alberdi's stagist strategy; they doubted that the "possible republic" was the path that would lead Argentina to the "true republic".

So, by questioning the "republic that actually existed" in the name of the true — democratic and federal — republic, they were impugning the separatism of Buenos Aires behind the nation's back, the presidentialist political system that operated in the guise of federal republicanism, the central government's interventionism over the provinces, the retention of the death penalty, electoral fraud and political assassination, and the policy of conquest of the land and repression of the aboriginal peoples carried out in the name of Civilization.

The "Popular Library" of Victory y Suárez

In the year 1864, nearly three years after the Battle of Pavón, the Argentine nation finally appeared to be unified, although the *montoneras* (insurgent rural militias) in the interior, weapons in hand, resisted the centralist policies of President Mitre. In spite of the conflicts and the fact that the country was preparing for an unpopular war against Paraguay, the national territory was being settled by immigrants at the same time as the telegraph, railways and wire fencing were spreading apace.

Although numerous press media were stirring up political life in the cities, it would be another decade at least before a stable press system would be formed. Even though the printing houses in the largest cities, such as Buenos Aires and Cordoba brought out newspapers, periodicals and books, true publishing houses had still not emerged in the country. The bookshops were stocked mainly with books published in Spain and France. Ernesto Quesada was still bewailing the fact in 1882, that "aquí no hay —con excepción de rarísimos ejemplos— editores que puedan llamarse propiamente así" (here — apart from the rarest of exceptions — there are no publishers who are truly worthy of the name). It was no coincidence, then, that one of the earliest Argentine publishing experiences undertaken in the city of Buenos Aires should bear the title *Biblioteca Popular* (Popular Library). The life and work of Bartolomé Victory y Suárez (1833–1897), whose brainchild the Popular Library was, seem to follow the course taken by the worker's movement; this son of 1840s and 1850s Catalan labour culture — he was a typographer, a member of that nineteenth-century "labour aristocracy" — went into exile in Buenos Aires in 1857, linking up with the incipient mutualist and associationist movement (the Buenos Aires Typographical Society), at the same time as he won a place for himself as publisher, journalist and even administrator among the educated Buenos Aires elite. The product of popular culture, he carved a niche for himself among the elites from which to launch his own publishing project, which he did not hesitate to baptize the "Popular Library". Sergio Pastormerlo observed that when the cycle of popular culture was complete round the 1920s, "the popular" overlapped with "the commercial". In1864, though, "Popular Library" meant only one thing: a collection of books chosen for the instruction of the People.

As far as we know, the Library consisted of only five titles that appeared between September 1864 and 1865. It is worth reviewing them one by one, noticing briefly their graphic characteristics and especially the system of publisher's notes. Once the titles have been introduced, we shall then consider the political and intellectual coherence of the collection.

The Popular Library was inaugurated in 1864 in Buenos Aires, with a thick 19-cm. high volume of a Spanish translation from the French of Alexis de Tocqueville's *Democracy in America*. On its cover, it said:

Popular Library
Democracy in America
by Alexis de Tocqueville
Translation of the tenth French edition
Buenos Aires, Central Press of Bartolomé Victory y Suárez Publisher
454 Calle de Rivadavia 458
1864[13]

This would be the first edition of this classic of nineteenth-century political thought in South America and the second in Latin America. The work had been originally published in two parts in Paris: the first in 1835 (two volumes) and the second in 1840 (another two volumes). In 1836, the first part had already been translated into English and Spanish (in Paris). The first complete translation into Spanish was the one carried out in Spain in 1842; there is a Mexican edition of 1855 and then the one published by Victory y Suárez in Buenos Aires in 1864.

On the first page, dated 1 September 1864, "The Publisher" addresses "the Reader" to justify the publication of this work in the new and dramatic context of the American Civil War (1861–1865). For Victory, this book would not only make it possible for readers in Argentina to understand the dynamics of the Civil War, but also the fact that the abolition of slavery in the North and the triumph of the Union would mean the implantation of a Federal Republic based on democracy, a system that henceforth would have to have "más seguridades de estabilidad en doquiera que exista o se proclame" (more assurances of stability wherever it existed or was proclaimed).

The publisher is careful to clarify on the title page from which edition (the tenth French edition of 1848) the work had been translated, although he does not supply the name of the translator (more than likely because he took a version that already existed, the Spanish one). However, after comparing the Spanish edition of 1854, on which it is based, with various French editions, Victory y Suárez points out that a series of notes, paragraphs and even articles have been suppressed. He considers that "in a free country like this" such mutilation is "unforgivable" and therefore he puts them back in at the end of the work, in an "Appendix", under the heading "Additional notes to the work". He then adds, on page 52 of the Appendix, an article, which "has been copied from the *Spanish democratic socialist* newspaper, *La Discusión* (Madrid, 11 July 1864) and the notes in it have been added by the author of this edition."[14]

A small engraving repeated on all title pages served as the logotype for the "Popular Library". It was an open book, propped upon another one, acting as its lectern; one book on another book.

The second book he issued was:

Popular Library
Second Publication
Communism by Étienne Cabet
Lawyer, public writer, former Member of the Chamber of Deputies, former Public Prosecutor and Founder of the Icarian Colony
Translated and augmented, with quotations and notes integrated in the text.
Buenos Aires, Central Press of Bartolomé Victory y Suárez Publisher
454 Calle de Rivadavia 458
1864[15]

This is the second edition of Cabet's text in Spanish and the first edition on the American continent. Cabet was not unknown in republican and democratic circles. His *Voyage en Icarie* (Voyage to Icaria) had made him very popular from the 1840s onwards, not only in Europe but also throughout America. The *Voyage* was translated into Spanish in Barcelona, in 1848,[16] and copies of that edition were sold in the bookshops of Buenos Aires, Santiago de Chile and Montevideo, and also figured in popular library catalogues at the end of the century. The Catalan followers of Cabet had, in that same year, also translated two of Cabet's pamphlets: "De qué manera soy comunista" (Why I am a Communist) and "Mi Credo Comunista" (My Communist Credo).[17] Cabet's pamphlet, reissued many times in France — entitled *Comment je suis communiste et mon Credo communiste* (Paris, March 1841) in its original edition —

and translated and divided into two parts in 1848 in Barcelona, was now published for the first time in America. Through the edition that Victory entitled *El Comunismo de Esteban Cabet*, the Argentinian reader would have had the first biographical information about the author and more specific details about his doctrine. We shall refer later to the notes that Victory y Suárez inserted into this edition.

His third publication was:

Popular Library
Third publication
Inquisitorial Jurisprudence or Inquisitorial Manual
by Nicholas Eymerich
Augmented edition with notes and historical documents
Translated by José Marchena
Buenos Aires, Central Press of Bartolomé Victory y Suárez Publisher
454 Calle de Rivadavia 458
1864[18]

This was doubtless taken from the abbreviated version of the Inquisitorial Manual, translated and with a commentary by Don José Marchena y Ruiz: *Manual de Inquisidores, para uso de las Inquisiciones de España y Portugal, o compendio de la obra titulada Directorio de Inquisidores, de Nicolás Eymerico. Traducida del francés en idioma castellano por J. Marchena; con adiciones del traductor acerca de la Inquisición de España* (Inquisitorial Manual, for the use of the Inquisitions of Spain and Portugal, or compendium of the work entitled *Directorium Inquisitorium*, by Nicholas Eymerich. Translated from the French into the Spanish language by J. Marchena; with additions by the translator about the Spanish Inquisition).[19]

According to the study by Jaume de Puig i Oliver, this was the eleventh edition of the *Manual* and the first in Spanish. A re-issue (the 12th) appeared in Montpellier in 1821 and a new edition in Madrid in 1822 (the 13th). The fourteenth edition was the one published by Victory y Suárez in Buenos Aires.

The Abbé Marchena (1768–1821) was a Spanish journalist and translator with liberal leanings who lived a good part of his life in exile in France, fleeing, as it happens, from the Inquisition. This enabled him to take part in the events of the revolution of 1789. It is to his translations that we owe the dissemination of authors such as Lucretius, Voltaire, Rousseau, Volney, Molière and Montesquieu in Spanish.

Nicholas Eymerich (Gerona, *c.*1320–4 January 1399) was a Roman Catholic theologian and also Inquisitor General of the Crown of Aragon during the second half of the fourteenth century. His most important work, the *Directorium inquisitorum*, composed in 1376, was a famous manual that initially circulated in manuscript copies and was used for centuries by the inquisitors of Spain and Portugal as a handbook of procedure and penal code. Eymerich set out to define different types of "witchcraft" as heresy, as well as a series of methods for identifying all types of heretical practice. He also described various ways of extracting a confession, including primitive forms of psychological manipulation, not excluding torture, which he himself practised.

It was the freethinkers and then their left-wing heirs who took pains to disseminate this sinister piece of inquisitorial thinking; it was even re-issued in the twentieth century, especially in Spain and Portugal. In 1982, for example, the left-wing publishers, Fontamara of Barcelona, reproduced the old translation by the Abbé

Marchena with the title *Manual de Inquisidores* (Manual for Inquisitors) and a year later the Hispano-Argentinian, Jacobo Muchnik, reissued it in the French translation by Francisco Martín, adding an introduction, a translation from Latin into French and notes by Luis Sala-Molins (*El Manual de los Inquisidores*).

The fourth volume was:

Constitutions
of the
United States
and of the
State of New York
Buenos Aires, Central Press of Bartolomé Victory y Suárez Publisher
N° 454 — Calle de Rivadavia — 458
1864[20]

This was a separate print run of the texts of the two constitutions that Victory had already published as part of the de Tocqueville edition.

Finally, the fifth and apparently last volume, published in 1865, was Ceferino Tresserra's *El derecho democrático* (Democratic Law):

Popular Library
Democratic Law
Buenos Aires, Central Press of Bartolomé Victory y Suárez Publisher
454 Calle de Rivadavia 458
1864[21]

Its author, Ceferino Tresserra y Ventosa (Barcelona, 1830 — La Coruña, 1880), now forgotten, even in Spain, was a democratic politician, a federalist and a mason. A journalist and writer, he was the author of various socio-philosophical novels that enjoyed widespread popularity in his time. He was persecuted for belonging to the Democratic Party in Spain (1849–1869), a group that broke away from the Progressive Party, led by Pi y Margall, who was a contributor to his newspaper, *La Discusión*.

Initially he played a part in propagating the ideas of utopian socialism in Spain, and founded a secret society in 1858 modelled on Italian Carbonarism, which spread throughout Catalonia and Andalusia and was finally dismantled by the police, as a result of which he ended up in prison that same year. In 1859, he published his famous *Cuadro sinóptico del derecho democrático* (Synoptic picture of democratic law) "which was reported and withdrawn by royal order at the behest of the Bishop of Barcelona after the legal term stipulated by the law, despite which it continued to be sold surreptitiously."[22] This experience of being held in jail was the inspiration for *Los misterios del Saladero: Novela filosófico-social* (1860) (The Mysteries of the Saladero: a Socio-philosophical Novel), a work that was very close to those of Eugenio Sue in the depiction of the lower classes and a life of crime. Tresserra was also the author of numerous political works, of which his *Cuadro sinóptico de la democracia española* (Synoptic Picture of Spanish Democracy) was outstanding.[23] According to Rodríguez Solís:

> Thousands and thousands of this sheet were sold, which many considered to be the Gospel of the people, and there was not a social club or workers' society in Catalonia that did not possess one, keeping it hidden from strangers, but was discovered and studied and learnt by heart by all the sons of the people, by all democrats.[24]

During the First Republic, Tresserra was the civil governor of Soria and Palencia. He was behind the creation of a collective volume of Catalan writers entitled *El libro del obrero* (1862) (The Worker's Book), which was a tribute to Anselmo Clave, the founder of the "choral societies" made up of working-class people.

Having now presented the five titles and their authors, the question arises what is the common thread that unifies these titles published as a collection? And an even more meaningful question must be where is the political unity of a series comprising a classic of liberal political thought, such as that by de Tocqueville, the Constitutions of the United States and of the State of New York, an inquisitorial manual, a lampoon by a radical democrat like Tresserra, and some texts by a Utopian like Cabet?

Little is known of the life and thought of Bartolomé Victory y Suárez, his education, readings and correspondence with his old companions from the world of Spanish typography and journalism to enable us to answer these questions definitively. However, a brief review of his life, paying specific attention to his formative period in Barcelona between 1840 and 1850, will provide us with a valuable clue to grasping the intrinsic ideological unity of a project such as the "Popular Library", which might otherwise appear to be an assorted jumble of books and authors.

The life and work of Victory y Suárez

Bartolomé Victory y Suárez was born in Mahon, on the island of Minorca, on 2 August 1833, into a working-class family. We know from his own testimony that when he was "only seven years old, he went to a modest elementary school, run by an old schoolmaster in poor health and rather prone to impatience".[25] In 1846, at the age of thirteen, he left Mahon, never to return; he went to live in Barcelona, where his father, José Victory, a typographer and mason, set up a printing house. There he worked as a typographer and took part, with his father, in the early stages of socialist agitation and Catalan working-class organization, which enjoyed a period of rapid growth between 1854 and 1856. In 1855, he contributed to the Barcelona daily, *El Tribuno*. Enlisted in the ranks of the republicans, father and son took part in the armed resistance to the suppression of the general strike in Catalonia, called in the context of the so-called "reacción del 56" (reaction of '56). It was within the framework of the repression of the workers that José Victory had to dismantle his printing business and take his family into exile, to Argentina, docking in Buenos Aires on 9 October 1857 on board an English merchant ship. In Buenos Aires, Bartolomé set up the "Imprenta Central de B. Victory y Suárez Editor" (Central Press of B. Victory y Suárez, Publisher) at 454/458 Rivadavia Street, from where he launched his "Popular Library".

For reasons of health, he had to leave the printing trade around 1865 and devote himself to journalism. During 1863, he edited the newspaper *El Artesano*, a pioneer of working-class socialist journalism in Argentina, which published texts by other European and Latin American *quarante-huitards* (Forty-Eighters), such as Francisco Bilbao, Amadeo Jacques and Alejo Peyret. Besides wanting to set himself up as a sort of "encyclopaedia" for artisans and small industrialists, he sought at a very early stage to legitimize the "social question" in the eyes of public opinion as part of the Nation building project without presenting them as in conflict with each other. Next he edited three newspapers in quick succession: *El Pueblo Español*, in 1865, *La Iberia*, in 1867 and *Crónica del Progreso*, also in 1867, the last one being masonic in orientation. Between 1867 and 1873, he was a regular contributor to *La República*, the daily newspaper edited by Manuel Bilbao, where he promoted industrial and agricultural development, secular education, the equality of women and democracy.

On the basis of his modern nation project, Victory y Suárez promoted associations for agriculturalists, industrialists, artisans and manual workers. He advocated cooperative and mutualist models for organizations, supported by "protective labour laws". He argued that, as well as the "political rights" enshrined in Argentina's Constitution, it was necessary to fight for "social rights". Social rights would only be included in the Constitution of Argentina in 1949, almost a century after Victory was writing. The democratic electoral reform of the Argentine political regime was approved in 1912, although a stable democratic system only became normal in 1984. How should we understand this and other examples of Victory being ahead of his time?

Our author called for the forces of labour to organize with a view to creating artisans' associations, grouped by skill, trade and profession, a sort of consumer cooperative, supplying associates with goods and services at direct cost and without the need for commercial intermediaries: "Work is the source of all wealth. Let us associate our labour then; let us associate by skill, trade and profession. Let us arrange it so that *money is not indispensable* for supplying us with what we need".[26] Victory y Suárez has been called "utopian" and "a utopian socialist" but, for him, "no hay utopía, todo es realizable, todo es fácil y sencillo, solo requiere fuerza de voluntad" (there is no utopia, everything is achievable, it is all easy and simple, it just requires willpower):

> Let all the skills, trades and professions be constituted as societies. Let the societies pay for only half the labour in money and the other half in vouchers. Let each society have a warehouse for selling articles of their craft and let them hand over to their associates anything they might need in exchange for vouchers at cost price. . . . Let the associates of whatever art or craft be able to dress and be shod, eat, and so on, in exchange for vouchers.[27]

Appealing for vouchers to replace money may seem naive to us today, but we should remember that this first outline of cooperativism in Argentina was written in 1863, when the earliest experiences of the English pioneers were just about starting to be known in other European countries.[28] In that same year, the Spaniard, Fernando Garrido, made those earliest experiences known to the Spanish-speaking world in his *Historia de las asociaciones obreras en Europa* (History of Workers' Associations in Europe).[29]

Several of his friends gathered the journalistic articles together in the 1873 volume, *Cuestiones de interés público* (Questions of Public Interest). In 1870, he figured as a supporting member of the Typographical Society of Buenos Aires and was a regular contributor to its 1870–1871 newspaper *Anales*. According to various sources, it was Victory y Suárez who facilitated the contact between the typographers' mutual benefit society and the *Asociación Internacional de los Trabajadores* (International Workers Association). So it was that the organ of the *Consejo Federal de la Región Española*

(Federal Council of the Spanish Region), *La Fraternidad* (Fraternity) reproduced in its pages one of Victory's articles from *Anales*. In those articles, he defended cooperativist principles, and disseminated the experience of the Rochdale Pioneers.[30]

Finally, turning his full attention to masonic propaganda, he published a series of pamphlets and, between 1873 and 1875, he edited the *Revista Masónica Americana* (the South American Masonic Review). He also contributed to *El Estudiante* (The Student) in 1867, and to the *Anales de Agricultura de la República Argentina* (Agricultural Annals of the Republic of Argentina) between 1873 and 1876. He became secretary of the Asociación Española de Socorros Mutuos (Spanish Association for Mutual Aid) and manager of the Ferrocarril Oeste (Western Railway). In addition to his work as a journalist, he promoted the modernization of agriculture in Argentina; he worked for seven years as a manager of the Sociedad Rural Argentina (Argentine Rural Society) and was treasurer and one of the promoters of the Exposición Nacional de Córdoba (Cordoba National Exposition), inaugurated in 1871. He was the editor of the *Boletín de la Exposición Nacional de Córdoba* (Bulletin of the Cordoba National Exposition), which was published from 1869 to 1873. Emilio Frers said of him "Fue como el alma secreta que ponía en movimiento todos los resortes sociales, desde la administración del local y las exposiciones, hasta la redacción de los *Anales*" (He was like the secret soul that set all the social springs in movement, from administering the premises and exhibitions to editing the *Anales*). In addition, the Province of Buenos Aires commissioned him to produce a Rural Code. He was editor of *El noticiero agrícola* (The Agricultural News Bulletin) in 1880 and the *Revista popular de la Exposición Rural Internacional* (Popular Review of the International Rural Exhibition) between 1884 and 1886. In 1881, he was one of the founders of the Liberal Club. He died in Buenos Aires on 10 May 1897. Some days after his death, on 15 May 1897, the socialist weekly *La Vanguardia* bade farewell to him as follows: "Ha fallecido el domingo pasado en esta Capital el fundador del primer periódico obrero en la Argentina" (The founder of the first workers' newspaper in Argentina passed away last Sunday in this Capital).[31]

The ideological universe of Victory y Suárez

The doyen of historians of anarchism, Diego Abad de Santillán, maintained that Victory y Suárez arrived in Argentina already influenced by the Fourierist-inspired radical democratic ideology of the Spaniard, Fernando Garrido:

> During his time in Spain, Victory y Suárez did not come across the ideas of the International, because they only began to be propagated after he had arrived in Argentina; his humanitarian socialism seems to be influenced by reading the books of Fernando Garrido, especially the study he wrote of the working classes in Europe [*Historia de las asociaciones obreras en Europa*], whose explanation of the Rochdale experience seemed to have interested him very much and influenced his social conceptions.[32]

However, even though the work of Garrido must have been among the required readings of Victory y Suárez, the two volumes of *Historia de las asociaciones obreras en Europa* appeared in Barcelona in 1863, when Victory had already been in Buenos Aires for six years. There is no doubt that shortly afterwards the work arrived in the

Buenos Aires bookshops where Garrido was read. Victory, however, as we know from the most specific biographical sketch that we have of him, was already a political activist in Barcelona in the 1850s and it was his militancy that forced him into exile to Argentina:

> He works as a typographer. He starts to write. In 1855, he is a contributor to El Tribuno in Barcelona. The political position is adopted. Father and son will defend republican ideas with weapons in their hands. And for that reason, both will have to flee Spain, hidden in an English merchant ship that leaves Barcelona for Buenos Aires on 24 August 1857, and where they arrive on 9 October.[33]

Even in its brevity, Olivier's line "Father and son will defend republican ideas with weapons in their hands", gives us a valuable clue that connects directly, as we shall see, with the titles in the "Biblioteca Popular" and the articles in *El Artesano*.

In 1846, Victory y Suárez, father and son, both typographers, arrived in a skilled working-class Barcelona that was a hive of intellectual activity and trade union organization. In the following decade, Catalonia would witness intense socialist propaganda and workers' agitation that would conclude with violent government repression.

According to Francisco Mora, one of the pioneers of socialist and working-class history in Spain:

> In 1840, on the initiative of a worker called Munts, the Society of Handweavers was founded in Barcelona, which in a short space of time, came to have more than two thousand members, and which is regarded as the first Society in Spain resistant to capital. This society also served as a charitable fund for helping disabled workers.[34]

Mora added that "the success enjoyed by this society encouraged workers in different trades to found others of a similar kind and, as mechanical industrialization continued to develop, so working-class societies were also created in Barcelona and other places in Catalonia."[35] So, if societies of weavers, bleachers, dyers, painters, spinners, hosiers, printers, makers of canvas sandals and so on, already existed in Catalonia in the 1840s, it is reasonable to suppose that the young Victory y Suárez and his father were already affiliated to a printers' society.

These working-class societies were particularly receptive to the universe of "associative" ideas. The first attempts to build a federation of societies, such as the "Union of classes", arose in that same decade, and the "Three Classes of Steam" in the following decade. In parallel "se fundaron también algunas Cooperativas de producción y consumo" (some producers' and consumers' cooperatives were also founded) in cities such as Barcelona, Cadiz and Valencia as well as "Sociedades obreras de instrucción y recreo" (working-class societies for instruction and recreation) like the Working Man's Cultural Association in Barcelona, the Reus Reading Centre, and the Cultural Association of the Working Class in Villanueva and Geltrú.[36] It is in this context in Barcelona that the young Victory y Suárez could have learned of the incipient attempts at cooperatives, whose programmes he would later advocate in the Argentinian press.

Later, these working-class societies languished for a few years, to resurface in the mid 1850s at the peak of the workers' conflicts in Catalonia and the start of the government reaction: "Coinciding with the July Revolution in 1854, which gave power to the progressives, the workers' movement was reborn and new working-class societies were created; in Barcelona and other localities in Catalonia, there was stiff opposition to the introduction of automated machines, known as *selfactinas* (from the English term 'self-acting'."[37] A Workers' Committee from the spinning mills, led by José Barceló, ordered a boycott of the machines. In January 1855, a Central Board of Directors of the Working Class was established. In June of the same year, however, the government reacted with vigorous repression, dissolved the workers' associations and worker members were dismissed. On 2 July, for the first time in Spanish territory, a general strike was declared throughout Catalonia. The strike failed, many workers' leaders were thrown into prison and others emigrated. A military tribunal sentenced Barceló to death by garrotting.

Nevertheless, a year later when the textile industrialists demanded of their workers an extra half hour on Saturday in weeks when there was a public holiday, the workers refused to comply and a new conflict broke out (June 1856). To make matters worse, these events coincided with the return of the "moderates" to government in Spain and the beginning of a new period of repression against labour organizations that would be remembered by the workers as "the reaction of '56".[38]

However, in order to obtain a more accurate idea of that incipient labour movement prior to the establishment in Spain of the *Asociación Internacional de los Trabajadores* in 1869, it is necessary to stress it was a communicating vessel with federalist republicanism. Francisco Mora observed that the workers' associations were "hotbeds of revolutionary ideas, which followed with interest, and supported with effectiveness all the preparations for the Glorious Revolution of 1868." Mora added that "the Spanish working class, not of their own initiative, and completely unaware of their destinies, in good faith followed the advice of the bourgeois parties that most flattered their own unconscious aspirations of social improvement".[39]

According to another classic of Spanish socialist history, "la masa obrera de las grandes poblaciones es progresista, primero, y luego demócrata, y más tarde republicana federal" (mass labour in the large towns is first of all progressive, and then democratic, and ultimately federal republican).[40] The progressive, especially the democratic, press — *El Republicano*, *La Soberanía Nacional*, *La Democracia*, and *La Discusión* — was read by the most politicized sector of the Spanish working class, starting with those who created it, the typographers. Morato went on to say that "De todos los hombres de los partidos republicanos, tres conquistan sobre todo la adhesión y el cariño de las masas, y estos hombres son Abdón Terradas, Sixto Cámara y Francisco Pi y Margall" (Of all the men in the republican parties, three in particular have won the support and affection of the masses, and those three men are Abdón Terradas, Sixto Cámara and Francisco Pi y Margall).[41] He explained that:

> It cannot, in all justice, be said that Terradas was a socialist. His newspaper, *El Republicano*, is read by the masses, mainly and almost exclusively in Catalonia, and its editor is held by them in high regard, but at the root of Terradas' doctrines, there is nothing more than a desire for political equality and a longing for taxes to fall on the rich.[42]

With his daily newspaper *La Soberanía Nacional* and pamphlets, Sixto Cámara went a step further, even as far as questioning individual property:

> And the first person to talk about the iniquity of capital was Pi y Margall in 1854, in a manifesto published in *El eco de la Revolución* (The Echo of the Revolution), a document that *Iberia* called incendiary . . . In 1864, when he took over as editor of *La Discusión*, and the historic polemic with Castelar's La Democracia started up, Pi y Margall went so far as to declare land ownership to be illegitimate and anti-social.[43]

Morato concluded that "the republican party deserved, then, — or some of its men did — the considerable popularity that it enjoyed among the workers. As a result, then, when the International was founded, a large number of industrial workers were republicans."[44] This is confirmed by Diego Abad de Santillán, who stated that "the socially inspired republican movement acquired considerable vigour for having been linked to strongly working-class neighbourhoods."[45]

More than half a century after the first attempts at working-class history, academic historiography confirmed this viewpoint:

> The workers and their weak, impotent, sporadic organizations acted in conjunction with the middle class in all the major political upheavals of the nineteenth century. We see them deserting the workplace in 1820, collaborating with the industrialists of Barcelona in the triumph of Riego and constitutionalist liberalism, at the street barricades during the progressive uprising of 1854, and in the innumerable riots, disturbances and uprisings in Madrid, Barcelona and other provincial capitals. They fought in the republican rebellions in Andalusia and Catalonia and contributed decisively to the overthrow of the Bourbons. But they never collectively formulated an opposing plan, nor even one that was just different from the one that the liberals, progressives, democrats or republicans were contemplating.[46]

In parallel to this, both the progressive left wing, as well as the workers' associations thrived on socialist ideas that originated in France. So Pi y Margall would translate *Du Principe Fédératif* (The Principle of Federation) and other works by Proudhon. According to Santillán, "en las sociedades obreras clandestinas se difundieron las doctrinas de Cabet" (the doctrines of Cabet were disseminated in clandestine workers' societies).[47] French socialist ideas also found their way into the incipient workers' press. Narciso Monturiol, influenced by the ideas of the author of *Voyage en Icarie*, brought out the newspaper *La Fraternidad* in Barcelona (1847–1848) and after that *El padre de familia* (1849–1850). Monturiol, together with the novelist and historian, Francisco J. Orellana, translated Cabet's *Voyage en Icarie* into Spanish in 1848 and published it in the pages of *La Fraternidad* before it appeared in book form. In 1850, another Cabetian, the musician, José Anselmo Clavé, founded the "choral societies", made up primarily of workers.[48] According to González del Rivero's unpublished doctoral thesis, Ceferino Tresserra formed part of the *Fraternidad* group, and might even have been the author of some of the translations from French:

> Tresserra, who was seventeen years old at the time, according to Iris Zavala's research, formed part of this small group of utopian socialists. The same researcher also attributes the translation of some of Cabet's writings to our author [Tresserra], but does not specify which ones. Elorza cites Orellana and Monturiol as certainly being the translators of *Viaje por Icaria*, although he mentions another two works by Cabet of 1848, *De qué manera soy comunista* and *Mi credo comunista* but does not tell us anything about their translator, who may have been Tresserra (1970: 102). Among the handful of members of the group of Cabetists that Elorza

does list, Monturiol, Terradas, Orellana, Paula Coello, Clavé and Suñer y Capdevila, there is no mention of our novelist (1970: 101). This list, which he warns was not exhaustive, does however coincide with the core group that shortly afterwards founded the Spanish Democratic Party, and which certainly included Tresserra.

On 12 December 1847, the Catalan Cabetists founded *La Fraternidad*, a newspaper which, from its inauguration, 'presented itself to the reader as the organ of the Spanish socialist party, the repository of social truth, just as Cabet's writings had set it out'.[49] Zavala regards Tresserra as one of the co-founders of the communist publication, but contributes no more details about his participation (1971: 140).[50]

According to Termes, Monturiol and the whole of "Catalan utopianism [was] closely linked to democratic currents of a Jacobin complexion". Monturiol, Clavé, Tresserra and other Cabetists joined the Democratic Party after the failure of the Icarian expedition to the United States in 1848. A left wing calling itself "socialist" — as distinct from the "individualists" — emerged within the party, the same left wing that published the newspaper, *La Discusión*. Termes specifies the scope of the term "socialism" as used by the men at *La Discusión*:

> They wanted to temper social differences, they were trying to mediate in the brutal struggle between the various social classes, and improve the conditions of the worker's life, because they had confirmed the deficiencies of the capitalist system of free competition. In summary, they were in favour of State intervention in the social and economic areas and desired legislation that would protect the worker, regularize the work of men, limit that of women, and suppress that of children.[51]

They were "socialists" in so far as they were opposed to economic individualism. This is therefore a transitional moment between the old craftsman-worker corporations where the boss was still "the master", and the workers' trade unions of the end of the nineteenth century for whom the boss would be "the bourgeois", and the class enemy who enjoyed the usufruct of other people's labour.[52] Even in the 1840s and 1850s, the dynamic of resisting the introduction of modern machinery, the wage claims, as well as an early desire to organize themselves drove the workers into confrontation with the Catalan industrial bourgeoisie, although that trend was counterbalanced by the political, intellectual and moral influence that the republican left wing still exerted over the workers.

The birth of the International in Spain in 1869 and the hegemony of the anarchists inaugurated a new cycle in the Spanish workers' movement, but until that time, workers' associationism was actively politicized; in their ideological universe, Cabet's utopianism combined with reformist and associationist socialism, cooperativism and the cult of study and enlightenment as the stage preceding social emancipation, republicanism, federalism and democracy. This and no other was the political and intellectual universe of Bartolomé Victory y Suárez.

A socialist in the Buenos Aires of 1864

The library of this ideological universe contained publications from the enlightened tradition that the Abbé Marchena had been producing since the beginning of the nineteenth century; it also had the 1843 and 1854 versions in Spanish of de Tocqueville's *Democracy in America*, as well as Spanish translations of the French romantic socialists, such as Lamennais (translated by Larra), Cabet (translated into Spanish by Monturiol) and Proudhon (translated by Pi y Margall). In Spain itself, this library included distinguished authors such as Emilio Castelar, Ramón de la Sagra, Narciso Monturiol, Ceferino Tresserra, Fernando Garrido and Pi y Margall himself.

This was the world of works and authors that Victory was steeped in when he went into exile to Argentina in 1857, at the age of twenty-four, taking with him in all probability his printing trestles with their trays, letter types, borders, fleurons, proofing press, press and typometer, and certainly the pamphlets and books from this socialist republican library, as well as his experience in the emergent Spanish workers' movement.

Now, although contemporary political traditions have been built on the opposition between liberalism and socialism, de Tocqueville, Tresserra and Cabet formed part — not without tensions, of course — of the library of the radical republican of 1848 or 1855. Social processes like the formation of an intensive network of institutions that made up civil society in the United States, which de Tocqueville studied thoroughly, earned their attention, study and plaudits, as did the Icarian movement experiences (also in the United States) or cooperative associations like the Rochdale Equitable Pioneers Society in England. The de Tocqueville that Victory y Suárez was so bent on publishing in 1864 was not so much the theorist of liberalism, which is how we think of him today, but the brilliant political essayist who saw the struggle for equality as the driving force of modern history, who renounced the aristocracy in order to announce the irreversible advent of democracy, the *député* of 1848, the Vice President of the National Assembly in 1849, the opponent of the coup d'état of 1851.

Victory y Suárez, as did, in their way, other Spanish republicans of his time, clung to a socialism that was defined by a will to form associations as a counterweight to the risks of free market individualistic liberalism. Why then would he publish a "communist" like Cabet in Buenos Aires in 1864? A "Foreword" by the publisher, signed in Buenos Aires on 1 November 1864, begins with what, at first sight, appears to be an implicit reference to the *Communist Manifesto*, "Communism has been one of the spectres used to frighten weak spirits through ignorance or self-interest."[53] Victory wanted Communism to be recognized as one of the "great conceptions of the world" aimed at remedying social evils:

> Communism, however, is one of the endless theories born of the study of social hardships and of the means that reason teaches in order to eradicate them. The system may be more or less complete or more or less in accordance with the nature of Man; so, as long as these two points are debated, Communism will remain on the high ground occupied by the great conceptions of thought; it will not descend to the level of farce as long as it is not shown to be unattainable or unstable.[54]

The publisher pointed out the practical projection of what seemed to be a nebulous utopia, by stressing the still experimental nature of the Icarian proposal:

> Cabet has proved that Communism can be a practical reality. Whether organizing it would offer a guarantee of long-term stability can neither be confirmed nor denied, because

> neither one nor a hundred frustrated attempts implies that the stability of Communism is impossible.[55]

Once again, Victory comes to the defence of what, from the perspective of order, is discredited as "anarchic" or "utopian".

> The hypothesis that the *Comunismo de Cabet* is unworthy of occupying the intelligence of man simply because its first trial was a failure, or because such and such people of intelligence have argued against the possibility of it being realized, and have in addition fought against it because they consider it anarchic and a disruption of order would be offensive and would not stand up to examination either; furthermore, what new idea is not anarchic before it becomes a reality? What reform does not entail disorder? What is the most insignificant governmental action that does not upset the established order that it acts upon? Can anyone cite a single step forward taken by humanity on the path to progress that has not first stirred things up and then caused disruption before the idea that germinated as the precursor of progress later became established in practical life? Is not human life one continuous revolution?[56]

Victory y Suárez here is echoing the final page of Cabet's brief treatise, in which the Frenchman argues that each and every one of humanity's innovations is always dismissed as "impossible".

Victory added an appendix to Cabet's text with the title "Una explicación" (An explanation). In this text, he reproduced an article of his own, "La Verdad social" (The Social Truth), which first appeared on 9 May 1863 in his newspaper *El artesano*, where he had professed his socialist faith. In a brief page of particular interest, however, he justifies the republishing of the article in that text, in order to establish the difference between Cabet's communism and his own non-communist, but democratic socialist conception of it. Victory clarifies his view:

> I found out that I have been labelled a *Communist* because I have published this book. This label would not concern me at all if I was not already certain that it was malicious and involved the idea of discrediting me in the eyes of a certain class of person. I say this because those who call me a communist harbour the twisted belief that Communism is the social religion of lost people . . . the riffraff! I believe therefore that I have the duty to demonstrate that I am not a Communist, but at the same time, faced with such an outrage to this System, I put forward a reading of this same book and a reminder of the communist teachings of Christ and his Apostles. I am not a Communist, but I am a socialist; I am not a supporter of Cabet's system, and even less of the monastic Communist system, but I am of another one.[57]

It is very unlikely that Victory y Suárez fully shared Cabet's programme, particularly his conception of a "community of goods", whether it was the monastic version or a modern one. In the same book published by Victory, Cabet made a distinction between the "community of goods" based on industrial progress and mechanization, which he postulated, and the "equality of poverty" that would turn men into "monks" and make society a "convent".[58]

Victory had proclaimed that he was a socialist a year before, in the article that he would subsequently transcribe as proof of his beliefs.

> I demonstrated it in *El Artesano*, when I was its editor. I demonstrated I was a socialist precisely by writing an article with the objective of separately defining two ideas that are generally confused in political discussions and which confusion necessarily stems from empty words, humbug, chaos. I am referring to the separate definitions for the political idea and the social idea, the first represented by the word republic, and the other by the word Democracy.[59]

The article did not go as far as Cabet, as far as the foundation of "Icarias" or the "community of goods", but it was undoubtedly inspired by his principles and values. Let us not forget that, in the end, Cabet's communism was not revolutionary but reformist and democratic; it was not classist, but addressed the complete spectrum of "the popular".

The utopia of social rights

In his journalistic campaign of the 1860s, Victory introduced the distinction between *republic* and *democracy*, political constitution and social constitution, political rights and social rights. The republic, he tells us, has conquered political rights: the right to vote, to free expression, to religious freedom. Democracy came and added social rights: the right to a State education that is secular and free of charge; the right to an old-age or invalidity pension; the right to housing; the right of the immigrant worker to receive land and tools of work from the State; the right to free justice for the worker; the right to protected industry, and so on. The first, the Republic, produces individuals; democracy, persons.

The self-taught Minorcan typographer here attains the stature of a social constitutionalist: "Si queremos ser, pues, consecuentes, si queremos dar al pueblo lo que es el pueblo, es preciso que nuestra República sea democrática" (If we want to be consistent then, if we want to give the citizenry what the citizenry is, it is necessary for our Republic to be democratic). For this, Victory y Suárez postulated — in 1863! — adding social rights to the Constitution of Argentina. He was aware that his proposal was a bold move and ahead of its time, and there was no naivety in the way he set it out:

> We know that if tomorrow we were to call for the declaration of social rights [to be included] in the constitution of the republic, we would be subjected to ridicule, at the very least; we know that if we were to ask for this, accompanied by all those who believe it just and necessary, we would risk being rewarded by having our political privileges taken away from us, but . . . we have made it our duty to record the rights of the people in writing, and we are doing that here. Time will do the rest.[60]

Utopian? Maybe, but only in so far as the realities of today are the utopias of yesterday:

Utopia has recognized the human personality and declared it sacred; utopia allows us to cross thousands of leagues in just a few hours; utopia enables us to converse in thirteen minutes from one world to another; utopia has changed the face of laws, customs, and even religions themselves.[61]

Notes

1 Marta Bonaudo, ed., *Liberalismo, Estado y orden burgués*, vol. 4 *(1852–1880)* (Buenos Aires: Sudamericana, 2003), pp. 13 and ff.

2 *Ibid.*, pp. 16–17: "La Constitución sancionada en 1853 afirmó el criterio de la soberanía del pueblo y colocó a la figura del ciudadano en la base de toda legitimidad. Sin embargo, a partir de las prácticas de poder concretas que emergieron y se desarrollaron durante estos treinta años, las elites violaron sistemáticamente aspectos fundamentales del ideario que estaba en la base de su legitimidad, lo que no impidió la consolidación de una trama de legalidad que apuntaló la construcción del Estado-nación."

3 Jürgen Habermas, *Historia y crítica de la opinión pública* (Barcelona: Gustavo Gili, 1982).

4 Pilar González Bernaldo del Quirós, *Civilidad y política en los orígenes de la Nación Argentina. Las sociabilidades en Buenos Aires. 1829–1862* [1999] (Buenos Aires: Fondo de Cultura Económica, 2007), pp. 174 and ff; Roberto Di Stefano, "Orígenes del movimiento asociativo: de las cofradías coloniales al auge mutualista", in *De las cofradías a las organizaciones de la sociedad civil. Historia de la iniciativa asociativa en la Argentina. 1776–1990*, edited by Roberto Di Stefano, Hilda Sabato, Luis Alberto Romero and José Luis Moreno, compiled by Elba Luna and Élida Cecconi (Buenos Aires: Gadis, 2002).

5 By 1855, Héctor Varela's *Almanaque comercial y guía de forasteros para el Estado de Buenos Aires, publicado por La Tribuna* confirmed the existence of eleven bookshops, ten printing houses, and two lithographic printers. In 1858, this figure, according to data from the Register of Printing and Publishing Activity in the City of Buenos Aires had increased to fifteen bookshops, twelve printing houses and two lithographic printers. By the 1870s, using an analysis based on advertisements appearing in the press, eighteen bookshops were accounted for, concentrated in and around the Buenos Aires City Hall. Alejandro Eujanian, "La cultura: público, autores y editores", in *Liberalismo, estado y orden burgués, 1852–1880*, edited by Marta Bonaudo, vol. 4 of the Nueva Historia Argentina collection (Buenos Aires: Sudamericana, 1999), p. 559.

6 Oscar Oszlak, *La formación del Estado argentino* (Buenos Aires: Belgrano, 1982).

7 José Carlos Chiaramonte, *Formas de sociedad y economía en Hispanoamérica* (Mexico: Grijalbo, 1984), pp. 63–64: "En Argentina, luego de la llamada generación del '37, la misión de los intelectuales deja de pensarse en términos de revolución. Los escritos de Esteban Echeverría y algunos de sus contemporáneos son los últimos en que los intelectuales del siglo XIX conciben su tarea como revolucionaria. A partir de allí, y los trabajos posteriores de varios de los integrantes de esa generación también testimonian el cambio, los intelectuales argentinos parecen resignarse a aceptar ciertos aspectos de la sociedad como inmodificables. En el conjunto de Iberoamérica, las perspectivas de brillante expansión económica que ofrece la división internacional del trabajo tiende a disminuir el énfasis en las desigualdades sociales, de manera que el problema central será el del orden social, que llegará a constituirse en la consigna central de gobiernos finiseculares, tanto en Argentina como en Brasil o México. Este fenómeno no es una simple moda, no es solo fruto de la *influencia* positivista —en cuanto el positivismo acepta el orden social existente y pone el acento en el orden para el progreso—, sino que constituye un verdadero cambio de mentalidad en las élites políticas latinoamericanas, inducido por las perspectivas de conexión con la economía mundial y por los propios cambios en la estructura social de estos países."

8 Tulio Halperin Donghi, "Prologue" to Domingo F. Sarmiento, *Campaña del Ejército Grande Aliado de Sudamérica* (Mexico/Buenos Aires: Fondo de Cultura Económica, 1958), p. xxvii.

9 Marcelo Segall, "En Amérique Latine: Développement du Mouvement Ouvrier et Proscription", *International Review of Social History* 17, 1 (1972): 325– 326.

10 Francisco Bilbao, *Revista del Nuevo Mundo* (Buenos Aires) no. 2 (July 1857): 44.

11 Alejo Peyret, *Cartas sobre la Intervención del Gobierno Federal a la Provincia de Entre Ríos.*

Por un estranjero (Buenos Aires: Imprenta Litografía y Fundición de Tipos de la Sociedad Anónima, 1873), p. ii.

12 Roberto Schwarz, "Las ideas fuera de lugar" (1973), in *Absurdo Brasil. Polémicas en la cultura brasileña,* edited by Adriana Amante and Florencia Garramuño (Buenos Aires: Biblos, 2000).

13 " Biblioteca Popular / *La democracia en América* / por Alejandro de Tocqueville / Traducción de la X edición francesa / Buenos Aires, Imprenta Central de Bartolomé Victory y Suárez Editor / 454 Calle de Rivadavia 458 / 1864." A 19-cm. volume of 535 + XLII + 56pp. The figures in Arabic numerals correspond to the pages as numbered in the original, and the Roman numeral to the number of paratexts. This book has three sets of figures probably because the editor decided to add an appendix at the last minute, which he renumbered in Roman numerals.

14 "se ha copiado del periódico *democrático socialista español La Discusión* (Madrid, 11 julio 1864) y las notas que lleva han sido añadidas por el autor de esta edición" (The italics are in the original).

15 "Biblioteca Popular / Segunda publicación / *El Comunismo* / de Esteban Cabet / Abogado, escritor público, exdiputado, exprocurador general / y Jefe de la Colonia Icariana. / Traducido y aumentado con citas y notas integradas en el texto. / Buenos Aires, Imprenta Central de Bartolomé Victory y Suárez Editor / 454 Calle de Rivadavia 458 / 1864" (A 157-page volume).

16 Étienne Cabet, *Viage por Icaria,* trans. Francisco J. Orellana and Narciso Monturiol (Barcelona: Imprenta y Librería Oriental, 1848).

17 Antonio Elorza, *Socialismo utópico español* (Madrid: Alianza Editorial, 1970) p. 102.

18 "Biblioteca Popular / Tercera publicación / *Jurisprudencia inquisitorial o Manual de inquisidores* / de Nicolas Eymerich / Edición aumentada con notas y documentos históricos. / Traducido por José Marchena / Buenos Aires, Imprenta Central de Bartolomé Victory y Suárez Editor / 454 Calle de Rivadavia 458 / 1864" (A 192-page volume).

19 Montpeller [sic], Imprenta de Feliz Aviñón, 1819: XII + 159 pages. The figure in Roman numerals corresponds to the paratexts. There are examples with a title page from Bordeaux.

20 "*Constituciones / de los / Estados Unidos / y del / Estado de Nueva-York* / Buenos Aires, Imprenta Central de Bartolomé Victory y Suárez Editor / N° 454 — Calle de Rivadavia — 458 / 1864" (A volume of forty-four pages).

21 "Biblioteca Popular / Ceferino Tresserra / *El derecho democrático* / Buenos Aires, Imprenta Central de Bartolomé Victory y Suárez Editor / 454 Calle de Rivadavia 458 / 1865" (A volume of sixteen pages).

22 Fernando Garrido, *Historia del reinado del último Borbón de España. De los crímenes, apostasías, opresión* (Barcelona: Librería de Antonio de San Martín, 1869), part 3, p. 861: "que fue denunciado y recogido de real orden, a la instancia del Obispo de Barcelona después del término legal que marcaba la ley, pero a pesar de esto se vendía subrepticiamente."

23 *Porvenir de las asociaciones de la clase obrera: origen y estado actual de la cuestión del trabajo en Cataluña* (Barcelona: Imprenta de Narciso Ramírez, 1855); *Tablas del derecho democrático* (Barcelona: Manero, 1859); *Cuadro sinóptico de la democracia española* (Barcelona: Librería de Salvador Manero, 1865). (This took the form of a folding 73 x 53 cm sheet of paper with portraits of Confucius, Plato, Christ, Gutenberg, Galileo and Franklin, a sort of genealogy of radical thought, which was stuck on the walls of houses, shops and businesses. No copies are preserved in Argentinian libraries, although it is cited in various works); *Cuadro sinóptico del Derecho democrático* (Banned by the Bourbon Governments) (Barcelona: Librería de Salvador Manero 1869); *¿Los anarquistas, los socialistas y los comunistas son demócratas?* (Barcelona: Librería de Salvador Manero, 1861); *Catecismo democrático-republicano* (Madrid: Manuel Galiano, 1868); *Catecismo de la Federación Republicano-democrática* (Madrid: Molino y Cía, 1870); *¿Hay Dios? Estudio crítico-filosófico de la cuestión de*

las cuestiones según el racionalismo puro 3rd ed. (Barcelona: Est. Tipográfico de Narciso Ramírez y Compañía, 1873).

24 Pablo Ramos González del Rivero, *Las armas de la república europea de las letras: propaganda y pedagogía democráticas en la narrativa popular decimonónica: Ceferino Tresserra* (PhD diss. Autonomous University of Madrid, 2008), p. 88: "Miles y miles se vendieron de esta hoja que muchos consideraron como el Evangelio del pueblo, y no había casino ni sociedad obrera de Cataluña que no lo poseyera manteniéndole oculto para los desconocidos, pero descubierto y estudiado y aprendido de memoria por todos los hijos del pueblo, por todos los demócratas."

25 "solamente siete años de edad, frecuentaba una humilde escuela dirigida por un pobre dómine anciano, achacoso y algo dado a la impaciencia."

26 B. Victory y Suárez, "Asociémonos", *El Artesano* no. 9, 25 April 1863, 2: " "El trabajo es la fuente de toda riqueza. Asociemos, pues, el trabajo; asociémonos por artes, oficios y profesiones. Arreglémonos de modo que *el dinero no sea indispensable* para proporcionarnos lo que necesitamos."

27 *Ibid.* "Que todos los artes, oficios y profesiones se constituyan en sociedad. Que las sociedades paguen solamente la mitad del trabajo en dinero y la otra mitad en bonos. Que cada sociedad tenga un almacén de venta de los artículos de su arte y entregue a los asociados todo lo que precisen en cambio de bonos al precio de costo . . . Que recíprocamente los asociados de cualquier profesión o arte que sean, puedan vestirse, calzarse, comer, etc. en cambio de bonos."

28 Although very much focused on experiences in England, Germany, France and Belgium, the classic study by Gromoslav Mladenatz, *Historia de las doctrinas cooperativas* (Buenos Aires: Intercoop, 1969) offers a concise history of European cooperativist thinking.

29 The classic work by George Holyoake, *Self-help by the People: The History of Co-operation in Rochdale (The Society of Equitable Pioneers)* had been published in part in the *Daily News* in London in 1857 and appeared a year later as an independent volume. Even though it was immediately reprinted in New York and in various reissues in England, it took several years before it was disseminated in other languages. The first French edition was in 1881, although various advance passages in *Le Progrès* in Lyon, achieved a very wide circulation. In the Spanish-speaking world, the first and main disseminator was Fernando Garrido, who included Holyoake's work in his two-volume, 443-page work *Historia de las asociaciones obreras en Europa ó Las clases trabajadoras regeneradas por la asociación* (Barcelona: Salvador Manero, 1864).

30 Ricardo Falcón, *Los orígenes del movimiento obrero (1857–1899)* (Buenos Aires: CEAL, 1984), Horacio Tarcus, *Marx en la Argentina. Sus primeros lectores obreros, intelectuales y científicos. 1870–1910* (Buenos Aires: Siglo XXI, 2007).

31 For Bartolomé Victory y Suárez, see Ángel Giménez, *Precursores del socialismo en la República Argentina* (Buenos Aires: La Vanguardia, 1917); Ernesto Olivier, "Bartolomé Victory y Suárez, propagador de la cooperación en la Argentina", in B. Victory y Suárez, *Cuestiones de interés público. La asociaciones cooperativas. 1870* (Buenos Aires: Círculo de Estudios Cooperativos de Buenos Aires, 1970); Juan Antonio Solari, "Un precursor", in *Recuerdos y anécdotas socialistas* (Buenos Aires: La Vanguardia), 1976, p. 50; Horacio Tarcus, *Marx en la Argentina.*

32 Diego Abad de Santillán, *El movimiento anarquista en la Argentina. Desde sus comienzos hasta el año 1910* (Buenos Aires: Argonauta, 1930), pp. 12–14: "Victory y Suárez no conoció en España las ideas de la Internacional, que comenzaron a propagarse después de estar él en la Argentina; su socialismo humanitario parece influenciado por la lectura de los libros de Fernando Garrido, sobre todo del estudio escrito por este sobre las clases obreras de Europa, cuya exposición de la experiencia de Rochdale parece haberle interesado mucho e influenciado sus concepciones sociales."

33 Olivier, "Bartolomé Victory y Suárez, propagador de la cooperación", p. 6: "Trabaja

como tipógrafo. Empieza a escribir. En 1855 colabora en *El Tribuno* barcelonés. La posición política está tomada. Padre e hijo defenderán las ideas republicanas con las armas en la mano. Y por ese motivo ambos deberán huir de España escondidos en un barco mercante inglés que abandona Barcelona el 24 de agosto de 1857 rumbo a Buenos Aires, adonde llegan el 9 de octubre."

34 Francisco Mora, *Historia del Socialismo Obrero Español* (Madrid: Imprenta de I. Calleja, 1902), pp. 44. "En 1840 se fundó en Barcelona, por iniciativa de un obrero llamado Munts, la Sociedad de Tejedores a la mano, la cual en poco tiempo llegó a contar más de 2000 socios, y que se considera como la primera Sociedad de resistencia al capital constituida en España. Esta sociedad tenía también el carácter de Montepío para el socorro de inválidos del trabajo."

35 *Ibid.*, p. 45: "El éxito alcanzado por esta sociedad animó a los obreros de otros oficios a fundar otras de la misma especie, y a medida que la industrialización mecánica fue desarrollándose, fueron también creándose Sociedades obreras en Barcelona y otros puntos de Cataluña".

36 *Ibid.*

37 Joseph Termes, *Anarquismo y sindicalismo en España. La Primera Internacional (1864–1881)* (Barcelona: Ariel, 1972), p. 20: "Coincidiendo con la revolución de julio de 1854, que dio el poder a los progresistas, el movimiento obrero renació y se crearon nuevas sociedades obreras; se produjo en Barcelona y otras comarcas catalanas una aguda oposición a la introducción de máquinas automáticas denominadas selfactinas."

38 Termes, *Anarquismo y sindicalismo en España*, p. 22.

39 Mora, *Historia del Socialismo Obrero Español*, p. 46: "semilleros de ideas revolucionarias, en los cuales se seguía con interés y se apoyaba con eficacia todos los preparativos de la Revolución del 68 . . . La clase obrera española, sin propia iniciativa, y desconociendo por completo sus destinos, seguía de buena fe los consejos de los partidos burgueses que más halagaban sus inconscientes aspiraciones de mejora social."

40 Juan José Morato, *El Partido Socialista Obrero* [1918] (Madrid: Ayuso, 1976), p. 41. The author is referring to the trend towards the division and recomposition of the Spanish political parties, the Progressive and the Democratic.

41 *Ibid.*

42 *Ibid.*: "Con justicia no puede decirse que Terradas fuese socialista. Su periódico *El Republicano* es leído de las masas, principal y casi exclusivamente en Cataluña, su director es estimado de ellas, pero en el fondo de las doctrinas de Terradas no hay más que un deseo de igualdad política y el anhelo de que los tributos recaigan sobre los ricos."

43 *Ibid.* "Y quien primero habla de la iniquidad del capital es Pi y Margall el año 54, en un Manifiesto publicado en *El eco de la Revolución*, documento que *Iberia* calificó de incendiario . . . En 1864, cuando se encarga de dirigir *La Discusión* y se entabla la histórica polémica con *La Democracia* de Castelar, Pi y Margall llega a declarar ilegítima y antisocial la propiedad de la tierra."

44 Morato, *El Partido Socialista Obrero*, pp. 42–43: "Merecía, pues, el partido republicano —o lo merecían algunos de sus hombres— la considerable popularidad de que gozaba entre los obreros. Tenemos, pues, que al fundarse la Internacional gran número de obreros industriales eran republicanos."

45 Diego Abad de Santillán, vol. 1 of *Contribución a la historia del movimiento obrero español* (Mexico: Cajica, 1968), p. 65: " El movimiento de inspiración republicana y social adquirió un vigor considerable por haberse vinculado con fuertes núcleos obreros."

46 Termes, *Anarquismo y sindicalismo en España*, p. 27: "Los obreros y sus débiles, impotentes y esporádicas organizaciones actuaron conjuntamente con la clase media en todas las grandes conmociones políticas del siglo XIX. Les vemos desertar del trabajo en 1820 colaborando con los industriales de Barcelona en el triunfo de Riego y del liberalismo constitucionalista; en las barricadas callejeras durante el alzamiento progresista de 1854; y

en los innumerables motines, asonadas y alzamientos de Madrid, Barcelona y otras capitales. Lucharon en las sublevaciones republicanas de Andalucía y de Cataluña y colaboraron decididamente en el destronamiento de los Borbones. Pero nunca formularon colectivamente un plan opuesto, ni tan solo distinto, del que planteaban liberales, progresistas, demócratas o republicanos."

47 Abad de Santillán, *Contribución a la historia del movimiento*, p. 64.

48 Termes, *Anarquismo y sindicalismo en España*, p. 17.

49 Iris M. Zavala, *Masones, comuneros y carbonarios* (Madrid: Siglo XXI, 1971), p. 102.

50 González del Rivero, *Las armas de la república europea de las letras*, pp. 42 and 43: "Tresserra, que por entonces contaba diecisiete años, según los trabajos de Iris Zavala, formó parte de este reducido núcleo de socialistas utópicos. La misma investigadora atribuye a nuestro autor la traducción de algunos escritos de Cabet, aunque no especifica cuáles. Elorza cita a Orellana y Monturiol como traductores seguros de *Viaje por Icaria*, aunque menciona otras dos obras de Cabet de 1848, *De qué manera soy comunista* y *Mi credo comunista*, y no informa sobre su traductor, que quizá fue Tresserra (1970: 102). Entre el puñado de miembros del grupo de cabetistas que enumera el mismo Elorza, Monturiol, Terradas, Orellana, Paula Coello, Clavé o Suñer y Capdevila, no aparece nuestro novelista (1970: 101). En todo caso, este elenco, del que advierte de su carácter no exhaustivo, coincide con el núcleo que muy poco después fundará el Partido Demócrata español y en el que es seguro que se halló Tresserra.
El 12 de diciembre de 1847 los cabetistas catalanes fundaron *La Fraternidad*, periódico que desde su aparición "se presentó al lector como órgano del partido socialista español, depositario de la verdad social, tal y como la misma quedara expuesta por los escritos de Cabet". Zavala considera a Tresserra uno de los cofundadores de la publicación comunista, aunque no aporta más datos sobre su participación (1971: 140)."

51 Termes, *Anarquismo y sindicalismo en España*, pp. 16–17, 27–28: "que deseaban suavizar las diferencias sociales, que pretendían mediar en la brutal pugna entre las diversas clases sociales, y mejorar las condiciones de vida del obrero, porque habían comprobado las deficiencias del sistema capitalista de libre competencia. En síntesis, eran partidarios de la intervención del Estado en el campo social y económico, y deseaban una legislación que amparase al obrero, que regularizase el trabajo de los hombres, limitase el de las mujeres y suprimiese el de los niños."

52 Morato, *El Partido Socialista Obrero*, p. 40.

53 B. Victory y Suárez, "Advertencia" [Foreword] to *El comunismo de Esteban Cabet* (Buenos Aires: Imprenta Central de Bartolomé Victory y Suárez Editor, 1864), p. 1: "El comunismo ha sido uno de los fantasmas con que se ha querido aterrar a los espíritus débiles por ignorancia o por egoísmo." If Victory knew the *Communist Manifesto*, it was at a later date, after the Spanish edition of 1886 or the Argentinian one of 1893. The "spectre of communism" is not the creation of Marx and Engels, but, as Gareth Stedman Jones has shown, was already being raised by the literature of reaction in the 1840s, which Victory knew very well, see Gareth Stedman Jones, *El Manifiesto comunista de Karl Marx y Friedrich Engels* (Madrid: Fondo de Cultura Económica, 2007), p. 19.

54 *Ibid.*: "El Comunismo es, sin embargo, una de las infinitas teorías nacidas del estudio de las miserias sociales y de los medios que la razón enseña para extirparlas. El sistema puede ser más o menos completo y más o menos conforme con la naturaleza del hombre; siempre que se discutan, pues, estos dos puntos, el Comunismo conservará el puesto elevado que ocupan las grandes concepciones del pensamiento; no se rebajará al nivel de la farsa, mientras no se demuestre que es irrealizable y que no es estable."

55 *Ibid.*: "Cabet ha probado que puede el Comunismo ser un hecho práctico. Que su organización ofrezca garantías de estabilidad por mucho tiempo, es lo que ni se puede negar ni se puede afirmar, porque la frustración de uno ni cien ensayos, no implica la imposibilidad de la estabilidad del Comunismo."

56 *Ibid.*: "Ofensiva y raquítica por demás sería la hipótesis de que el *Comunismo de Cabet* es indigno de ocupar la inteligencia del hombre por el solo hecho de haber fracasado en su primer ensayo, o porque tales o cuales personas de talento han argumentado en contra de la posibilidad de su realización y lo han combatido además por considerarlo anárquico y perturbador del orden: mas, ¿cuál es la idea nueva que no es anárquica antes de ser un hecho? ¿cuál es la reforma que no entraña desorden? ¿Cuál es la más insignificante acción gubernativa que no trastorna el orden establecido sobre el que acciona? ¿Se puede citar un solo paso dado por la humanidad en la senda de su progreso, que no haya agitado primero, perturbado luego, para establecer después en la vida práctica la idea que germinaba como precursora del progreso? ¿La vida humana no es una revolución continua?"

57 B. Victory y Suárez, "Una explicación", in Cabet, *El comunismo*, p. 151: "He sabido que se me ha calificado de *Comunista* por el hecho de publicar este libro. Ninguna importancia daría yo a esta calificación si no me constara que es maliciosa y que envuelve la idea de hacerme desconceptuar ante cierta clase de personas. Digo esto, porque los que me llaman comunista, abrigan la torcida creencia de que el Comunismo es la religión social de la gente perdida... de la chusma! Por esto me creo en el deber de manifestar que no soy Comunista, pero al mismo tiempo, contra semejante ultraje a este Sistema, opongo la lectura de este mismo libro y recuerdo las doctrinas comunistas de Cristo y de sus Apóstoles. No soy Comunista, pero soy *socialista*; no soy partidario del sistema de Cabet y mucho menos del sistema Comunista monacal, pero lo soy de otro."

58 Cabet, *El comunismo*, p. 10.

59 Victory y Suárez, "Una explicación", in Cabet, *El comunismo*, p. 151. "Lo he manifestado en *El artesano*, cuando estaba bajo mi dirección. Precisamente manifesté ser socialista escribiendo un artículo que tenía por objeto definir separadamente dos ideas que generalmente se confunden en las discusiones políticas y de cuya confusión nace necesariamente la palabrería hueca, la farsa, el caos. Me refiero a la definición separada de la idea política y de la idea social: la primera representada por la palabra república, y la otra por la palabra Democracia."

60 B. Victory y Suárez, "La verdad social", *El Artesano*, no. 11, 9 May 1863, pp. 1 and 2. Reproduced by the author as an appendix in *El comunismo de Esteban Cabet* (Buenos Aires: Imprenta Central B. Victory y Suárez Editor, 1864), pp. 152–157: "Sabemos que si mañana aconsejáramos la declaración de un derecho social en la constitución de la república, seríamos cuando menos motivo de ridículo; sabemos que si lo pidiéramos acompañados de todos los que lo creen justo y necesario, nos expondríamos a recibir en premio un desafuero político, pero... nos hemos impuesto el deber de consignar el derecho del pueblo, y lo consignamos. El tiempo hará lo demás."

61 *Ibid.*: "La utopía ha hecho consagrar el reconocimiento de la personalidad humana; la utopía nos hace cruzar millares de leguas en pocas horas; la utopía nos hace conversar en trece minutos desde uno a otro mundo; la utopía ha cambiado la faz de las leyes, de las costumbres y hasta de las mismas religiones."

7 The Utopia of the "Latin Race": Michel Chevalier, Victor Considerant and Public Debate in Spain Concerning the Intervention in Mexico (1861–1867)

Nere Basabe

France responded to the announcement made by the liberal government of Benito Juárez in 1861 that Mexico's foreign debt payments had been suspended by leading a military expedition to that country, supported by British and Spanish troops, which ended in armed conflict. This chapter in history, known as the Second French Intervention in Mexico, culminated in the establishment of the Second Mexican Empire under Archduke Maximilian of Habsburg and generated heated controversy in French and Spanish public opinion. Set in the context of the new race for empire, an attempt was made to justify this interference with a rhetorical discourse that appealed to the supposed ideals of brotherhood and revolved around a central concept that quickly spread: the Latin race. The aim of this chapter is to explore the more utopian dimensions of the concept of "the Latin race" by means of an analysis of its meanings and the ways it was employed in contemporary intellectual debate, first of all in France, where Saint-Simonian and Fourierist authors launched into arguments over the objectives of the intervention and the destiny of the so-called Latin race, then in the press and in Spanish parliamentary debate. It is of interest because, apart from its ideological justification of imperial domination, the discourse on the Latin race harboured many classic utopian aspirations: as a project for social regeneration and a new form of universal association.

America as an "irrevocable decree of destiny"

As Edgar Quinet would recall, "Fourier and the other visionaries taught us that Mexico is the natural capital of the world."[1] The Fourierist newspaper was called *Le Nouveau Monde*, and with good reason, for the New World as a whole, and not only Mexico, had always been the territory par excellence for utopia, which had finally ceased to be a "no-place".

It had been such since the age of the discoveries and the conquistadors, when the American continent was introduced as a "large blank slate waiting to be filled with meaning,"[2] or a "gift to human life", according to the Enlightenment definition by Antonio de Alcedo in his famous geographical and historical dictionary of the West Indies, the *Diccionario geográfico-histórico de las Indias Occidentales* (1786–1789).[3] By the nineteenth century, the consolidation of the various processes of emancipation and the birth of the new nations turned America once more into a continent where anything was possible. The space was now being opened up to republics in which fairer models of society could be constructed, with vast territories waiting to be populated. It became a privileged field for experimentation and, consequently, the driving force for criticism of the Old World with a strong performative mission, in which the old was opposed to the new, the past to the future, and discourse became prophetic and emancipatory. In the words of the Mexican historian, Guillermo Zermeño, "America refers to a land, a geography, a climate, and a demand for a future in the face of 'Old' Europe."[4] Once the new theories of Rousseau's good savage had triumphed over the discourse of inferior civilizations, there was a shift towards acclaim for the virtues of America, especially its unprecedented uniqueness, and a search for a form of civilization appropriate to the "American genius". The theories conceived in the Old World would therefore be unsuitable for the needs of the New World, as was repeatedly stated by Juan Manuel de Rosas, leader of the regime in Buenos Aires (1835–1852), who developed a project for an "American System".[5] The same assertion was made everywhere: "Spanish America is unique; its institutions and government should be unique, as should the means of establishing both of them. Either we invent, or we lose our way".[6] It was a call for a "political mission", one that was ready even to set "an example for the world" and also to bring about transformation in Europe.[7] The American continent was set up pre-eminently as a project that looked to the future, a desideratum of collective liberation. In the words of the Liberator, Simon Bolivar, "America . . . is a sovereign, irrevocable decree of destiny,"[8] fertile ground for utopias of every stamp.

Charles Fourier was never in America. His leading disciple, Victor Considerant, however, arrived there in 1852 and tried to set up "the New World of Fourier adapted to the geographical New World."[9] In 1855, he founded the colony of La Réunion on the banks of the Red River in newly independent Texas, which had been Mexican territory until 1836. Within a year, though, few of the initial three hundred colonists were still there, and by 1863, it had been completely abandoned. It was not the only one. In Mexico, numerous phalansteries, utopian colonies and projects were set up. These included the *Champ d'Asile*, or Field of Refuge, a colony set up by Bonapartist exiles who contemplated rescuing the Emperor from the island of Saint Helena and offering him a throne in America;[10] the communes of Aguascalientes, founded by José María Chávez in 1850; the one that Rhodakanaty set up in Chalco (1865); and the Tobolobambo community, founded by an engineer from Pennsylvania, Albert Kimsey Owen, in the bay of Ogüira (State of Sinaloa), which survived until 1895. In 1848, a vanguard of followers of Étienne Cabet tried to establish their Icarian colony there, only to move shortly after to Illinois, while, twenty years earlier, Robert Owen himself had tried to persuade the Mexican government to cede the provinces of Texas and Coahuila to him so that he could put into practice there "a harmonious new society that would put an end to social, national, commercial and religious strife."[11]

Compared with these experiments, there seems to be little that is intrinsically utopian about the historical chapter of the French Intervention and the Second Mexican Empire, which has been described as that "great moment of collective delirium in which utopia blends with the sordid."[12] The interests in play there were undoubtedly economic and military ones, rather than benevolent or humanitarian. Nevertheless, it is instructive to take a look at the rhetoric that was deployed to justify the enterprise — specifically, the notion of "the Latin race", a term that was repeatedly brandished to vindicate the imperial project — because this was the moment when America ceased to be Spanish in order to become Latin, and to break once and for all with its colonial past. As we shall see, this reinvention owed a good deal to Michel Chevalier, who took a Saint-Simonian view of the Mexican question, a combination of scientific undertaking, typical of a political economist, and a vision of a shining future in a better world, where vast, fertile plains would be scored in all directions by canals and railway tracks.

Michel Chevalier and Victor Considerant: Two French utopians on the intervention in Mexico

After the aspirations of the 1848 revolutions had been dashed, the European utopian schools in the second half of the century placed all their hopes on the other side of the Atlantic. On many occasions, many of these formerly Europeanist movements now also adopted new supranational types of doctrine, while keeping their roots firmly planted in the prevailing nationalism. Pan-Latinism arose in response to the concepts of *Mitteleuropa* and Pan-Slavism. It was promoted to a high degree by the Second French Empire and gave rise to specific policies, such as the Latin Monetary Union, which was founded in 1865 by France, Belgium, Italy and Switzerland. The new term of "Latin America" was also coined in this environment and was promoted in France — like the term Pan-Latinism — by Michel Chevalier, an outlawed former Saint-Simonian who became a senator and palace advisor to the Emperor.[13] America, which had hitherto been only "Spanish" or "Iberian", now became "Latin" and a space where French ambitions could also be accommodated.

In the 1830s, Michel Chevalier, as editor-in-chief of the Saint-Simonian newspaper, *Le Globe*, had already launched various projects for establishing a European (and even world) confederation of states by means of a "Mediterranean system", an extensive infrastructure network of railways and canals that would bring the West closer to the Near East. In those articles, Chevalier preached the advent of an "Amphictyonic Congress of Europe" and a new "universal associationist civilization"[14] and also supported the independence movements in Spanish America. At the end of the 1850s and beginning of the 1860s, after a study trip round North America that took him as far as Mexico,[15] his thinking changed. He not only started to develop a theory of the new Pan-Latinism, but also to postulate the existence of a Latin America, based on the notion of "the Latin race", which he had frequently referred to in his earlier work. This concept would ultimately justify France's right to intervene in affairs in that part of the world.

Madame de Staël had used the idea of the Latin race — even before Chevalier did — when, at the beginning of her famous work *De l'Allemagne* (1813), she portrayed Europe as divided into three distinct races, which were the origin of the main nations of the day: the Latin, Germanic and Slavic races. The Latin race — which included, in her opinion, the French, Italians, Spaniards and Portuguese — had received its language and civilization from the ancient Romans; it was, therefore, the race with the oldest civilization, equipped for politics and domination as well as earthly pleasures, able to adapt to circumstances, originally pagan and not given to abstract thought, unlike those who resisted the Romans, the Germanic peoples.[16]

Somewhat obscured for almost half a century, the concept of the Latin race reappeared in the 1850s and was common currency again at the time of the joint French, British and Spanish intervention in Mexico. Some aspects of its conceptual content had changed since the original formulation by Madame de Staël. In addition to the historical heritage and common civilization, the Latin race was now characterized by a shared religion, Roman Catholicism (as opposed to Protestantism in Northern Europe), and was extended so as to encompass also all those American peoples with shared linguistic and religious roots. What was still missing, however, was any attempt to justify the term on ethnic or truly racial grounds, this at a time when racial studies and arguments were starting to proliferate (after all, Arthur de Gobineau had published his *Essai sur l'inégalité des races humaines* in 1853). In this context, therefore, the term "race" was merely a synonym for civilization and a justification for the supranational unity that they sought.

The Pan-Latinism of the former Europeanist, Michel Chevalier, the invention of Latin America and the supposed existence of a Latin race all served to lay the doctrinal foundations for the French to intervene in American territory and to justify their presence in Mexico. In 1861, taking advantage of the American Civil War, France and Great Britain decided to send an army to Mexico to depose the new government of Benito Juárez, who had decided to cancel Mexico's foreign debt payments. At the height of the race for empire and driven by economic interests, Bonapartist discourse adopted the idea of "the Latin race" as an antidote to, and a means of holding back, the robustly dynamic "Anglo-Saxon race" to the North. The historian Pierre-Luc Abramson notes that "without the help of Saint-Simonian ideology, and without Chevalier and his knowledge, there is no doubt that the emperor could neither have produced nor presented to the general public such forward-looking justifications for the 'Grand Design of the Reign'."[17]

Thus, Michel Chevalier went from defending the universal association of his master, Saint-Simon, to being the standard-bearer of Latin association. His trip to the United States and Mexico in 1835 marked the turning point in his thought. He followed the example of Alexander von Humboldt, who had also written about Mexico in his *Essai politique sur le Royaume de la Nouvelle-Espagne*, by extolling "this inexhaustible fund of wealth contained in the New World,"[18] while in an early letter to Madame Mathieu de Saint-Hilaire in 1835, he had already stated that what kept him in Mexico was the interest that the country offered and, furthermore, the interest that it was likely to offer within what he supposed would be a very short period of time.[19] Chevalier considered the possibility of cutting the canal that would later connect the two oceans through the Isthmus of Tehuantepec (rather than in Panama) and planned a whole series of social and agrarian reforms and transport links so that the Mexican fields "would be populated with a new race."[20] These reforms, nonetheless, would be backed by a good number of Europeans with technical qualifications who would have to go and settle in the North American country (in 1831–1832, in the *Globe* newspaper, he had similarly defended the conquest of Algeria under the pretext of the so-called civilizing mission that history had entrusted to the French).[21]

Chevalier thought that he had found the best example of the deterioration of the Latin race in Mexico, against the Anglo-Saxons. In 1842, in his *Cours d'économie politique*, he quoted the words of Jefferson, who predicted that Mexico would be conquered piece by piece and annexed by the United States; Chevalier believed that the most industrious race in the world would inevitably dominate the other. He hoped, however, that Mexico could also be a space for the "regeneration of the race", as long as the other Latin peoples mobilized to collaborate in appreciating its great natural wealth.[22] That opportunity seemed finally to have arrived when the French troops embarked for Veracruz and Chevalier became, what Abramson referred to as, the "rhapsodist of the expedition",[23] with a lengthy article in two instalments in *La Revue des deux mondes* in April 1862, in which he justified intervention, pointed out the advantages of a monarchical regime (although expressing his reservations about the origins of Maximilian of Habsburg, citing the incompatible characters of the Germanic and Latin races as his reason) and examined the human and economic resources of the country, offering practical advice for the success of the enterprise. The article concluded with him denouncing the Monroe doctrine and invoking France's role as the protector of the Latin nations.[24]

A fuller version of this work was published the following year in his two-volume study, *Le Mexique ancien et moderne*, in which he reviewed the history of Mexico from its pre-Columbian past and considered its potential for the future. Chevalier set out two political reasons for defending the French intervention. The first, which he considered to be "of universal interest" was to erect a barrier to prevent the imminent invasion of the whole of the continent by the United States; and the second, which drew more on French policy, was to protect and save "from irreparable ruin", not only Mexico, but also "the whole Spanish branch of Latin civilization in the New World." The expedition was not interference as such, but the point of departure for "the political regeneration of Mexico."[25]

Chevalier had still not forgotten his former dreams of a unified Europe — which he did by evoking Napoleon I, who had claimed that any European war was a civil war — but thought that the harmony and unity that should govern relations between the different European states was more necessary, and possibly more achievable, between the Latin nations, that branch of Western civilization marked by Catholicism, and whose soul and eldest sister was France.[26] The Latin races, however, were losing their dominant position in the world in the face of the Protestant advance; he also noted that it was fundamental for French politics to safeguard its role in the world, because France herself would lapse otherwise into international irrelevance:

> The destinies of France and the greatness of her authority are subordinate to the future prospects of the Catholic States in general and of the Latin races in particular. This is the most powerful argument that can be offered in support of the expedition to Mexico.[27]

Chevalier's thinking about Mexico, which was based on the idea of a "clash of civilizations" *avant la lettre*, swung therefore between imperialist zeal, the *Realpolitik* that it now seemed to have become, and echoes from his Saint-Simonian past. The whole vocabulary of the industrial classes and government by administrators is still there, as well as his great passion for railways and steamboats as "powerful instruments of civilization". The call for the unity of the Latin race was just another of the many forms of universal association, expressed in terms of the protection and guardianship that were necessary for a people that was consumed by anarchy and "would not know how to provide its beautiful country with the most essential elements of social order and public prosperity." This would justify the "civilizing mission" that he had so often advocated before, because, as he concluded, with an idea that any utopian would appreciate, "in Mexico, it all remains to be done."[28]

The fact is that Saint-Simonianism always paid more attention to technical issues (Chevalier's works are lengthy studies on economics, agriculture and demographics) and industrial development than to public or individual freedoms, so it was not so much that Chevalier was a utopian thinker whose thought became more politically pragmatic in the shadow of power, but that there persisted in him a strand of ideological continuity. Chevalier's thoughts on Mexico and the works he wrote, although forgotten and reviled by many today, contributed nonetheless to the emergence of a future "Latin" identity as the Uruguayan, José Enrique Rodó, claimed in his famous essay "Ariel".

Chevalier's decisive influence on the Mexican enterprise was certainly acknowledged at the time and he was attacked by some of his contemporaries who were opposed to intervention. The Republican, Edgar Quinet, for example, did not hesitate to mock the old Saint-Simonian and his idea of a Latin race. It was an alleged right based on kinship and elevated to a "new divine law" that granted authority and power to fight those who did not belong to the family and to keep the close relatives on a tight rein. It was an absurdity carried to extremes (for who was able to state with any certainty that he did not have a single drop of Latin blood in his veins or one word with Latin roots in his vocabulary?) and its only purpose was to impose Caesarism on the Spanish republics. Words like "civilization" were used to mask "un coup d'état contre les libertés du genre humain" (a coup d'état against the liberties of humankind).[29]

The Frenchman who was most familiar with the New World was without a doubt that other major utopian involved in the Mexican campaign, the Fourierist, Victor Considerant. From 1852, he resided in America, between the failed colony in Texas and his subsequent retreat in San Antonio, where he remained until 1869, when an amnesty was declared and he was able to return home to France. Fourier's leading disciple refined his master's theory and stripped it of its most eccentric features so that it could be put into practice. He was an early participant in phalanstery experiments in France, and also involved in the politics of his time as a parliamentary representative in the Second Republic of 1848. Then, following the coup of 2 December 1851 and because of his protests against Napoleon III's military expedition to Rome — a bone of contention also for Quinet, another Republican — he was imprisoned and forced to seek exile in Belgium. Considerant, like the Saint-Simonian, Chevalier, was a former student of the École Polytechnique; also like him, he devoted some of his early works to the project for a "universal democratic federation", a "Congress of Spherical Unity" and, more specifically, a European federation.[30] The democratic humanitarianism of Fourierism, however, with its doctrine of regeneration and social harmony, moved away from the more authoritarian theories of Saint-Simonianism, and after 1840, Considerant took up the baton of Fourierism to devote himself to the task of spreading utopia overseas. Until 1848, Considerant was heavily involved in the European question; after that date, America was uppermost in his mind. He believed that that was where the social difficulties and the great European question needed to be resolved. It was not a matter of abandoning the European mother country, but of preparing the ground for its salvation and that of the world.[31]

Once the Fourierist experiment of La Réunion had collapsed, Considerant and his wife took refuge in San Antonio between 1863 and 1869. For Considerant, this town was a "symbol of the conflict between the Mexican mestizos and the white Anglo-Saxons. . . . a privileged observatory of the political and military events . . . of the French Intervention." In point of fact, both the Fourierist, Victor Considerant, and Robert Owen, whose own project was referred to earlier, viewed the Texan frontier as an area of common ground, a land between two cultures that was ripe for socio-religious utopia.[32] From there, Considerant visited Nuevo León and Coahuila, assiduously read Francisco Zarco's newspaper, *El Siglo XIX*, and met one of the precursors of the Mexican revolution, Alberto Santa Fe, author of an ambitious plan for agrarian reform, as well as Santiago Vidaurri, the architect of the breakaway Republic of Sierra Madre. Considerant's wide knowledge of Mexican society, language and culture is captured in his *Quatre lettres au maréchal Bazaine* (Four Letters to Marshal Bazaine), published anonymously in Brussels.[33] The objective of these letters was none other than to persuade Marshal Bazaine, and through him, the Emperor, of the need to abolish the system of servitude among the Mexican peasantry, known as *peonazgo*, or forced labour. Although Considerant was ideologically closer to Benito Juárez than to Maximilian and did not shrink from expressing his reservations critical of the French intervention, he did nonetheless write to his "enemies" in order to set out his ideas. The traditional indifference of Fourierism to the nature of the political regime — since it wanted socialism to spread through its own example rather than be imposed by subversive means — led Considerant to concentrate on its social and humanitarian aspects and to propose the need for reform "from below", which was in tune with the idea of his master, Fourier, who had dreamt of the "annihilated races" taking their revenge on the conquistadors and turning Mexico into the *Magnat* of future humanity.[34] In this respect, Considerant's Mexican letters constitute a plan for social reform of the agrarian question that would involve "una visión utópica de México, tierra nueva donde todo es posible por hallarse poblada de hombres sencillos y rectos, cercanos a la naturaleza" (a utopian vision of Mexico, a new land where anything is possible because it is populated with simple, upright men who live close to nature).[35]

In the same way that Considerant inserted references here and elsewhere to Rousseau's good savage, so he also echoed the Owen project and lamented that the independent Republic of Texas had not survived as a neutral buffer against the incursions from the North. In the main, however, he paraphrased — although from the other end of the ideological spectrum — the ideas of Michel Chevalier about the Latin race. The goal of protecting the Latin races from Anglo-Saxon influence, the oft-repeated reason for the intervention, seemed to him to be "une vue élevée et de politique vraiment *humanitaire*" (a lofty, truly *humanitarian* political vision). Meanwhile, Maximilian's objective should be to pluck those races that spoke Latin tongues from the jaws of Anglo-Saxon individualism and help them fulfil their destiny, which was none other than Pan-American unity.[36]

So, Victor Considerant had also absorbed the prevailing racial discourse of the time, although with some slight differences. Rather than "Latin race", he preferred to talk about "races that spoke *Latin tongues*" (unlike Chevalier, Considerant privileged linguistic ties over religious ones, because he also sought freedom of worship for Mexico). More particularly, he favoured the term "Mexican race", even though in his philosophy, he was seeking to achieve universal brotherhood in which the Mexican race was already freed from the European yoke and open to Pan-American unity. In the *Quatre Lettres*, Considerant characterized the Mexican race as "une race inculte très cultivée" (an uneducated race that is very cultured) (p. 17), "[qui] possède une grande douceur . . . et une extrême docilité" (very gentle . . . and extremely docile) (p. 15) and naturally sociable, in contrast to the steely *self* that characterized the North American (p. 18). He even referred to them in the Prologue as "la race rouge" (the red race). Paraphrasing Fourier, he said that it would be a race with a *lesser* title that was not without *greater* capabilities (p. 17). It was, however, in danger of being swallowed up by the stronger race from the North, just as the buffalo had been, and so needed a bulwark to protect it (p. 81 and p. 86). The question of *peonazgo*, however, which was at the heart of the matter, was not a problem of race (as slavery was), but embraced a whole social class (p. 44), and that was why Considerant favoured a humanitarian, socialist policy rather than a civilizing approach.

Victor Considerant, unlike Chevalier, never abandoned his Fourierist activism, which he continued to vindicate in his work. The course of events, he tells us, was proving his master right and he ventured the prediction that the humanitarian policy that still provoked derision among pragmatic politicians would end up being the key question in future times. He never abandoned his commitment to the lot of the dispossessed Mexicans and declared that the true, pure Mexican was the poor Mexican (p. 14), and he ended his days, twenty-five years later, an eccentric old man who strolled through the Latin Quarter in Paris, mixing with the young students and wearing a serape and Mexican sombrero.

The "Latin race" in the Spanish press

Although the expression "Latin race" was a French invention, it very quickly spread to Spain, a country, we should remember, that was not a founder member of the Latin Monetary Union, joining it a few years after it had been established; what is more, the idea spread so successfully that it eventually eclipsed the notion of Spain's claim to its former colonies (albeit with some exceptions, as we shall see). The *Revue des races latines* [Journal of the Latin Races] (1857–1861), the French journal founded by the government of Napoleon III, had a Spanish counterpart, *Revista española de ambos mundos* [Spanish Journal of the Two Worlds] (1853–1855), whose first issue opened precisely with an article by Michel Chevalier, "Sobre el progreso y porvenir de la civilización" (On the progress and future of civilization), and which went into print again in 1874 with the new title of *Eco de ambos mundos: Órgano de la raza latina* [Echo of the Two Worlds: Organ of the Latin Race].

Spain, led by General Juan Prim, joined the Franco-British expedition to Mexico from the start. Prim, however, contrary to the opinion of the government, led by General Leopoldo O'Donnell, favoured negotiations under the auspices of Great Britain, rather than the monarchist, interventionist solution that Napoleon III was aiming for in the figure of Emperor Maximilian.[37] This issue generated a good deal of controversy, and the conflict was increasingly interpreted in Spanish public opinion as a clash between the Latin and Anglo-Saxon races. It was particularly well received among the conservatives of the Liberal Union party, led by O'Donnell, and those who supported a Mexican monarchy.

The expression "Latin race" first appeared in the Spanish press in that revolutionary year of 1848, the turning point in European history and in utopian dreams; it appeared, therefore, even before it became commonplace through the work of Michel Chevalier. In the 1850s it was frequently used with one outstanding feature: whenever it was mentioned, it would be in terms of the "degeneración de la raza latina" (degeneration of the Latin race), "decadencia" (decadence) or "abatimiento" (dejection).[38] When compared with the progress that the Teutonic races had made in both Europe and America, it marked the end of the influence of the Latin race and its loss of "virility". Newspapers such as the conservative *El Heraldo* — the decadence of the Latin race was a leading topic in the more conservative media, although not only there — attributed it to the excessive presence of and dependence on the State among the peoples of the south, as opposed to Anglo-Saxon individualism and *self-government*,[39] an idea repeated later by Victor Considerant. In short, the conclusion at that time could not have been more pessimistic: "we may as well start weeping like the *motherland* of today because of the danger that threatens the *Latin* race, which seems to us to be now very *Greek*."[40]

A single exception to this spate of lamentations before the intervention in Mexico was found in the Balearic newspaper, *El Genio de la Libertad* (The Spirit of Freedom), affiliated to the Liberal-Progressive party, where the expression "Latin race" came to enshrine the most advanced political values. So, they echoed the manifesto of the joint French–Spanish–Italian Democratic Committee, set up in Paris and headed by Lamennais, in which these three countries were presented as "the nucleus of the Latin race" and the vanguard of democratic, republican reform that was intended to reach every corner of Europe. The news that the Russian tsar had recognized the new Belgian government was also enthusiastically received, and interpreted as evidence of divine right losing ground before the "truth" of the Latin race (this "truth" being a constitutional monarchy elected by suffrage).[41]

Finally, another article on 8 May 1853 (p. 2), this time in the Liberal Monarchist newspaper, *La España,* provided an illuminating advance explanation of what the Mexican question and the confrontation of the races that was symbolically unfolding there was all about. The article, entitled "Sobre la situación de México" (On the situation in Mexico), sounded a warning against the growing dominance of the Anglo-Saxon race vis-à-vis the Latin one, since it upset the providential equilibrium of Western civilization. More particularly, readers were alerted to "la guerra de exterminio que Washington tiene jurada a México" (the war of extermination that Washington had sworn to wage against Mexico).

Use of the term the "Latin race" soared in the Spanish press in 1861, just as Prim was making preparations for the expedition to Mexico. By that time, its meaning had changed. It was no longer concerned with decadence and degeneration, but had become an open celebration of its virtues and possibilities. Many of these articles began with a characterization of the Latin race, or a clarification of what it meant. *La Discusión* (Discussion) for example, stated, "Unity, society, this is the eminently distinctive character of the Latin race. The distinguishing feature of the Anglo-Saxons is freedom [and] individualism."[42] This same newspaper (founded by the Democratic parliamentarian, Nicolás Rivero, whose parliamentary speeches, as we shall see, also dealt extensively with the topic of the Latin race), extolled the three major contributions of the Latin race to humanity as being, "el imperio, la sociedad universal, el catolicismo, la religión universal y la revolución francesa, que es la revolución universal" (Empire, the universal society; Catholicism, the universal religion; and the French revolution, which is the universal revolution). It also stressed the importance of understanding history, and not only the present day, in terms of races; in its controversy with other newspapers, such as *El Constitucional*, which was accused of denying the existence of such races, it became the leading left-wing exponent of this discourse on race. *La Discusión* pointed out that whereas Russia and Prussia had found strength in Pan-Slavism and Pan-Germanism, the Latin race, unlike those countries, would find unity, not by absorption, but by means of the free confederation of the various nations of which it was comprised.[43]

Only one newspaper dared to criticize the supposed values attributed to the Latin race at that time (in a text that appeared, oddly enough, in *La América* in 1861), taking up an argument from a decade earlier that had appeared in *El Heraldo*, namely, that the general tendency of the race "que se ha convenido en llamar raza latina" (that we have agreed to call the Latin race) was to obliterate the individual in favour of the government. Apart from this exception, however, the press, on the left and the right, was quick to embolden the Latin spirit and to call accordingly for the necessary unity of the Latin peoples so that those values would prevail in the Western hemisphere: "in this age of the rebuilding of nationalities and the grouping together of fraternal peoples, the union of the Latin race is not an empty phrase."[44] *La España* reported a grand scheme for an alliance between the Latin races and their resources, whose objectives would include the abolition of slavery overseas, the return of Gibraltar to Spain, the ending of England's maritime dominance, and making a variety of deals and agreements with the American republics to protect them from being absorbed by the United States.[45] A few months earlier, the view of *La Época* had been that an alliance between the three peoples of the Latin race (Spain, France and Italy) was the only "possible guarantee for the future in the West and that liberal Europe [would] applaud."[46]

This more or less utopian, more or less geostrategic programme for an alliance between the peoples of the Latin race focused its attention at that time on the military intervention in Mexico led by France. *La Iberia* repeated time and again that "Spain's dignity and interest oblige us to bring about the union of our race in America by every means at our disposal."[47] Spain, the bridge between the Old and New continents, was seen as the key to bringing the American peoples within the purview of the Latin race,[48] which was why the racial character of Mexico as Latin would continue to be asserted, even though, as *La Iberia* stated, it was a "raza degenerada sin duda pero a la cual restituiría su energía natural un gobierno bueno y firme" (degenerate race, certainly, but one whose natural energy would be restored by good, firm government). It was for that very reason that it was necessary to "intervenir en Méjico para salvar allí los grandes intereses de la humanidad, de la civilización y de la raza latina" (intervene in Mexico in order to save the major interests of humanity, civilization and the Latin race there).[49] *La Época*, a conservative monarchist newspaper and the organ of the Liberal Union party at the time, was the one that placed most emphasis on supporting the initiative of O'Donnell's government, appealing constantly to the unity of the Latin race as its motive. It stated explicitly that it did not do so in a display of rhetorical instrumentalization, but because that question was more important than lending rates:

> It is necessary for the European governments, especially those of the Latin race, to agree once and for all that they should intervene in the Spanish-American republics, not to rule over them or to impose their will on them, but to restrain them, protect them, in short, to prevent the total ruin that threatens them.[50]

However, not only the newspapers in favour of O'Donnell's government supported intervention in Mexico. So, and for identical reasons, did those of the opposition, such as *El Contemporáneo*, which supported the Moderate Party. The main reason, common to all of them, was to prevent "a powerful State not of our family" from dominating in Mexico.[51] *La Época* presented the union and strengthening of the Latin race, first and foremost, as the "last line of defence against the Anglo-Saxon invasion" and intervention was the only means of heading off the present anarchy and future conquest.[52]

This was how the Mexican question came to take the form of a "clash of civilizations", a battle of counter-concepts between two opposing races (a topic that Michel Chevalier had already developed at length, as we have seen) in a discourse that historically interpreted the 1815 treaties as "an agreement between the nations of the North to destroy the nations of the South; . . . an absolutist pact against freedom; the machinations of several races in order to finish off our Latin race."[53] *La España* claimed that "all the efforts of the Anglo-Saxon race have been directed at destroying the influence in America of the Latin race, which should, according to the noblest of ideals, carry the greatest weight in those countries."[54] *La América* expressed its view as follows:

> Our most sincere wish is that our brothers and sisters in South America should consolidate their freedom and organize their social and political life under the conditions that may have the most effective influence on their happiness; we wish those Republics . . . to become rich, peaceful, flourishing States. . . . But we shall never approve of these results being achieved as wards of the Anglo-Americans . . . , because the distinctive traits of the Latin race of the inhabitants of our former colonies are in stark contrast to those that are dominant in the race of their neighbours, and they would degenerate into a coarse copy and shapeless caricature if they lost their originality and adopted an outward appearance so contrary to their native character.[55]

The confrontation even reached such a pitch that *La Época* found itself rejoicing at the outbreak of the American Civil War: "con placer hemos visto la profunda escisión de aquel pueblo que, enemigo de nuestra raza latina, trataba de lanzarnos de las Américas, cambiando la religión, las costumbres y la lengua de nuestros hermanos" (it is with pleasure that we have seen the deep division within that nation — the enemy of our Latin race — which was trying to drive us out of the Americas, changing the religion, customs and language of our brothers and sisters).[56]

Nonetheless, although most of the newspapers seemed to be unanimous in their support for the intervention in Mexico, based on the claim of supposed Latin solidarity, controversy broke out subsequently over the details of the intervention and the possible solutions that could be applied. The first concerned whether the tripartite nature of the intervention was the right one and the fact that Spain had surrendered the initiative — which some claimed should have been theirs for historical reasons — to France. Reactionary newspapers, such as *El Reino*, or the Catholic-monarchist *La Esperanza*, expressed the humiliation they felt because Spain had not occupied that prime position, "because we, the representatives of the Latin race in America, those most threatened by the ambition of the United States, we, . . . should have tried to be the leading, though not the only actors, there in that great undertaking."[57] *La Época*, on the other hand, did support the tripartite intervention and the government's action in the matter, and took the view that unilateral intervention would have been "fatal

. . . for the destiny of the Latin race in America."[58] Apart from the role of France, another thornier question was the role played by England, "enemiga de nuestra raza latina" (the enemy of our Latin race), which was bitterly attacked in the pages of *La Época*. So, they raised the situation of Gibraltar and asserted that "talking to England about . . . the regeneration of the Latin race is to talk to her in a language that the Foreign Office does not understand."[59] The columns of the *Contemporáneo* were also critical of Britain's policy. By leaving France to its own devices in Mexico and putting pressure on Spain to withdraw from the conflict as well, Britain had "achieved another objective . . . to thwart the plan to unite the Latin race that would bring France, Italy and Spain together in a common policy."[60]

The arguments intensified particularly when the time came to propose a candidate for the Mexican throne and France put forward the name of the Austrian Maximilian as Emperor. Moderate and democratic newspapers alike, all opposed to O'Donnell, virulently attacked the French aspirations: "France's policy in America should not be to stir up trouble if it does not want the Latin race to lose its present dominance there or the major interests bound up with it to suffer irreparable damage."[61] It was for that very reason that *La España* proposed an alternative candidate for the throne of Mexico. Its opinion was that Isabella II's sister was the most suitable: "from the point of view of Mexican interests and Spain's transatlantic ones; from the point of view that the cause of liberalism wins if she triumphs; from the point of view of how desirable it is for the preponderance that the Latin race must acquire there where the Anglo-Saxon race is trying to assert its dominance."[62]

Meanwhile, the attitude of *La Época* (the official organ of the government) vis-à-vis the Mexican throne remained ambiguous. While the newspaper supported the plan for a monarchy, which had, it said, been a Spanish initiative, it lamented that it was being consolidated with a candidate that was opposed to the Latin race since Spain would end up losing influence in the territories that it had created with its laws and its children. It reiterated constantly that "the salvation of the Latin race" depended on the stance that Spain took on this issue.

> We do not want to harbour the regret of having missed perhaps our only opportunity to have contributed to our union with Europe, to the salvation of the Latin race and Spanish influence in the New World. . . . A constitutional throne raised and consolidated in Mexico . . . is, above all, the European barrier against the Latin race being absorbed by peoples with interests in [Spanish] America that rival our own.[63]

The conservative newspaper *La Esperanza* (the unofficial mouthpiece of Carlism) was in favour of a monarchy and reacted angrily to the letter that General Prim addressed to the French government, pointing out that it was of the utmost interest to the glory of Spain to save "the Latin race, Catholicism and the language of Cervantes" in America.[64] Such was the success of the notion of the Latin race that it spread, not only through the ranks of the more utopian democrats and socialists, but also to the most reactionary and traditionalist elements.

Reality, however, finally prevailed. *La América* pointed out that, in contrast to that imperial dream of the Anglo-Saxon race lying prostrate on the ground, while the Latin race rose up powerfully in the shadow of the monarchy that was to be established in Mexico, the paramount influence of the Latin race on South American soil was on the wane, because, as it reminded its readers, "en América no hay elementos monárquicos" (there are no monarchist elements in America). The expedition led by Prim withdrew from Mexican territory with scarcely anything to show for it, and the Austrian Emperor Maximilian imposed by the French did not last long either. Meanwhile, the Mexican government under Juárez was prepared to reach an agreement after all with the United States, leaving the ultraconservatives of *La Esperanza* lamenting yet again that "the possession of Mexico by the Anglo-Americans means the definitive triumph of the Anglo-Saxon race over the Latin race in America."[65]

The debate about the question of the Latin race in the context of the intervention in Mexico took place not only in the pages of the newspapers, but was also a major topic of parliamentary debate during those years. The Minister for Grace and Justice at that time, Santiago Fernández Negrete, defended the intervention with the following words: "There is a nation in America that is not of Spanish origin, the United States, whose circumstances make it a rival of our race, and which believes and says that the Latin race is to be conquered by force in America."[66] General Prim, addressing the Senate, also based his arguments on the idea of the interests of the Latin race, but precisely in order to defend the opposite viewpoint; leadership of the race could not be conquered with cannon fire.

> Stop there, gentlemen, stop there. What will you gain by taking your armies to Mexico? Nothing, neither honour, nor glory. The only thing that you are going to achieve is to destroy the influence over the Latin race that Spain should exercise, now and forever, and influence is not imposed with cannon fire. Do not lose sight of the fact that the United States, at the head of the Saxon race, is advancing with every day that passes. Let us not, therefore, by employing Spanish arms, facilitate their eagerness to invade.[67]

The Mexican diplomat assigned to Paris, José Manuel Hidalgo y Esnaurrizar, also published, for the benefit of the French public, a selection of debates held in the Spanish Cortes on the Mexico question, outstanding among which were the speeches of Bermúdez de Castro, the Marquis of Havana and Ríos Rosas. In his introduction, Hidalgo used that notion of the Latin race on numerous occasions and predicted an impending "struggle between races".[68] In the session on 13 January 1863, Ríos Rosas delivered a speech to Congress, which he began by recalling the presence of two European races in America, which had taken civilization to its limits over there, so presenting the conflict once again as a clash of races.

In the parliamentary arena, however, the man who most vigorously upheld the existence of the Latin race and asserted the need for its unification was Nicolás María Rivero, the leader of the Democratic Party and founder of the newspaper, *La Discusión*. Other newspapers of the time, as well as *La Discusión*, reported (on 12 March 1861) one of Rivero's speeches, in which the parliamentary representative called for the unity and confederation of the Latin race. His proposal was backed by other progressive members of the Cortes, including Salustiano Olózaga, and even Sagasta of the Liberal Party, who later served as Prime Minister on a number of occasions. The opposing benches, however, accused them of wanting to resurrect Napoleonic projects. The most hostile to Rivero's proposal was undoubtedly the conservative, Cánovas del Castillo, the future prime minister, who replied that he did not believe in the unity of races, and that it was not, in any case, the business of government to talk about races, let alone suggest that the Spaniards could be referred to as "Latin" since, he quipped, the only thing that they had preserved of Latium were the ruins of Saguntum and Numantia.[69] General José Gutiérrez de la Concha, Marquis of Havana, expressed himself in a similar vein a year later (on that occasion, in a reply to Prim himself, "the Marquis of Castillejos"), equating liberal policy with "racial policy" and dismissing it as "unworkable" (in other words, utopian).[70]

It is clear, therefore, that, in spite of the constant references to the Latin race in the political discourse of the time, and regardless of the enthusiasm of those who thought that race was a "law of history and biology" — like the columnists of *La Discusión*, mentioned at the beginning of this section — there were, nonetheless, a number of voices here and there expressing their scepticism of the new concept. The sceptical point of view was eloquently laid out in an article that appeared on 9 May 1862 in *La Época* (p. 2), in which the contradictions inherent in "esta ruidosa cuestión de razas con que diariamente nos atruenan los oídos" (this much discussed question of races with which our ears are deafened daily) were pointed out, those for example that led to an Austrian emperor being endorsed as an appropriate choice for a nation of the Latin race. After all, the writer wondered, what is the Latin race? Is it a question of religion, language, or blood? On what is that unity based? What right has Napoleon III to claim such representation for himself? The author of the article deflates that "sueño o manía predilecta de algunos de nuestros políticos" (dream or favourite obsession of some of our politicians), by reminding his readers that those nations, supposed brothers in race, instead of behaving fraternally, had historically waged war against one another, made alliances with nations from outside their own alleged race and had never obeyed any reason other than the reason of state, which was contrary to any utopia.

Conclusion: a project of social regeneration and political unity for the New World

Of all the historical experiences and experiments in Latin America — a territory that has lent itself to the conquest of all kinds of dreams and ideals and in which so much blood has been shed — the second European intervention in Mexico in the 1860s, the overthrow of Juárez's liberal government and the French occupation do not appear to warrant a chapter in the history of utopias, apart from the political oddity of placing a monarch from the House of Habsburg, however fleetingly, on the throne of a so-called Mexican Empire. The reasons for this foreign intervention (the suspension of debt payments by the revolutionary government, the defence of the rights of the clergy and the status quo, the threat of invasion from their Anglo-Saxon neighbours to the north, and the race to empire of the European powers) seem even less idealistic.

And yet this enterprise with interests that were so patently less than noble made use of a rhetorical charge whose range far exceeded the bounds of self-interest to appeal to sentiments of a more altruistic nature. The use and misuse of such a distinctive concept as the "Latin race" is an outstanding example. Some of the most prominent French utopian thinkers of the second half of the nineteenth century were directly involved in its formulation and dissemination, which spread even to the most extreme reactionary groups. In France, conceptualization of the supposed existence of a Latin race served to create a new ideology that justified imperialist expansion in Latin America, one that Spain made use of also in its attempt to maintain its post-imperial ascendancy in those territories.[71] And if, for the Europeans, it represented a first-rate intellectual instrument that could be used to rearticulate relations with their now independent former colonies, the Latin Americans (Bilbao, Torres Caicedo), for their part, also made use of it to lay claim to their own identity in opposition to Old Europe and the giant to the north. In the face of the real advance of the influence of the United States and the Monroe doctrine, which claimed the whole continent as its "exclusive sphere of influence", the "Latin race" would turn out to be an illusory invention, almost naïve in its impotence, and its success, ephemeral and merely rhetorical.

Agents of extraordinarily diverse ideologies were able to exploit the notion of the Latin race, because the very definition of it was always so vague. It did not imply a biological sense, so much as a cultural or "civilizing" one, which was understood in terms of religion, language, a shared history and supposed moral characteristics. What it implied, however, in most cases — which made it an excellent utopian tool — was the commonly projected idea of social regeneration and supranational political unity.

Latin America had always been conceived of as a place "destinado a la regeneración de la humanidad" (destined for the regeneration of humanity), where palingenetic aspirations joined hands with the Bolivarian utopia of the unity of the New World,[72] and the idea of the "Latin race" was certainly more than enough to embrace both these historical and utopian aspirations. It is no coincidence, therefore, that convinced former Europeanists like Michel Chevalier or Victor Considerant — who fought on behalf of European unity and universal association and harmony until the middle of the century — should, in the 1860s, join in the battle for the Latin race (much more reduced in scope, of course, but for that very reason more feasible and concrete).

With the revolutionary dreams of emancipation and brotherhood in the Old Continent dashed after 1848, many of those utopians turned their gaze towards America. The very idea of a free and united America was not new and not even enjoying its best moment given the consolidation of the newly independent autonomous states, but the longing for unity and regeneration expressed in the ideology of the Latin race helped revitalize it. A long time before this invention, the Argentinian revolutionary and promoter of independence in Latin America, Bernardo de Monteagudo, had already appealed (in 1823) to these affinities between Spanish-American nations (political affinities, he stressed, but also physical and moral ones), which ought to lead to the creation of a single nation. In a draft plan to achieve this, drawn up at the behest of Simon Bolivar himself, he encouraged the establishment of a great common congress:

> A spotlight that will light up [Spanish] America; to create a power that unites the forces of fourteen million individuals; to strengthen the relations between the [Spanish] Americans, uniting them with the great bond of a common congress, so that they learn to identify their interests and literally form a single family.[73]

With or without the bonds of kinship, the idea of universality and association as a constant of every enlightened or utopian project has continued to surface from time to time, even down to the present day, although always with a certain ambivalence, because, ultimately, it was never certain whether the project for uniting the Latin race concealed an imperialist plan for domination or the promise of regeneration and the emancipation of brothers. Even the French republican, Edgar Quinet, who was as critical of his country's interference in Mexican affairs as he was of the notion of the

"Latin race", concluded his short work on the expedition to Mexico by calling for the establishment of a great American confederation that would also include the powerful democracy of the United States, which would put a stop to all attempts at Old World Caesarism, since, as he pointed out, the New World "est encore l'espérance de tous les amis de la liberté" (is still the hope of all the friends of freedom).[74]

Notes

1 Edgar Quinet, *L'expédition du Mexique* (London: Imprimerie de J. Taylor, 1862), p. 10: "Fourier et les autres visionnaires nous ont enseigné que Mexico est la capitale naturelle du monde."

2 Luis Ricardo Dávila, "América: Venezuela", in *Diccionario político y social del mundo iberoamericano*, compiled by J. Fernández Sebastián (Madrid: Fundación Carolina, Sociedad Estatal de Conmemoraciones Culturales y Centro de Estudios Políticos y Constitucionales, 2009), p. 166: "gran página en blanco en espera de ser llenada de sentido."

3 Antonio de Alcedo, *Diccionario geográfico-histórico de las Indias Occidentales o América*, 5 vols. (Madrid: Imprenta de Benito Cano, 1786), 1:68: "regalo de la vida humana."

4 Guillermo Zermeño, "América: México", in *Diccionario político y social*, Fernández Sebastián, p. 134: "*América* refiere a un suelo, una geografía, un clima y un futuro reivindicado frente a la 'vieja' Europa."

5 *Gaceta Mercantil* (Buenos Aires), 20 January and 23 January 1844.

6 Simón Rodríguez, *Sociedades Americanas* (Caracas: Monte Ávila, 1992), p. 39 (the original citation is in the 1828 Arequipa edition, p.151): "la América española es original: originales han de ser sus instituciones y su gobierno, y originales los medios de fundar uno y otro. O inventamos, o erramos."

7 *El Telégrafo de Lima*, 29 April 1833.

8 Simón Bolívar, "Carta a Francisco de Paula Santander, Guayaquil, 6 de agosto de 1823", in *El mundo según Simón Bolívar*, edited by Carlos José Reyes (Bogota: Icono, 2006), p. 26: "América . . . es un decreto soberano, irrevocable del destino."

9 Pierre-Luc Abramson, *Las utopías sociales en América Latina* (Mexico: Fondo de Cultura Económica, 1999), p. 33: "la adecuación del Nuevo Mundo fourierista al Nuevo Mundo geográfico."

10 This *Champ d'Asile* was established in Texas on the orders of General Rigaud in 1815. It was constituted as a "Napoleonic confederation" with a militia of nine hundred men that called themselves the "Independent Soldiers of Mexico" whose aim was to fight the Bourbons in America and free Napoleon from his prison on Saint Helena. They were finally defeated in 1818 ["Plan pour la Confédération napoléonienne" cited by Jacques Penot, *Les relations entre la France et le Mexique de 1808 à 1840*, 2 vols (Lille/Paris: Atelier de reproduction des thèses de l'Université de Lille III — Librairie Honoré Champion, 1976), 1:169].

11 Carlos M. Rama, *Utopismo socialista (1830–1893)* (Caracas: Biblioteca Ayacucho, 1977), p. 185. The original note with Owen's text is kept in the Mexican Archivo General de la Nación, Exp. H/554: "una nueva sociedad armónica que acabase con las disensiones religiosas, mercantiles, nacionales o sociales."

12 Abramson, *Las utopías sociales*, p. 57: "gran momento de delirio colectivo donde la utopía se mezcla con lo sórdido."

13 Michel Chevalier was one of the leading disciples of Count Henri de Saint-Simon and a prominent representative of the Saint-Simonian school, based in the Ménilmontant commune. A graduate, like so many utopians, of the École Polytechnique in Paris, his work was devoted to social and economic reform, and to developing railways, canals, finances and colonization. After spending six months in prison when the Saint-Simonian commune was closed down (1832), a trip to the United States changed the direction of his work. Although most researchers have credited Chevalier with the neologism [Aims McGuiness,

"Searching for 'Latin America': Race and Sovereignty in the Americas in the 1850s", in *Race and Nation in Modern Latin America*, edited by Nancy Appelbaum et al. (Chapel Hill: University of North Carolina Press, 2003)], another version maintains that the expression "Latin America" was first used by South Americans, such as the Colombian, José Maria Torres Caicedo (in the poem "Las dos Américas", 1856) or the Chilean, Francisco Bilbao (in a lecture "Iniciativa de la América. Idea de un congreso federal de las repúblicas" delivered in Paris in 1856, or in *La América en peligro*, 1862, a dissertation against the French intervention in Mexico), although both authors at that time were residing in Paris and wrote their works there [see Walter Mignolo, *The Idea of Latin America* (Oxford: Wiley-Blackwell, 2005) and Joao Feres and Flavio Alfredo Gaitán, *La historia del concepto "Latin America" en los Estados Unidos* (Santander: Universidad de Cantabria, 2008)].

14 *Le Globe*, 12 February 1832, etc. Press articles gathered together in Michel Chevalier, *De la religion saint-simonienne. Politique européenne* (Paris: Au bureau du Globe, 1832) and Michel Chevalier, *De la religion saint-simonienne. Le Système de la Méditerranée* (Paris: Au bureau du Globe, 1832).

15 Michel Chevalier travelled to the United States between February and May 1835, with a commissioning order from the then Minister of Public Works, Adolphe Thiers, to study the transport links in North America. In Mexico, he visited Vera Cruz and the Federal District, where he took copious notes that he would use later for his many writings on Mexico, to the extent that he became the official specialist in France on Mexican questions, with works such as *Le Mexique* (1851) and *Le Mexique ancien et moderne* (1863).

16 Anne-Louise-Germaine, Madame de Staël, *De l'Allemagne* (Paris: Charpentier, 1839), pp. 9ff.

17 Abramson, *Las utopías sociales*, p. 54: "sin el auxilio de la ideología sansimoniana, y sin el de Chevalier y sus conocimientos, es indudable que el emperador no habría podido elaborar ni desarrollar, ante la opinión pública, las justificaciones progresistas del 'gran designio del reinado'." In the 1840s, Napoleon III had been an enthusiastic reader of Enfantin and Bazard, since he had been interested at the time in the proposed inter-oceanic American canal. It comes as no surprise that the Latin American idea germinated in the industrial Saint-Simonian environment; many such as Olinde Rodrigues and the Pereire brothers, founders of the first modern credit bank in Spain (1856), showed early interest in doing business in the Iberian peninsula and in the possibility of replacing the power of the long-standing decadent metropolis of Madrid in America with the influence of Paris.

18 Michel Chevalier, *Cours d'économie politique fait au collège de France, année 1841–1842* (Paris: Capelle, 1842), p. 234: "cet inépuisable fonds de richesses que renferme le Nouveau-Monde". There he already gives a foretaste of most of his reflections on Mexico (the quotation is reminiscent of "la richesse du sol, l'abondance des subsistances, la beauté du climat" in Alexander von Humboldt, *Essai politique sur le Royaume de la Nouvelle-Espagne*, 4 vols (Paris: Renouard, 1825–1827), 4: 266).

19 Letter from Michel Chevalier to Madame Mathieu Saint-Hilaire, 28 February1835 (kept in the Prosper Enfantin Collection, Bibliothèque de l'Arsenal, Sign. 7706).

20 Chevalier, *Cours d'économie politique*, p. 298: "se peupleraient d'une race nouvelle."

21 Osama W. Abi–Mershed, *Apostles of Modernity: Saint-Simonians and the Civilising Mission in Algeria* (Stanford: Stanford University Press, 2010), pp. 27ff.

22 Chevalier, *Cours d'économie politique*, p. 226, and "Le Mexique", in *Encyclopédie du XIXe siècle* (Paris: Maulde et Renou, 1851).

23 Abramson, *Las utopías sociales*, p. 53.

24 Michel Chevalier, "L'Expédition européenne au Mexique. I. Les révolutions mexicaines depuis l'indépendance" and "L'Expédition du Mexique. II. Des ressources et de l'avenir du pays. Des motifs et des chances de l'expédition", *Revue des deux mondes*, 38 (1 and 14 April 1862): 513–561 and 879–918.

25 Michel Chevalier, *Le Mexique ancien et moderne*, 2 vols. (Paris: Librairie de L. Hachette,

1863), p. 479: "le rameau espagnol tout entier de la civilisation latine dans le nouveau monde." And p. 474: "la régénération politique du Mexique."

26 *Ibid.*, p. 588. In Chevalier's opinion, the distinguishing characteristic of the Latin race was religion, not language, which presented him therefore with a conflict, since, at the same time, he looked upon the Church of Rome as an "enemy of modern civilization", which is why the prior reconciliation between Catholicism and "progress, liberalism and civilization", constituted the greater challenge: "le plan de régénération (du Mexique) . . . ne pouvant avoir d'autres fondements que les principes libéraux" [the plan for the regeneration (of Mexico) . . . can have no other foundations than liberal ones].

27 Chevalier, *Le Mexique ancien et moderne*, p. 508: "Les destinées de la France et la grandeur de son autorité sont subordonnées aux chances d'avenir des États catholique en général et les races latines en particulier. C'est le plus puissant argument que soit possible de faire valoir á l'appui de l'expédition du Mexique."

28 *Ibid.*, p. 518: "ne saurait procurer à leur beau pays les éléments les plus indispensables de l'ordre social et la prospérité publique." And p. 539: "tout est à faire au Mexique."

29 Quinet, *L'expédition du Mexique*, pp. 10–15 and 36–38.

30 Victor Considerant, *De la politique générale et du rôle de la France en Europe* (Paris: Bureau de la Phalange, 1840) and *La dernière guerre et la paix définitive en Europe* (Paris: Librairie Phalanstérienne, 1850), in which he calls for social democracy and the fraternity and association of nations.

31 Victor Considerant, *Au Texas* (Paris: Librairie Phalanstérienne, 1854), quoted in Silvio Zavala, "Victor Considerant ante el problema social de México", *Historia Mexicana*, vol. 7, 3 (1958): 312.

32 Cited in Abramson, *Las utopías sociales*, pp. 59–60: "símbolo del conflicto entre los mestizos mexicanos y los blancos anglosajones. . . . un observatorio privilegiado de los acontecimientos políticos y militares . . . de la Intervención francesa". And pp. 41–42. The failure of Owen's 1828 project, which set out to make Texas an independent buffer state between Mexico and the United States, failed, according to his own testimony, as a result of the Mexican government's opposition, because of the requirement for religious freedom in the emancipated "regenerated" territories.

33 Victor Considerant, *Quatre lettres au maréchal Bazaine* (Brussels: C. Muquardt, 1868). The letters are dated 1865 and 1867 (the last one). The first of these is franked in Concepción. Marshal Bazaine was in command of the second French intervention in Mexico and was regarded as suspect because of his political attitudes that were critical of Emperor Maximilian and close to those of the Mexican resistance. Prior to this attempt to find a more propitious interlocutor (although it is not known whether the letters actually reached the Marshal), Considerant also wrote a lengthy letter to Emperor Maximilian (1865), the unpublished draft of which is kept in the École Normale de Paris [quoted in Russell M. Jones, "Victor Considerant's American Experience (1852–1869)", *The French-American Review* (1976 and 1977): 65–94 and 124–170].

34 Charles Fourier, *Théorie des quatre mouvements* (Paris: J.-J. Pauvert, 1967), p. 150, cited by Quinet, *L'Expédition du Mexique*, p. 10.

35 Abramson, *Las utopías sociales*, p. 63; although Considerant was an enemy of the Bonapartist Empire and any monarchist project that attacked Mexican independence, he also thought that the liberal reformist programme of Juárez did not go far enough, because it did not initiate the redistribution of land or abolish *peonazgo*. This practice would not be officially abolished until 1910 (Silvio Zavala, "Victor Considerant ante el problema social de México").

36 Considerant, *Quatre lettres au maréchal Bazaine*, p. 81, and Zavala, "Victor Considerant ante el problema social de México", pp. 320–321.

37 Manuel Ortuño Martínez, "La intervención en México del general Prim (1861–1862)", *Trienio: Ilustración y liberalismo* 58 (2011): 21–60.

38 See for example, *El Heraldo*, 14 October 1852, p. 1 and 21 December 1852, p. 3; *La España*, 12 May 1852, p. 3, and *El Clamor Público*, 26 June 1852, p. 2; *La Esperanza*, 22 September 1849, p. 3.
39 *El Heraldo*, 9 November 1852, p. 1.
40 *La Esperanza*, 21 September 1849, p. 4: "estamos por ponernos a llorar como la *Patria* de hoy por el peligro que amenaza a la raza *latina*, que nos parece ya muy *griega*."
41 *El Genio de la Libertad*, 7 September 1851, p. 1 and 2 March 1853, p. 1.
42 *La Discusión*, 26 August 1862, p. 1: "La unidad, la sociedad, es el carácter profundamente distintivo de la raza latina. La libertad, el individualismo, es el rasgo de los anglo-sajones."
43 *La Discusión*, 10 May 1861, p. 1.
44 *La Época*, 13 November 1861, p. 3: "en esta época de reconstitución de las nacionalidades, de agrupamiento de los pueblos hermanos, la unión de la raza latina no es una vana frase."
45 *La España*, 13 June 1861, p. 3.
46 *La Época*, 7 January 1861, p. 2: "garante posible del porvenir en Occidente y que la Europa liberal aplaudirá."
47 *La Iberia*, 8 March 1861, p. 1: "la dignidad y la conveniencia de España nos obligan a procurar por cuantos medios estén a nuestro alcance la unión de nuestra raza en América."
48 "En esta gran crisis humana, en esta gran evolución de los pueblos, España debe tener el primer puesto. Siendo el lazo de unión entre Europa y América, podrá representar en ambos continentes el papel de jefe de la raza latina y hacerse respetar como tal" [In this great human crisis, in this great evolution of the peoples, Spain should occupy the leading position. Being the bond that unites Europe and America, [Spain] will be able to represent on both continents the role of head of the Latin race and be respected as such for it] *La Iberia*, 15 November 1861, p. 1.
49 *La Época*, 12 and 13 September 1861, p. 2.
50 *La Época*, 5 September 1861, p. 2: "Es necesario que los gobiernos europeos, sobre todo los de raza latina, se convengan una vez por todas de que deben intervenir en las repúblicas hispano-americanas, no para dominarlas ni imponer allí su voluntad, sino para moderarlas, protegerlas, para impedir, en fin, la total ruina que las amenaza."
51 *El Contemporáneo*, 10 September 1861, p. 1: "un Estado poderoso que no sea de nuestra familia."
52 *La Época*, 12 September 1861, p. 2: "última barrera opuesta a la invasión anglo-sajona."
53 *La Iberia*, 8 March 1861, p. 2: "un convenio de las naciones del Norte para destruir las naciones del Mediodía; . . . el pacto del absolutismo contra la libertad; la inteligencia de varias razas para acabar con nuestra raza latina."
54 *La España*, 7 April 1861, p. 3: "todo el trabajo de la raza anglo-sajona se ha dirigido a destruir en América el influjo de la raza latina, que por los más altos conceptos debiera ser preponderante en aquellos países."
55 *La América*, 24 February 1861, p. 2: "Deseamos sinceramente que nuestros hermanos de la América del Sur consoliden su libertad y organicen su vida social y política con las condiciones que más eficazmente puedan influir en su ventura; deseamos que aquellas Repúblicas (...) lleguen a convertirse en Estados ricos, pacíficos y florecientes. . . . Pero no aprobaremos jamás que estos resultados se consigan bajo la tutela de los anglo-americanos . . . , porque los caracteres distintivos de la raza latina que poseen los habitantes de nuestras antiguas colonias, están en abierta oposición con los que sobresalen en la raza de sus vecinos, y degenerarían en copia grosera e informe caricatura si perdiesen su originalidad y adoptasen exterioridades tan opuestas a su índole nativa."
56 *La Época*, 26 September 1861, p. 1.
57 *La Esperanza*, 16 May 1862, p. 3: "porque nosotros, representantes de la raza latina en América, los más amenazados por la ambición de los Estados Unidos, nosotros . . . debíamos haber tratado de ser allí los primeros, ya que no los actores únicos, en esa grande obra."

58 *La Época*, 29 March 1862, p. 2: "fatal . . . para el destino de la raza latina en América."

59 *La Época*, 13 September 1862, p. 2 and 15 December 1862, p. 2: "hablarle a Inglaterra de . . . regeneración de la raza latina es hablarla un lenguaje que es incomprensible para el *Foreign Office*."

60 *El Contemporáneo*, 10 January 1863, p. 3: "conseguido otro objetivo, . . . estorbar el plan de unión de la raza latina que reuniría en una política común a Francia, Italia y España."

61 *La España*, 26 February 1862, p. 2: "La política de Francia en América no debe ser perturbadora, si no quiere que la raza latina pierda allí su actual preponderancia, y los grandes intereses con ella enlazados sufran daños irreparables."

62 *La España*, 22 March 1862, p. 2: "desde el punto de vista de los intereses mejicanos y los trasatlánticos de España; de lo que con su triunfo gana la causa del liberalismo; de cuán conveniente es a la preponderancia que la raza latina debe adquirir allí donde la anglo-sajona intenta dominarla."

63 *La Época*, 26 and 28 April 1862: "No queremos abrigar el remordimiento de haber perdido acaso la ocasión única de haber contribuido a nuestra unión con la Europa y a la salvación de la raza latina y de la influencia española en el nuevo mundo. . . . Un trono constitucional alzado y consolidado en Méjico . . . es sobre todo el dique europeo opuesto a la absorción de la raza latina por pueblos que tienen en América intereses rivales a los nuestros."

64 *La Esperanza*, 28 April 1862, p. 3: "la raza latina, el catolicismo y la lengua de Cervantes."

65 *La Esperanza*, 24 June 1862, p. 3: "La posesión de Méjico por los anglo-americanos significa el triunfo definitivo en América de la raza anglo-sajona sobre la raza latina."

66 Reported by *El Clamor público* and other newspapers on 23 November 1861: "Hay en América una nación que no es de origen español, el pueblo norteamericano, cuyas circunstancias la constituyen en rival de nuestra raza, y el cual cree y dice que la raza latina le ha de estar supeditada en América."

67 *Le général Prim, le Sénat, les Cortès et la presse espagnole dans la question du Mexique* (Paris: E. Dentu, 1863), p. 4: "Deteneos, señores, deteneos. ¿Qué ganaréis llevando vuestros ejércitos a México? Nada, ni honor, ni gloria. Lo único que vais a lograr es destruir la influencia que España debe ejercer, ahora y siempre, sobre la raza latina, y la influencia no se impone a golpes de cañón. No pierdan ustedes de vista que los Estados Unidos, a la cabeza de la raza sajona, avanzan cada día más. No les facilitemos por tanto, mediante las armas españolas, su avidez de invasión" (the date of the debate on the Mexican question in the Senate: 1858).

68 *La question du Mexique devant les Cortès d'Espagne: discours prononcés au Sénat et au Congrès / par MM. Bermudez de Castro, Concha, Mon et Ríos y Rosas; traduits de l'espagnol et précédés d'une introduction par J. M. Hidalgo* (Paris: A. Lainé et J. Harvard, 1863), p. 9.

69 *La Discusión*, 23 December 1862, p. 2.

70 *La question du Mexique devant les Cortès d'Espagne*, p. 197.

71 Joseba Gabilondo, "Genealogía de la raza latina. Para una teoría atlántica de las teorías raciales hispanas", *Revista Iberoamericana* 75, 228 (2009): 803.

72 Abramson, *Las utopías sociales*, pp. 345 and 348.

73 Bernardo de Monteagudo, *Ensayo sobre la necesidad de una federación general entre los estados hispano-americanos y plan de su organización* (Guatemala: Imprenta del estado por J. González, 1825), p. 22: "Un foco de luz que ilumine a la América: crear un poder que una las fuerzas de catorce millones de individuos: estrechar las relaciones de los americanos, uniéndolos por el gran lazo de un congreso común, para que aprendan a identificar sus intereses y formar a la letra una sola familia."

74 Quinet, *L'expédition du Mexique*, pp. 17 and 15.

8 Rhodakanaty in Mexico

Carlos Illades

Very little is known about Plotino Constantino Rhodakanaty. It is said that he was born in Athens on 14 October 1828 and that his father, a doctor and writer, fought in the Greek War of Independence against the Turks (1821–1829), dying before the London Protocol of 1830. After his father's death, his mother took him to Austria to live with his grandparents. However, when Rhodakanaty registered his own son, Plotino Nefi in 1881, he was forty-seven years old according to the records, which means that he must have been born in 1833 or 1834, four or five years after the date stated by Valadés and repeated by all of us who have followed the enigmatic life story of this Greek socialist. This would also modify the record that we have of his education, at least with regard to the dates. Based on this, Plotino Constantino's father could not have died during the Greek War of Independence either, but maybe a few years later.[1]

While still young, Rhodakanaty began studying medicine at the University of Vienna, which was one of the oldest institutions of higher education in Central Europe (founded in 1365) and with a strong tradition. At that time, the Austrian capital was a bastion of European culture in which one of the most robust of socialist traditions on the continent would flourish. Possibly in the early months of 1848, Rhodakanaty left for neighbouring Budapest — which had been shaken by the democratic revolution of Lajos Kossuth — and would have resumed his medical studies later in Berlin, where his family may have taken up residence in 1848. He could have remained there perhaps until 1857. Rhodakanaty probably visited Paris in 1850 in order to meet Pierre-Joseph Proudhon in person.[2] It is also possible that seven years later the young Greek moved to the City of Light for good, where it is presumed that he published a pamphlet entitled *De la Naturaleza*.[3]

In the French capital, Rhodakanaty would have gone further into his philosophical studies and learnt other languages, Spanish among them. According to his own records, it was in Paris that he joined the ranks of the socialists, and it was possibly there that he also became aware of the decree of 1 February 1856 favouring the establishment of agrarian colonies in Mexican territory. This document extended the rights of foreign residents and also extended the circumstances under which naturalization would be granted.[4] This opportunity was worth considering, and so, at the end of

1860, Rhodakanaty travelled to Barcelona, later embarking for Mexico. He arrived in Veracruz at the end of February1861 and shortly after, in the capital. Practically unknown, probably with little money, but having his medical and philosophical knowledge to hand, as well as his ability in both classical and modern languages (Greek and Latin, Spanish, German, French, English and Italian), Rhodakanaty started to earn his living and put the finishing touches to his plans for founding a study circle and an agricultural colony. That was a time of intense political conflict, when political and religious ideas emanating from Europe and the United States were circulating while at the same time the national press was growing. All this, and perhaps some contact with Paris, helped Plotino Constantino to integrate quickly.

Changing times

Rhodakanaty's *Cartilla Socialista o sea el catecismo elemental de la escuela de Carlos Fourier: el falansterio* [Socialist Notebook or the Elementary Catechism of the School of Charles Fourier: the Phalansterian] (Imprenta de Vicente García Torres, 1861) suggested that the supreme objective of human reason was to bring about the universal association of peoples and that only an order based on association could put an end to conflict that caused individuals to fight among themselves and that set families, peoples and classes against each other. Rhodakanaty's short tract considered that man was naturally good, but that flawed social institutions had altered the course of his behaviour, so favouring conflict. It would therefore be necessary to rebuild the institutions in order to establish a fair and equitable social contract that would bring different groups into harmony instead of causing social disagreement. A contract of this kind ought to benefit from the natural attributes of individuals by bringing them together in such a way that they would give of their best. Work, as well as being varied, would alternate with play and recreation. There would be a mix of sexes and ages to make the most of the intrinsic virtues of each. Isaiah Berlin points out that, for the romantics, what mattered was

> not the common base but the differences, not the one but the many; for them the craving for unity — the regeneration of mankind by recovery of a lost innocence and harmony, the return from a fragmented existence to the all-embracing whole — is an infantile and dangerous delusion: to crush all diversity and even conflict in the interest of uniformity is, for them, to crush life itself.[5]

Simple conviction based on concrete experience would be enough for humanity as a whole to recover its providential purposes as a guide to individual and collective behaviour. The social model sketched out in the pamphlet considered abolishing the privileges of the aristocracy, in particular putting an end to idleness, and promoting the work of the whole community, by sharing out material and spiritual wealth in an equitable manner. With all members having products and services to exchange, there would be no point in using money as a medium. The result would be universal harmony of the community and passions, in other words, fraternity realized in full. A social experiment on this scale could not be accomplished by breaking existing laws: conviction and persuasion would be the way to carry it forward. The resources to achieve this would include written and oral propaganda, and practical exercises. When the collective as a whole was convinced that this was the best way to organize the life of the community, attention could then turn to transforming the political regime, which would happen in a natural way, as the validity of the experiment could be tested and verified at the local level.[6]

After a didactic presentation of Fourier's doctrine, Rhodakanaty sought to disseminate his ideas directly. In 1863, he set up a free school that some referred to as "El Falansterio", or the Phalanstery, in which, "por principios se enseñaba al pueblo los derechos y prerrogativas de su soberanía nacional" (as a matter of principle, the people were taught the rights and prerogatives of their national sovereignty) and also to be wary in religious matters of any authority that was not "reason and common sense", and spreading awareness among the working classes of "los principios más puros y luminosos de la moral universal" (the purest and most luminous principles of universal morality).[7] It may have been at that time that he invited some young men who were interested in philosophy and concerned about social issues to join his circle; they included Francisco Zalacosta, Hermenegildo Villavicencio, Juan B. Villarreal and Santiago Villanueva. He also published a second short tract entitled *Neopanteísmo, consideraciones sobre el hombre y la naturaleza* (Neo-pantheism: Considerations Concerning Man and Nature) (Imprenta de Rivera, 1864).[8]

Possibly in March 1868, Rhodakanaty left the capital to settle in Chalco with the intention of founding an agricultural colony and putting the "agrarian law" into practice. He failed in his first intention, but he was successful in setting up the "free school" where he spread his philosophical and social ideas. One of those who attended was Julio López, a farm labourer who led a peasant uprising that gave the Mexican landowners no respite that year. López was probably taken prisoner in San Nicolás del Monte, district of Yautepec, and executed by firing squad in Chalco on 9 July1868. It is said that Rhodakanaty and Zalacosta tried to join the peasant uprising, but the former was apprehended in Huamantla and threatened with the death penalty, which was commuted to banishment from the region. After this failed rebellion, the Greek socialist went to the state of Morelos to spread Fourier's societary doctrine.[9]

Upon his return to Mexico City, Rhodakanaty worked on creating an organization that would lead to social regeneration. Founded by him on 20 March 1871, the *Social* envisaged the affiliation of its members as free and voluntary and that its operation would be the responsibility of a "Central Organizing Committee", consisting of seven delegates elected by majority, with its headquarters in the federal capital and a newspaper entitled *La Internacional*, which was the responsibility of Zalacosta. The organization proposed ending the exploitation of labour, recognizing women's rights and rehabilitating the indigenous population. Rhodakanaty prioritized its objectives: in the first place, it would try to put an end to pauperism and prostitution by encouraging public wealth, health, hygiene and morality; later, it would fight for the transformation humanity through beauty, virtue and science. Rhodakanaty conceived the association as a "sociocratic party" that would inform the "proletarian class" of its rights and obligations, and be constituted as the "órgano oficial del pueblo mexicano" (official organ of the Mexican people). It would also be an "internationalist party", not to mention providential, given that it had to implement evangelical charity. It represented the impoverished of Mexico and the world, taking the fight to the aristocracy, despotic governments and false churches, in other words those that made resignation the habitual sentiment of the poor. Poverty, unemployment, war, prostitution and corruption constituted the worst scourges of modern civilization, and they could only be overcome with socialism, a stage in human development where there would be bread, work and education for all.[10]

According to Rhodakanaty, natives, workers and women made up the race, class and gender oppressed by modern society. All social reform projects should include reparation of their rights and bring them closer to the rest of the community with more advantageous conditions. The emancipation of women and workers would set the standard for social regeneration. There would be a verifiable improvement in the working class provided that certain conditions were in place for them, such as organization at local, national and worldwide levels, if wages were increased, productive activities were financed and cooperatives formed, and if the factors of production of capital, labour and talent were harmonized. The "agrarian law", which included abolishing the monopoly of the *latifundistas* (owners of large estates) and regularizing individual property, was the solution to the problem involving the natives. Divorce was one answer to the disadvantageous situation of women and not even the Gospels opposed it. When the Legislature began to discuss the matter, Rhodakanaty came out in favour of accepting civil divorce because it guaranteed the freedom of the spouses and their right to marry again. His argument was also grounded in the law since it stressed the obvious contradiction between individual freedom — the natural right of man — and the fulfilment of a contract — marriage — that was indissoluble and lifelong, since divorce was prohibited by the Constitution of 1857. He was not against the family, but rather in favour of adapting it to the present time; he vindicated free love, "Christian marriages", although in cases of immoral behaviour on the part of the parents, the State would be entrusted with the care of the children.[11]

A new social contract had to be formulated, but totally different from the one proposed by Rousseau. According to Rhodakanaty, for Rousseau "todos los hombres le aparecen como otras tantas fuerzas o individualidades separadas, no solamente iguales, sino idénticas, que no pueden ser unidas en nada más que por contrato" (all men appear to him as so many separate forces or individualities, not only equal but identical, that cannot be united in anything except by contract), the consequence of which was the uniformity of the contracting parties, making them homogeneous instead of "semejantes" (equals, fellow beings). Formal equality, which was what the contract was based on, was for Rhodakanaty a subterfuge that left out natural inequality and led to individualism and selfishness. Furthermore, ceding individual sovereignty to an outside entity, the State, seemed to him to be an outrageous waste of human freedom. In Rousseau's contract, the citizen pledged everything. Not even in classical antiquity did they go that far, since some rights were reserved for citizens that were not subject to state arbitration. Subordinating the minority to the will of the majority inevitably led to slavery and despotism, since, unlike Rousseau — for whom "la voluntad constante de todos los miembros del Estado es la voluntad general; por ella son ciudadanos y libres" (the constant will of all members of the State is the general will; that is why they are citizens and free) — Rhodakanaty thought that, in the social contract, man was free only "in having his say and a vote", for "once the law is passed, he is a slave" of the majority who crush those who think differently.[12]

Rhodakanaty proposed creating sovereign federated bodies, which, when added together, made up society itself. Within the neighbourhood or municipality, the interests of individuals, families and classes could be expressed in a harmonious way. The core element comprised individuals and, as with any agreement in a harmonious society, it ought to be free, voluntary and revocable. At a later time, it might even be necessary to suppress the State, replacing it with a social pact endowed with political functions. With the elimination of governments and abominable politicians, public administration would be managed by the "social contract". The meaning that Rhodakanaty gave to the term was different from Rousseau's, since for the former, it had to do with civil society, not political society, as Rousseau thought, or, at any rate, a synthesis of the two that operated in the socio-economic sphere: "the organization of economic forces" would eventually "replace public powers", in other words, "the commonwealth of labour" would be born.[13] Politics would then lose its autonomy and be reabsorbed by intermediate federated bodies, which, after crossing national boundaries, would make up the universal association of peoples.

Rhodakanaty, an 1848 revolutionary, after all, defended democracy because it enshrined universal rights for all humankind. This could not function fully without socialism, so the question was, "What would be the use of enjoying our inalienable political and natural rights [...] if, on the other hand, we have no guarantee that ensures our livelihood, providing us with bread and work, the two elements necessary for our physical existence?"[14] He also attached greater weight to "social democracy" over representative democracy or, in other words, to "direct government by the people". If the ruler breaks the social contract by turning into a tyrant, citizens who are organized as a people acquire the "natural" right to rebel: "at that moment, undoubtedly, the anarchist revolutionary principle of Proudhon, which has up till now been looked upon as an exaggerated utopia, would be justifiable."[15]

"Garantismo Humanitario", published in instalments in *El Socialista* between 1876 and 1877, conceived of future society as the recreation of a Golden Age, the return of man to his original "state of nature", to "Edenism". In Fourier, Rhodakanaty identified various stages of development that human societies would pass through: Edenism, savagery, patriarchate, barbarism, civilization, socialism, guarantism and universal harmony. The present age corresponded to civilization, albeit an imperfect one because it was ravaged by hunger, poverty, prostitution, war and despotism. If the existence of the preceding historical phases was recognized, the feasibility of future stages could be deduced, given that the assumption that the present could not be transformed meant denying that there was a Divine Plan.[16]

The historical phases to come would synthesize the best of the preceding periods, and wrongs and outdated institutions would gradually disappear. Every historical stage had its own defining characteristics: in the case of Edenism, freedom of the passions and harmony, as well as its universality and the absence of sickness and geological cataclysms. During the phase of savagery, the prevailing characteristic was the unrestrained independence of individuals, an indomitable warlike spirit, cruelty, oppression and slaughter; slavery also appeared. Patriarchate, although despotic, increased the rights of legally recognized wives and was implemented in the Orient. Barbarism was obscurantist, rejected science and favoured serfdom, institutionalizing it under feudalism. Civilization produced slavery in freedom, poverty in plenty, ignorance in encyclopaedism, the exploitation of labour by capital and of woman by man. Progressing beyond civilization was only possible by means of association, the element that connotes socialism. Under guarantism, marriage would survive, abject poverty would be reduced, political revolutions would decrease as social and natural balance increased. Meanwhile, universal harmony would lead to the human passions being in perfect equilibrium and to the fullness of the natural order, with individual interests coinciding with collective interests. It would be an enhanced return to primitive

Edenism, since humanity would have benefited from what it had learned during the course of its history, and through enlightenment it would recognize its providential purposes. Given the universality of movement and progress, the Earth would reach its optimal position in space. Climates would improve and natural catastrophes would disappear. Arriving at this goal would be inevitable, because history was governed by a natural inexorable law created by God.[17]

God

In the 1870s, Rhodakanaty wrote articles about religion for *La República* and was the editor of *El Craneoscopio, periódico frenológico y científico* (*The Cranioscope: Scientific Phrenological Weekly*), but earned his living by teaching homeopathic medicine and languages.[18] He saw churches as a useful tool for social reform, since they were spaces where groups of people assembled and so places where he could explain his ideas and engage in political proselytism. All types of circles and schools, whether socialist or religious, whether they discussed philosophy or talked about social calamities, as well as use of the religious press were all seen as compatible with the societary doctrine and as methods of spreading its principles, in as much as it saw persuasion as the ideal instrument for winning followers and all means were valid for disseminating its doctrine. Except for the Roman Catholic Church, all the other Christian congregations practised an austere morality that Rhodakanaty found persuasive, as well as being suitable for shaping the minds of his followers; they would be highly valuable, therefore, given the process of secularization that the country had gone through.

Rhodakanaty was certain that the intolerant Roman Catholic Church had distorted the Gospel, perverted the faith and encouraged hypocritical and sexually deviant behaviour among the ministers of Catholic worship. In accordance with what he had learnt in the Orthodox Church of his native Greece, he considered it pressing, therefore, to end the celibacy of priests, as well as to adopt austerity in the practice of religion and daily life as preached by the Mormons. The ability to freely consult the Holy Scriptures, postulated by the Protestants, was both indispensable and urgent. The goal was to return to primitive Christianity. The separation of Church and State was crucial, not only because it corresponded to the theological division between the earthly and the divine, but also because it had generated innumerable conflicts throughout history. On the temporal plane, ministers of worship would have to be subject to civil laws and authorities. According to Rhodakanaty, socialism meant updating the message of Jesus Christ and the possibility of restoring the natural order that had been lost, put off or corrupted by the ambition and selfishness of the powerful. The process was inevitable because, from his point of view, all human societies were moving naturally towards progress and perfection.

Rhodakanaty was not a deist, nor was he an atheist; on the contrary, he considered that external worship was a justifiable expression of the Christian faith; consequently, he did not coincide with the Protestants, who were too strict on this point. This emphasis on the manifestations of faith did not distance him from Baruch Spinoza (1632–1677), of whom Rhodakanaty considered himself to be a follower, since the Jewish Dutch philosopher did not reject revealed religion as such, but the adulterations that were the work of theologians, which is why he is acknowledged as a precursor of biblical criticism. The geographical spread of Protestantism, according to the Greek socialist, corresponded to the thoughtful positivist character of northern Europeans, but not to the warm sentimental temperament of Latin peoples and there was no doubt that religious convictions played a role in the destiny of societies. Accordingly, while Protestantism accommodated social inequality, Catholicism harboured intolerance and obscurantism. Midway between the two positions, the Orthodox Church was the legitimate heir to the purest Christianity of the first Christian martyrs, and corresponded to a society that was democratic, austere and also practised the necessary external worship of God. Greece and Russia were its most important bastions; Turkey its limit. One of its most appreciable characteristics was its marked emphasis on persuasion and conviction, resources that Rhodakanaty regarded as essential and which harmonized easily with Fourier's societary doctrine.[19]

Towards the end of the 1870s, Rhodakanaty taught Greek and philosophy courses in the seminary of the Church of Jesus and also helped out in their shelter. It was said that he received twenty pesos a month for his classes.[20] He even managed to publish a booklet on the correct pronunciation of Greek, on which he was still an acknowledged authority at the turn of the century:

> To clarify this most interesting point on Greek pronunciation, let us see what Professor Rhodakanaty says about Erasmus' phonetic system. 'Classical Greek pronounced as it is taught in schools and universities in Europe [in Italy, France, Spain, Germany and England] as well as in the United States and here in Mexico, that is, according to the system of pronunciation *invented* by Erasmus of Rotterdam [for the single purpose of making it easier for foreigners to study] and more or less modified later in each country in accordance with the special characteristics of their respective languages, is just as intelligible today to modern Greeks, as it would be equally to ancient Greeks, since they had the same pronunciation, which has been preserved without interruption by a centuries-old tradition, and whose refusal to move on certain matters is such a particular feature of our race [...]'[21]

The shelter for young girls of the Church of Jesus had belonged to the convent of San Antonio Abad (St. Anthony the Great), which was nationalized by the Liberal government. María Josefina Hooker, who ran the institution, invited the Greek socialist to visit the premises, where Jacinto Hernández, one of his disciples, coordinated religious worship. Rhodakanaty was attracted by the austerity and the atmosphere of the orphanage and by Mrs Hooker's strict moral leadership. The fact that forty girls of different classes and nationalities could live together under the same roof clearly demonstrated that "la igualdad cristiana ha sabido nivelar bajo su fraternal disciplina" (Christian equality had managed, with fraternal discipline, to place on the same level) those whom society had unacceptably separated on the basis of their wealth. In a way, it represented an ideal community that ought to multiply its influence in society as a whole. Rhodakanaty, as he usually did, talked about social regeneration and the pressing need to improve the conditions of the poor inhabitants of the country and of the world. Nor did he pass up the opportunity to point out how urgent it was to create social institutions aimed at achieving women's emancipation. At that time, Rhodakanaty also wrote for *La Verdad*, modifying his view of Protestantism, showing greater benevolence towards the theological assumptions of the Church and its rites. He started from the consideration that Roman Catholicism and Christianity were not synonymous, so that it would be possible to be a Christian without being a Catholic. He would now characterize it as "pure Christianity", equivalent to the Orthodox religion, the Egyptian Copts, the Christians of Syria or the Armenians, in other words, all the "dissident communions", given that God had revealed himself in different ways and forms to all the peoples of the globe.[22]

In the middle of 1874, Brigham Young, President of the Church of Jesus Christ of Latter-day Saints, thought that the time was ripe to take the Gospel to the Mexicans, which is why he instructed Melitón González Trejo and Daniel W. Jones to translate the *Book of Mormon* into Spanish, publishing 1,500 copies of a selection from the sacred text, which they carried out themselves using a fund of 500 dollars received in donation. Shortly afterwards, the first mission was organized that would travel to Mexico; it comprised Jones himself (accompanied by his son), Helaman Pratt, James Z. Stewart, Anthony W. Ivins, Robert H. Smith and Ammon M. Tenney. Of these, only Jones and Tenney spoke Spanish. The evangelists set off from Nephi, Utah, in September 1875 and entered Mexico by present-day Ciudad Juárez, formerly Paso del Norte, just as winter was beginning. They spent some time in the city of Chihuahua from where, with the assistance of some supportive postal workers, they sent out hundreds of copies of *Trozos selectos del Libro de Mormón* (Choice Selections from the Book of Mormon) to prominent men across the country; one of these fell into the hands of Ignacio Manuel Altamirano and another into Rhodakanaty's. Daniel W. Jones records that the writer from Guerrero, Altamirano, sent him a letter of thanks, while Rhodakanaty's missives informed him that he had "una visión que le mostraba la verdad del *Libro de Mormón*" (a vision that showed him the truth of the *Book of Mormon*).[23]

In the early months of 1878, Rhodakanaty wrote to Trejo asking for more information about the Church, and John Taylor, its president after Young's recent death, sent him various pamphlets.[24] Although the Greek did not defend polygamy, he was rather closer to the approach of members of the Mormon Church on matters of agricultural colonization, collectivism, "united order" and concern for the indigenous population. Furthermore, he possessed the "gift of tongues", much esteemed by the Mormon rite. On November 15, Rhodakanaty declared that he was sympathetic towards it:

> After reading the Spanish translation of the Mormon Bible, which has reached these latitudes, I am convinced of the truth and purity of the Mormon faith, and am anxious to become a member of its Church but, given that there are no Mormon missions here, I am asking you to send one, with full authority to preach and convert.[25]

On 15 December, Rhodakanaty and some followers that he used to meet with on Sundays — at its height, the group increased from twenty to seventy people[26] — communicated to the Mormon hierarchy that they considered themselves to be members of that faith. According to the account by historians of the Church, the socialist Rhodakanaty published a monthly periodical entitled *La Voz del Desierto* in the Mexican capital in order to spread the faith.[27] The letter, sent to Salt Lake City, stated that:

> Having been convoked to a private meeting in the home belonging to Dr. Plotino Constantino Rhodakanaty, managing promoter of the same Church, for the purpose of organizing a small circle or congregation of religious and social persuasion in this capital city, said gentleman read to us for such purpose a work entitled Choice Selections from the Book of Mormon,

translated into Spanish by the reverend Elder Meliton G. Trejo and Daniel W. Jones, the mystical and highly transcendent meaning of which was later elucidated to us by the same Doctor, who proved unto us and fully convinced us of the evidence of the divine origin of such a precious book, and of the lofty mission that its doctrine has to accomplish in the world causing through its entirely providential and divine influence a complete humanitarian palingenesis or transformation in the religious as well as in the moral, social, and political orders.[28]

Dr Plotino Constantino Rhodakanaty signed as "propagador gerente de la Iglesia" (managing promoter of the Church), and Domingo Mejía, Darío F. Fernández, Miguel Enríquez, José Cleofas G. y Sánchez, Luis G. Rubín and Félix Rodríguez also signed the letter.[29] In January 1879, Taylor replied to the missive from Rhodakanaty, who responded on 27 February, finalizing details about the scope and composition of the new Mormon mission:

I very much hope that a month at most does not pass before the arrival of the missionary appointed by the decision of that centre. It should be noted that said missionary should be able to speak some Spanish and if he is accompanied by gentlemen with the same objective, it would be better for spreading the word to bring books too. It is best to take the Veracruz route because it is safe and the rail transport goes to the capital. I stress the need for the agent of the mission to come invested with broad powers, so that when I confer with him, we can settle some quite important business that I have to communicate to you.[30]

The "Quorum of the Twelve Apostles" of the Church of Jesus Christ of Latter-day Saints accepted Rhodakanaty's request. Despite Rhodakanaty's impatience, the second mission, composed of James Z. Stewart, Melitón G. Trejo and Moses Thatcher, embarked in New Orleans and did not arrive in the port of Veracruz until 14 November 1879, on board the steamer *City of Mexico*. Two days later, they put up at the Hotel Iturbide in the centre, in calle Francisco I. Madero, called calle de San Francisco at that time. On 16 November, the Mormon envoys received a visit from "Doctor Plotino Constantino Rhodakanaty, who was glad to see us, and we immediately had a good impression of his frank warm hearted manners and his frank and intelligent appearance."[31] Thatcher was struck by the number of natives that were still in the country, the opportunities available for agricultural colonization and the religious tolerance written into the Mexican laws.[32]

The mission settled into the Hotel San Carlos and later at No. 7 calle de San Francisco. After intense dialogue lasting four days, they baptized Rhodakanaty and Silviano Arteaga, and also confirmed them as members of the Church on Thursday, 20 November:

In company with Elders Stewart and Trejo, I visited Doctor Plotino Constantino Rhodakanaty at 8:30 this morning at his residence, and being there together with two believers we all knelt and offered prayers.[33]

They then moved on to Tacuba and, in a private garden near the tree of the Night of Sorrows, they located

a rustic dressing-room, circular in shape, of perhaps some three metres in diameter [which they entered] through a narrow bathroom door with a pool of some two and a half by three and a half metres, and one metre thirty centimetres or a metre and a half deep; [it was surrounded by a wall and railings], so that we were completely protected from the sight of anyone in the garden.[34]

Thatcher stepped into the pool, followed by Rhodakanaty,

With the watery element surrounding our bodies from just below the arm pits and with the clear blue sky of heaven above us on which the sun was shining brightly and while all nature was smiling around us and I believe angels were rejoicing above [I baptised] in the name of the Father Son and Holy Ghost the first person ever baptised in this part of the world by the authority of the Holy Priesthood since the days of the Nephites.[35]

Seven more people were baptized in those days of November. Three of those were ordained within the Melchizedek Priesthood and made elders, including Rhodakanaty who was also appointed as the president of the Mexican branch. Becoming a Mormon cost him his job, a wage of 20 pesos a month and caused a rift with the Church of Jesus.[36] Elder Thatcher noted that:

Brother Rhodakanaty has received news of his dismissal from the Episcopal Academy where he has been employed as a teacher of Greek. The bishop informed him that he could defend Catholicism or the doctrines of any of the Protestant sects, but not those of Mormonism, and that, through its [the church's] newspaper, he had to retract what he had said in *La Voz del Desierto*[37] in favour of those perverse persons, the Mormons, and that he had to deny all connection with and sympathy for those persons and that, if he did not, he would have to leave the Academy.[38]

To strengthen their evangelization campaign, the Mormons published the first Spanish edition of *Voz de la amonestación e instrucción al pueblo o sea la introducción a la fe y doctrinas de la Iglesia de Jesucristo de los Santos de los Últimos Días* (*A Voice of Warning and Instruction to All People, or, an Introduction to the Faith and Doctrine of the Church of Jesus Christ of Latter-day Saints*), written by Parley P. Pratt and translated by Trejo, Stewart and Rhodakanaty; this work kept the doctor busy from the middle of December 1879. By then, the relationship between the Mormon missionaries and Rhodakanaty had already deteriorated because of his attempts to persuade the faithful to join in his socialist project, assuring the Elders "que el pueblo aceptaría el Evangelio a cambio de escuelas, hospitales y colonización" (that the people would accept the Gospel in exchange for schools, hospitals and colonization). Rhodakanaty and some of his followers later broke with the Church of Jesus Christ of Latter-day Saints, although, according to Thatcher, the problem was a different one: Rhodakanaty was poor both in "spirit" and poor "economically" and when he was offered work, he would not accept it. Besides, given the straitened circumstances in which he was living, propagation of the Mormon faith was little more than a job for him, an attitude that was condemned by the congregation, which considered this materialistic motivation a spurious one when it was a question of spreading the doctrine. Furthermore, the Elder accused him of having published an article critical of the Mormon Church in *La Reforma Social* (2 May 1881), perhaps to put pressure on it. As proof, the congregation translated the text into English and sent it to Salt Lake City. Finally, on 28 August, Elder Wilcken read out in church Rhodakanaty's resignation from leadership of the Mexican branch, which was accepted by all those present, and he was excommunicated as an apostate.[39]

But man does not live by religion alone, and Rhodakanaty was sufficiently preoccupied with other things. The Greek socialist warned that the social revolution would happen suddenly if there was not a greater levelling of the social classes by attending to the problems of those most in need, who could turn the *status quo* upside down at any moment. The Paris Commune and the International Association of Workers were setting the trend in Europe, and in Mexico, sooner or later "the commune" would be paying us a visit. This revolution was as necessary as it was urgent for the "poor classes" who were without the basics. It also concerned the "reasonably comfortably off classes", because while their situation was not precarious, it was uncertain. Both were victims of "la anómala e irregular manera de ser de la propiedad" (the anomalous and irregular nature of property), which enables "un número muy reducido de personas" (a very small number of people) to own much more than other classes and live well above the level that was necessary. To combat these social problems it was essential to ignore the "teorías economistas, que son las que matan al pobre pueblo en nombre de la ley" (economist theories, which are what are killing poor people in the name of the law).[40] The vice at the root of this insensitive, unfair order was none other than the material inequality of the society that it was based on. Faced with this state of affairs, communism represented one of the most worrying "dangers for the future", given that although this "government of all, by all" obeyed the noble ideal of redeeming the poor, it would stir up disorder and hatred, making social harmony impossible. Rhodakanaty was not mistaken; the Rebellion of the *Pueblos Unidos* of Querétaro and Guanajuato soon broke out in which "General" Félix Rodríguez, one of the members of the *Social* baptized by the Mormons, played a prominent role.[41]

Transcendental philosophy

Even though philosophy occupied a central place in Rhodakanaty's intellectual discourse, perhaps from when he lived Berlin, the philosophical texts that he actually published correspond fundamentally to the final stage of his Mexican journey. He called his doctrine rationalism or transcendental philosophy and he began to disseminate it in the study circle that he formed in Mexico City. He went back to it in 1880, when he put on a course, open to the public, on the subject of positivism, the theoretical paradigm of the time, but which he rejected out of hand; however, he also published essays in which he presented both the philosophy of nature and Spinoza's pantheism. His phase of proselytism and practical action was coming to an end, and he returned to reflection, which became ever more abstract.

Rhodakanaty became familiar with Romanticism as a young man in Vienna and Berlin. In Central European circles, it was still considered to be a valid intellectual current, with its interplay between dreams, madness and the harmony of the human soul and the cosmos. The exploration of eastern religions and the occult sciences also formed part of Romantic hermeneutics. Karl-Gustav Carus and Eduard von Hartmann were to develop the notion of "the unconscious", linking it to either the universe or the individual psyche. For von Hartmann, even "the manifestations of the universal will are purposive and conscious, but without knowing themselves to be so". There would remain traces of a link between the past, the present and the future because human behaviour preserves traces of past behaviours belonging to a "golden age" that would possibly be discovered again in the future.[42]

Rhodakanaty's philosophical texts have more in common with Romanticism than with Kant's transcendental idealism or Hegelian dialectics, although he had great respect for both. With the spread of positivism, which banished God from the sphere of knowledge of nature and society, hostility towards pantheism — which Rhodakanaty identified with — was possibly increasing in Mexican academic circles. Even among spiritualists, who opposed positivism, there was no sympathy whatsoever for Fourier's societary doctrine and by the same token, as far as Rhodakanaty was concerned, spiritualism lacked a solid basis since it reduced "science to mere speculation" and also discarded "all experimental observation on the grounds that the idea is the true reality". In this, spiritualism gained very little, if anything, over positivist philosophy. In spite of the fact that the Greek doctor started from metaphysics, his intellectual horizons were complex, given that he included contemporary science and accepted a priori knowledge, although without dismissing empirical evidence, as is clear from his social engineering project, the Phalanstery, which would be empirically verifiable.[43]

Rhodakanaty thought of himself as a rationalist who embraced the philosophy of nature and pantheism. He believed, with Spinoza, that the universe, the world and man were the objectification of God in natural processes. Certain analytical and epistemological consequences followed on from this divine patent; the first was the assertion of the tendency towards perfection and harmony in all these orders, and the second was the necessary character of, on the one hand, events, processes and historical stages, and, on the other, of celestial and natural phenomena. Since nothing was arbitrary, everything could be known. Nonetheless, and this was fundamental to his philosophical viewpoint, science was one of the various ways of grasping reality. There were, rather, *a priori* truths, such as moral truths, which needed to be tested by experiment while others, such as physics, chemistry or social science, did not.[44]

Rhodakanaty had a wide-ranging philosophical culture. He made constant reference to Descartes, Spinoza, Pascal, Vico, Kant, Fichte, Schelling, Hegel, Feuerbach, Schopenhauer and von Hartmann in his manuscripts. He read Darwin and agreed with the principle of natural selection. He was also of the opinion that Wundt's psychology and the logic of Tiberghien were sound. He was well acquainted with the writings of Comte, Bain, Littré and Spencer and although he did not share the positivist ideas at their core, he thought the work of the English thinker was the richest and subtlest, while Comte's orthodoxy bordered on dogmatism. He abjured positivism's naïve belief in sensory knowledge, the value it attached to factual evidence as the criterion of truth, its rejection of *a priori* knowledge, the imperialism of the hard sciences over the human sciences, its denial of universal laws, its contempt for psychology (seen as a sort of metaphysics), as well as its vulgar materialism. According to Rhodakanaty, positivist morality was the morality of Anglo-Saxon utilitarianism; based on Hobbesian authoritarianism, its politics was driven by the law of "might is right", while its economic doctrine amplified social inequality.[45]

Rhodakanaty's controversy with the positivists began when Ignacio Mariscal, the Minister for Foreign Affairs, recommended him to Ezequiel Montes, responsible for the department of Justice and Public Education in the government of Manuel González, for a post (bear in mind that he had now severed relations with the Mormons). Apart from anything else, this enabled the good doctor to provide support for his "School of Transcendental Philosophy", which he set up in his surgery on 15

March 1880. He opened his seminar — given by him or one of his brightest followers — to students of both sexes (whether the classes were individual or group) on Mondays, Wednesdays and Fridays from 7 p.m. to 9 p.m. He charged four pesos per month, a sum that covered the costs of translating, printing, importing and publishing books and other items. This income, along with his consultations, the sale of homeopathic medicine and the classes in logic that he taught helped with Rhodakanaty's living expenses.[46] It is possible that he was still offering the Alkaheste Fluídico-Astral (universal panacea) to the public at 12 *reals* a flask, which was advertised as follows:

> This excellent medicine, the best of all those known until now, is the most efficacious restorer of the life force and a blood tonic and cleanser of ill humours, as experience has demonstrated, and as vouched for by the numerous authentic certificates from the sick, for whom three bottles at the most have sufficed to leave them completely healthy.[47]

Rhodakanaty stressed the importance of studying psychology and how useful it would be to include it in the syllabus of the Escuela Nacional Preparatoria (National Preparatory School) (ENP), as a prerequisite for understanding the Krausist-inspired work of the Belgian philosopher Guillaume Tiberghien, whose *Lógica. La ciencia del conocimiento* (Logic: The Science of Knowledge) — translated by José María del Castillo Velasco — had recently been adopted as a textbook in that renowned establishment. Rhodakanaty did not wholly agree with the Krausist line of panentheism, because he believed that it deviated from Spinoza's doctrine; nevertheless, panentheism itself lent weight his anti-positivist crusade. At the end of the day, Rhodakanaty did not achieve his objective (the first chair of psychology was instituted in the ENP in 1896) because in 1881 he became embroiled in an academic dispute with the naturalist Alfonso L. Herrera, the director of the institution, who, much to his surprise, met the suggestion of "such an important reform" with great reticence, and then hostility and open opposition.[48]

Rhodakanaty rejected the positivist approach according to which the object of psychology was the study of the laws governing the phenomena of the spirit, laws emanating from physiology that could be captured through empirical data perceived through the senses, empirical data systematized *a posteriori* by scientific discourse. He also argued that it was impossible to deduce the order of the mind from a biological foundation, since the mind constituted an *a priori* and obeyed an absolute abstract principle, like logic and mathematics (as Pythagoreans and rationalists showed), that could not be reduced to particular empirical manifestations, or be registered by the world of the senses or historical contingency. The soul, the object of psychology, could be neither measured nor weighed, which was why the method used to record spiritual processes would have to start from rationalist philosophy or metaphysics, since science was concerned with individual phenomena (whether natural or social), as well as accounting for the physical world, whereas the task of philosophy was to tackle the foundations of the world, absolute truths, universal values.[49]

In 1881, Rhodakanaty issued a short simplified version of Spinoza's *Ethics*, or a "pantheosophical catechism", for use as teaching material in his school, which merited an honourable mention at the Industrial Exhibition in Querétaro the following year. He credited Spinoza with having laid the foundations of the German school that followed by pondering, as did Fichte, on the "Absolute Ego", or the "transcendental subjective-objective identity" according to Schelling, or Hegel's "essential, logical evolutionary idea". Western philosophy was also indebted to Spinoza for the discovery of the one absolute universal substance, God, who included all things and had, at the same time, innumerable attributes, each one of them expressing "an infinite eternal essence", perfect, self-regulating, free and necessary, independent of any external factor and making self-determination possible. Since the universal essence was "a cause of itself", nothing was contingent and everything corresponded to "divine nature's need to exist and to work in a given manner". So, "second causes" — in other words, the manner in which empiricism tackled knowledge of the world of the senses by sidestepping the problem of God — were dismissed, in order to go back to the first cause, but not to the origin, because for pantheism nothing was created, everything was an extension of the Godhead.[50]

Rhodakanaty translated some passages from the *Theory of the Unconscious* (1868) by Eduard von Hartmann, a "respectable and very learned friend", and wrote a synthesis, published in instalments in *El Socialista* in 1884, as well as an essay, which he brought out the following year. The philosophical pessimism of von Hartmann, inspired by Schopenhauer, characterized the unconscious as a universal, omnipresent will that ruled the human psyche, an immanent rationality that acted purposively and consciously, "plastic energy", although without awareness of itself, like the universal Hegelian spirit. "But in rising above its individual striving, it also unites itself with the striving, teleological cosmic process as a whole. Recognizing the cosmic purpose, it can now selflessly identify with it."[51] Rhodakanaty saw von Hartmann as the modern interpreter of pantheism. It should be remembered that Spinoza included "intuitive knowledge" among the three kinds of cognition, that is, knowledge directed at grasping the "formal essence of certain attributes of God", whereas von Hartmann employed the "inductive method of the natural sciences" to penetrate the human unconscious. This introspection demonstrated the existence of a pre-established harmony, and, as the soul was indivisible, that harmony acted simultaneously in every part of the organism and even if this were cut up into pieces, its original attributes would remain intact.[52]

The final years

Rhodakanaty frequently changed his place of residence. In 1877, he was living at no.10, calle de la Amargura; in 1879, he lived and had his surgery and medical dispensary on the ground floor of no.1, Puente de Balvanera, and after that, possibly at no. 9, callejón de López; by 1880 he was in the callejón de Cuajomulco, where he also had his surgery. He subsequently occupied a house at no. 10, callejón de la Santa Veracruz. The Greek socialist constantly had problems earning a living, but it did not prevent him from carrying out his proselytizing activities and earning something of a reputation in the Evangelical churches, workers' organizations, local City Hall and the liberal circle. He also formed a family about which, as with everything concerning Rhodakanaty, we know little, and what we do know is confused. In the only one of his poems that we know of, from 1877, he addresses Ida as his soulmate, and recalls the moment "cuando tus bellos ojos lanzaron sobre los míos esa mirada fascinadora, en ella me comunicaste toda la fuerza de tu inteligencia, todos los encantos de tu amor" (When your beautiful eyes held me spellbound by their gaze, in it you conveyed to me all the force of your intelligence, all the charms of your love).[53]

Whether Ida was real or imaginary, it was not with Ida that he engendered a son, but with Soledad Girón. Plotino Nefi Rhodakanaty Girón was born in Mexico City in 1879. Rhodakanaty's "infant son" was blessed by the Mormon missionaries, who gave him the name of "Nephi Moses" within the congregation; from that moment on he would be welcomed by the Church and would live, as Moses Thatcher wrote in his journal, "in the grace of the Lord and [would] perform ordinances for the redemption of his forefathers".[54] A letter of 1880 gives an account of the straitened circumstances in which the "teacher of transcendental philosophy and languages" lived with his family:

> Sincerely wishing to contribute in some way to the enlightenment and morality of the masses of the town, I have established a People's School in this house (which is your house) at number 10 callejón de la Santa Veracruz, where, for a modest stipend, I teach various subjects of positive utility for the working and commercial classes, as you will see from the attached syllabus accompanying this request; however, having reached the critical unfortunate moment where I find myself currently devoid of all pecuniary resources, with the inevitable burden of sustaining my family, and without being able to continue paying the rent of the house that I am occupying as a consequence of finding myself, in addition, quite ill and for this reason prevented from raising the wherewithal to keep my establishment going [...].[55]

Apparently, the City Hall did not grant his request because when they registered Plotino Nefi, Rhodakanaty and Soledad Girón stated that they were still living at number 10 callejón de la Santa Veracruz. There are two birth certificates for the child, issued on 28 and 30 March 1881 respectively; the mother's name appears on the first one and Rhodakanaty's on the second; the witnesses are the same in both cases. Soledad Girón is said to be from Huichapan, of legal age and "casada amasiamente", that is, a common-law wife. And two days later, "Mr Plotino Rhodakanaty of Greece, 47 years old, a bachelor, homeopathic doctor, living at callejón de la Santa Veracruz, number 10, appeared and presented a little boy, alive, Plotino Nefi, who was born at No. 9 López on 8 eighth July 1879 at 2.00 am, his natural son."[56]

The financial hardships did not ease while Rhodakanaty persevered in his attempt to acquire a piece of land to set up his school and also move his home there. In December 1884, he reported to the city council the existence of "un terreno eriazo, que mide 30 metros de longitud por 20 de latitud, situado hasta la parte este de la garita de Peralvillo, entre estación Irolo y el Rastrillo núm. 4 del ferrocarril de Circunvalación" (a piece of waste ground, measuring 30 metres long by 20 metres wide, bordering the eastern part of the Peralvillo Customs House, between Irolo station and Rastrillo No. 4 of the railway loop). The legal process followed its course, so that on 6 March 1885, the local council agreed "se pregone por los periódicos dicha denuncia" (to announce said report in the newspapers). On 3 October, the council resolved to adjudicate "to citizen Plotino Rhodakanaty the piece of land he reported". Nevertheless, one problem remained to be solved; the city council set the price of the land at 228 pesos and 22 centavos, and received a counteroffer from him that "el honorable ayuntamiento se sirva conceder la referida adjudicación en el precio de cien pesos ($100) pagaderos, con hipoteca del mismo terreno, en 20 mensualidades, conforme al precedente general de las leyes de nacionalización" (the honourable city council should kindly concede the referred adjudication at a price of one hundred pesos ($100) payable, with the land itself standing as surety, in twenty monthly instalments, in accordance with the general precedent of the laws of nationalization"). In response to his request, on 24 November, the council reduced the sum to 154 pesos and 20 centavos.[57] This amount, however, was beyond Rhodakanaty's financial means and, in early 1886, he wrote as follows to the city council:

> Since it is absolutely impossible for me to pay a fixed sum every month, in view of the fact that what I can generate in my shaky profession as a homeopathic doctor is so intermittent, and which I [can only] exercise with a good deal of uncertainty, because of suffering from a chronic illness that prevents me from leaving my home unaccompanied, and on account of such a lamentable state, I beg of you, citizen president, to deign to allow me to make the payment of the amount referred to, or the value of the land, in the same period of one year that has been indicated to me, but by paying what I can when I can, solemnly committing myself to reimburse in said period, the full amount stipulated at once, which I certainly can do, provided that I can make whatever use I please of the land mentioned, in order to make it easier to expedite the payment of it.[58]

In March, the city council decided in favour of Rhodakanaty's request, without this being much use, since the doctor resolved to transfer the right to the property to another person. What could have been so urgent that made Rhodakanaty cede the rights to 616.0082 square metres of land for fifty pesos? It must have been something important. Furthermore, when he realized that he could not meet the commitment, he played the card of writing to President Díaz requesting the free gift of a building in the Centro del Panteón de Santa Paula, in view of "the pressing nature of my circumstances and the alarming seriousness of a chronic illness that I suffer from".[59]

The last piece of information that we have in this pitiful chronology indicates that on 14 June 1886, Helaman Pratt, accompanied by Elder Stewart, Silviano Arteaga and Amando Pérez, visited Rhodakanaty at his home (perhaps still in callejón de la Santa Veracruz), enquiring of him whether "he wanted to continue to be recognized as a member of the Church", as well as calling upon him to "attend to his duties" (of president of the branch and elder); his reply is not known. By then, the Greek socialist was suffering such economic hardship that he trying to survive by giving classes in the Instituto Monasterio, a Catholic school; furthermore, he was ill and it had been a year since he had published any of his writing in socialist newspapers.[60]

Nevertheless, in January 1886, the national press had reported the foundation — on 22 of that month at Rhodakanaty's home — of the literary society, the Arcadia Mexicana, under the motto of "honradez en la igualdad" (integrity in equality). The circle comprised Rhodakanaty himself, José Vera, Manuel Agoitia, José Muñoz, Carlos Servo, Felipe Torres, Eduardo Álvarez and José Monroy, these last two being secretary and president respectively. None of them were veterans of his political adventures. A year later the Arcadia Mexicana brought out a scientific and literary weekly magazine with the same name in which only Monroy appeared among the contributors. The Greek doctor no longer had the opportunity to return to Europe as Valadés and Hart speculated. Residing, very probably in abject poverty, in one room at number 5 callejón de Magueyitos (today Galeana, in the Guerrero Colony), Rhodakanaty died of pernicious fever at midnight on Sunday, 2 February 1890 in Mexico City, and was buried the following day. His remains lie in the Panteón Civil de Dolores.[61]

Notes

1 José C. Valadés, *El socialismo libertario mexicano (siglo XIX)*, prologue by Paco Ignacio Taibo II (Culiacan: UAS, 1984), p. 17; Registro Civil del D.F., 1881, fols 109, 111, MHMM [Museum of the History of Mormonism in Mexico], unclassified.

2 Valadés, *Socialismo libertario mexicano*, p. 18. There is no evidence of later contacts, since, in Proudhon's correspondence at least, no letters have been found addressed to Rhodakanaty. Pierre-Joseph Proudhon, *Correspondance*, préc. J.A. Langlois, 14 vols (Paris: Lacroix, 1875).

3 The reference to the pamphlet is from Valadés, *Socialismo libertario mexicano*, p. 18. After a thorough search of major libraries in Europe and the USA, I found no testimony to its publication.

4 Plotino C. Rhodakanaty, *Obras*, edition, prologue and notes by Carlos Illades, compilation by María Esther Reyes Duarte (Mexico: UNAM, 1998), p. 34; Manuel Dublán and José María Lozano, *Legislación mexicana o colección completa de las disposiciones legislativas expedidas desde la independencia de la República Mexicana*, 42 vols (Mexico: Imprenta de Comercio, 1876–1890), 8: 95.

5 Isaiah Berlin, *The Crooked Timber of Humanity: Chapters in the History of Ideas*, edited by Henry Hardy, foreword by John Banville (London: Pimlico, 2013), p. 46.

6 Carlos Illades, *Las otras ideas. Estudio sobre el primer socialismo en México 1860–1935* (Mexico: ERA/UAM, 2008), p. 137.

7 Carlos M. Rama, *Utopismo socialista, 1830–1893* (Caracas: Biblioteca Ayacucho, 1977), p. lix; Pierre-Luc Abramson, *Las utopías sociales en América Latina en el siglo XIX* (Mexico: FCE, 1999), p. 227; Rhodakanaty, *Obras*, p. 22, cited here.

8 A good deal of confusion surrounds Villanueva. While Valadés, *Socialismo libertario mexicano*, p. 22, portrays him as an active, vigorous young man, who died before his time, Rosendo Rojas Coria, *Tratado de cooperativismo mexicano*, 1952, 3rd ed. (Mexico: FCE, 1984), pp. 152–153 attributes the foundation of the philanthropic friendly society *La Gran Familia* in 1840 to someone of the same name, which means that, if it was the same individual, he would have been at the very least a mature man in the 1860s.

9 Marco Antonio Anaya Pérez, *Rebelión y revolución en Chalco y Amecameca, Estado de México, 1821–1921*, 2 vols (Mexico: Inherm/Universidad Autónoma de Chapingo, 1997), 1: 117; Rhodakanaty, *Obras*, p. 22; Valadés, *Socialismo libertario mexicano*, p. 170.

10 Plotino C. Rhodakanaty and Juan de Mata Rivera, *Pensamiento socialista del siglo XIX*, edition, prologue and notes by Carlos Illades, compiled by María Esther Reyes Duarte (Mexico: UNAM, 2001), p. 30; Carlos Illades, *De la Social a Morena: Breve historia de la izquierda en México* (Mexico: Jus, 2014), pp. 42–43. The former is cited. For a fuller, more detailed version of the *Social*, see Illades, *Las otras ideas*, chap. 5.

11 Carlos Illades, *Rhodakanaty y la formación del pensamiento socialista en México* (Barcelona: Anthropos/UAM, 2002), pp. 68–69; Rhodakanaty and De Mata Rivera, *Pensamiento socialista del siglo XIX*, p. 116.

12 Rhodakanaty, *Obras*, pp. 170, 171; Jean-Jacques Rousseau, *Contrato social*, 1969, translated by Fernando de los Ríos Urruti, 10th ed. (Mexico: Espasa-Calpe, 1992), p. 127.

13 Illades, *Rhodakanaty y la formación del pensamiento socialista*, p. 68; Rhodakanaty, *Obras*, p. 249. The latter is cited.

14 Rhodakanaty, *Obras*, p. 261: "¿de qué nos serviría solamente gozar de nuestros inalienables derechos políticos y naturales [...] si, por otra parte, carecemos de una garantía que nos asegure la subsistencia, proporcionándonos pan y trabajo, que son los dos elementos necesarios de nuestra existencia física?"

15 *Ibid.*, p. 240: "entonces, a no dudarlo, sería justificado el principio anárquico y revolucionario de Proudhon que, hasta ahora, se ha mirado como una exagerada utopía."

16 Illades, *Rhodakanaty y la formación del pensamiento socialista en México*, p. 57.

17 *Ibid.*, p. 58.

18 *La Democracia*, 1 December 1872.

19 Rhodakanaty and De Mata Rivera, *Pensamiento socialista del siglo XIX*, pp. 24–26; Rhodakanaty, *Obras*, pp. 216, 246.

20 "Journal of Helaman Pratt", MHMM, unclassified.

21 Cited in Jesús Díaz de León, *Curso de raíces griegas*, 7th ed., corrected and enlarged (Paris: Librairie de la Veuve de Ch. Bouret, 1903), p. 16: "Para esclarecer este punto tan interesante sobre la pronunciación griega, veamos lo que dice el profesor Rhodakanaty sobre el sistema fonético de Erasmo. "El griego clásico pronunciado como lo enseñan los colegios y universidades de Europa [en Italia, Francia, España, Alemania e Inglaterra] y también en los Estados Unidos y aquí en México, es decir, conforme al sistema de pronunciación *inventado* por Erasmo de Rotterdam, [con el único fin de facilitar su estudio a los extranjeros] y más o menos modificado después en cada nación según la índole especial de sus respectivos idiomas, es tan inteligible hoy para los griegos modernos como lo sería igualmente para los griegos antiguos, puesto que éstos tenían la misma pronunciación que se ha conservado por una tradición secular ininterrumpida, cuyo carácter de inmovilidad en ciertas cosas, es tan peculiar en nuestra raza [...]." Author's own emphasis. I am grateful to Rodolfo Suárez Molnar for the reference.

22 Rhodakanaty, *Obras*, pp. 30, 256. These more restrained writings with regard to Protestantism, as well as his collaboration with the Church of Jesus, have persuaded some historians that Rhodakanaty embraced Protestantism in 1877. Jean Pierre Bastian, *Los disidentes. Sociedades protestantes y revolución en México, 1872–1911* (Mexico: El Colegio de México/FCE, 1989), pp. 65–66. This seems unlikely if we take his earlier criticisms into consideration. At most, the episcopal affiliation of the Greek socialist lasted hardly any time at all, since, in 1878, he approached the Church of Jesus Christ of Latter-day Saints. According to Mormon historians, Rhodakanaty was already sympathetic to their Church from 1876. Raymundo Gómez González and Sergio Pagaza Castillo, *El águila mormón o el anarquista cristiano. Plotino Constantino Rhodakanaty primer miembro de la Iglesia de Jesucristo de los Santos de los Últimos Días en México* (Mexico: Museo de Historia del Mormonismo en México AC, 1997), p. 12.

23 F. Lamond Tullis, *Mormons in Mexico: The Dynamics of Faith and Culture/Los mormones en México. La dinámica de la fe y la cultura* (Mexico: Museo de Historia del Mormonismo en México A.C, 1997), pp. 172 and ff; Daniel W. Jones, *Memoirs*, n.d. MHMM, unclassified. The translation was done by the Museo de Historia del Mormonismo [Museum of Mormon History] in Mexico A.C. The latter is cited. See also Daniel W. Jones, *Forty years among the Indians: A True Yet Thrilling Narrative of the Author's Experiences Among the Natives* (Salt Lake City: Juvenile Instructor Office, 1890).

24 Tullis, *Mormons in Mexico*, pp. 179 and ff. Years earlier, Taylor had been on a mission to Paris where he made contact with a journalist named Krolokoski who was an admirer of Fourierism. Robert C. Webb, *The Real Mormonism: A Candid Analysis of an Interesting but Much Misunderstood Subject in History, Life and Thought* (New York: Sturgis & Walton Company, 1916), p. 119.

25 MHMM, unclassified. The letter is written in English. The Museum's own Spanish translation is cited here: "Después de haber leído la traducción española de la biblia mormona, que ha llegado hasta estas latitudes, he quedado convencido de la verdad y pureza de la fe mormona, y ansío llegar a ser un miembro de su iglesia pero, dado que aquí no existen misiones mormonas, les solicito que envían una, con amplia autoridad para predicar y convertir."

26 Agricol Lozano Herrera, *Historia del mormonismo en México* (Mexico: Zarahemla, 1983), p. 28. A figure of eighty sympathizers has also been suggested. Ygnacio Zárraga, "Editorial", *In Yaotlapixqui (El atalaya de México)*, September 1938. Amando Pérez talks of an initial group of twelve people that later grew to sixty. *Sendero Lamanita*, 1942.

27 B. H. Roberts, *A Comprehensive History of The Church of Jesus Christ of Latter-day Saints:*

Century I, 6 vols. (Provo, Utah: Brigham Young University Press, 1965), 5: 569; "The division of the Mexican mission", *Liahona*, July 1956. The information is taken from the journal of Moses Thatcher.

28 It has been rightly pointed out that, in the letter itself, the socialist conviction of the signatories is made explicit, a fact that was not sufficiently taken into account by those in the Church hierarchy. Bill L. Smith, "Impacts of the Mexican Revolution: The Mormon Experience, 1910–1946", (PhD diss.,Washington: Washington State University, 2000).

29 MHMM, unclassified. The museum has a copy of the letter.

30 MHMM, unclassified. The letter is written in Spanish. "Deseo mucho que no pase un mes a lo más sin que venga el misionero ya indicado según la determinación de aquel centro. Conviene poner que dicho misionero sepa hablar algo del idioma español y si viniese acompañado de señores con el mismo objeto sería mejor para la propaganda a la vez que provisto de libros. El mejor camino es el de Veracruz por la seguridad y el transporte del ferrocarril que conduce hacia la capital. Insisto en la necesidad de que el agente de la misión venga investido de amplias facultades para que al conferenciar yo con él podamos arreglar negocios bastante importantes que tengo que comunicarles."

31 "Registro de miembros e hijos de la Iglesia de los Santos de los Últimos Días perteneciente a la Misión Mexicana" [Register of the members and children of the Church of Latter-day Saints belonging to the Mexican Mission] book 1, MHMM, unclassified; "Paquete americano de Nueva Orleáns", *El Monitor Republicano*, 15 November 1879; "Letter from Moses Thatcher to John Taylor, Mexico D.F. 29 November 1879", MHMM, unclassified. The latter is cited.

32 *The Year of Jubilee. A full report of proceedings of the Fiftieth Annual Conference of the Church of Jesus Christ of Latter-day Saints, held in the Large Tabernacle Salt Lake City, Utah, April 6th, 7th and 8th, A.D. 1880. Also a report of the Exercises in the Salt Lake City Assembly Hall, on the Sunday and Monday just preceding the Conference, reported by George F. Gibbs and John Irvine* (Salt Lake City: Desert News Printing and Publishing Establishment, 1880), pp. 15 and ff.

33 "Letter from Moses Thatcher to John Taylor, Mexico D.F. 29 November 1879", MHMM, unclassified. "En compañía de los élderes Stewart y Trejo esta mañana visité a las 8:30 al doctor Plotino Constantino Rhodakanaty en su residencia, y estando allí junto con dos creyentes ofrecimos oraciones todos de rodillas."

34 *Ibid.*: "un rústico cuarto vestidor en forma circular de tal vez unos tres metros de diámetro [al que entraron] a través de una angosta puerta de baño con una pila de unos dos y medio por tres metros y medio, y con una profundidad de un metro treinta centímetros o de un metro y medio; [una pared y un enrejado la rodeaban] de tal manera que estábamos completamente protegidos de la vista de los que estuvieran en el jardín."

35 *Ibid.*

36 Illades, *Rhodakanaty y la formación del pensamiento socialista en México*, p. 104.

37 A monthly publication edited by Rhodakanaty himself according to the Mormon envoys. "Registro de miembros e hijos de la Iglesia de los Santos de los Últimos Días perteneciente a la Misión Mexicana", [Register of the members and children of the Church of Latter-day Saints belonging to the Mexican Mission] book 1, MHMM, unclassified.

38 "Letter from Moses Thatcher to John Taylor, Mexico D.F. 29 November 1879", MHMM, unclassified: "El hermano Rhodakanaty ha recibido noticia de haber sido despedido de la Academia Episcopal en donde ha sido empleado como profesor de griego. El obispo le informó que podía defender el catolicismo o las doctrinas de cualquiera de las sectas protestantes más no las del mormonismo, y que, por medio de su periódico, tenía que retractarse de lo que había dicho en *La Voz del Desierto* a favor de esas perversas personas de los mormones, y que tenía que desmentir toda conexión con y simpatía para estas personas y que, de no ser así, tendría que dejar la Academia."

39 "Journal of Moses Thatcher", MHMM, unclassified; Lozano, *Historia del mormonismo en*

México, p. 34; F. Lamond Tullis, "Early Mormon Exploration and Missionary Activities in Mexico", *Brigham Young University Studies* vol. 22, 3 (1982): 309–310; Tullis, *Mormons in Mexico*, p. 207.

40 Rhodakanaty, *Obras*, pp. 63,112. For Fourier, "la economía política invadía todo el dominio de la charlatanería" (the political economy was taking over the whole domain of charlatanism). Charles Fourier, *La armonía pasional del nuevo mundo*, introduction by Eduardo Subirats and Menene Gras (Madrid: Taurus, 1973), p. 142.

41 Rhodakanaty, *Obras*, p. 260; Carlos Illades, *Conflicto, dominación y violencia. Capítulos de historia social* (Mexico: Gedisa/UAM, 2015), pp. 73 ff. The former is cited.

42 Isaiah Berlin, *Las raíces del romanticismo*, edited by Henry Hardy (Madrid: Taurus, 2000), p. 27; John W. Burrow, *The Crisis of Reason: European Thought, 1848–1914* (New Haven and London: Yale University Press, 2000), p. 65. The latter is cited.

43 Rhodakanaty, *Obras*, p. 129; Illades, *Rhodakanaty y la formación del pensamiento socialista*, p. 36. The former is cited.

44 Illades, *Las otras ideas*, p. 139.

45 *Ibid.*, p. 138.

46 Rhodakanaty, *Obras*, pp. 68, 128.

47 Don Evaristo, "Plumadas", *El Combate*, 6 June 1878: "Esta excelente medicina, la mejor entre todas las que se conocen hasta ahora, es el más eficaz restaurador de la fuerza vital y depurativo de la sangre y de los malos humores, según lo ha demostrado la experiencia y lo acreditan los numerosos certificados auténticos de los mismos enfermos a quienes han bastado a lo más *tres botellas* para quedar enteramente sanos." Author's emphasis.

48 Rhodakanaty, *Obras*, p. 68. A detailed analysis of the translation of Tiberghien's book is in Antolín C. Sánchez Cuervo, *Krausismo en México* (Mexico: UNAM, 2004), pp. 140 and ff. Psychology would finally find a niche in the positivist system when Justo Sierra presided over the Segundo Congreso de Instrucción Pública (Second Congress of Public Education) in 1891. It was only in 1907, however, that Ezequiel Chávez and Justo Sierra, influenced by the American psychologist, James Mark Baldwin, opened the door to the Sociedad de Estudios Psicológicos (Society of Psychological Studies), Charles A. Hale, *La transformación del liberalismo en México a finales del siglo XIX* (Mexico: Vuelta, 1991), pp. 407–408; Mauricio Tenorio-Trillo, "Stereophonic Scientific Modernisms: Social Science Between Mexico and the United States, 1880s–1930s", *The Journal of American History* 86, no. 3 (1999): 1180.

49 Illades, *Las otras ideas*, pp. 141–142.

50 Celestino Díaz, *Memoria de la Primera Exposición Industrial de Querétaro, y lista de objetos presentados en la misma* (Querétaro: Imprenta de Luciano Frías y Soto, 1882), p. 86; Rhodakanaty and De Mata Rivera, *Pensamiento socialista del siglo XIX*, pp. 117, 169, 119, 126, 120, 124. The latter is cited.

51 Burrow, *Crisis of Reason*, p. 66.

52 Rhodakanaty, *Obras*, pp. 185, 186. Rhodakanaty and De Mata Rivera, *Pensamiento socialista del siglo XIX*, pp. 134, 135. Von Hartmann, considered to be a precursor of psychoanalytic theory, died in 1906. Eleven editions of his book were published during his lifetime.

53 Rhodakanaty, *Obras*, p. 74.

54 "México, bautismos: 1560–1950", MHMM, unclassified; Gómez González and Pagaza Castillo, *El águila mormón o el anarquista cristiano*, addendum 1.

55 "Carta de Plotino Rhodakanaty al presidente del ayuntamiento, México D.F., 22 de octubre de 1880" (Letter from Plotino Rhodakanaty to the president of the city council, Mexico D.F., 22 October 1880), Archivo Histórico del Distrito Federal [AHDF], Fondo Ayuntamiento, Instrucción Pública, vol. 2490, exp. 1480: "Deseando sinceramente contribuir de alguna manera a la ilustración y moralidad de las masas del pueblo, he establecido una Escuela Popular en esta su casa callejón de la Santa Veracruz núm. 10, donde por un

módico estipendio enseño varias materias de positiva utilidad para las clases obrera y mercantil, según verá usted por el adjunto programa que acompaño a esta solicitud: más habiendo llegado el momento crítico y fatal de encontrarme en la actualidad exhausto de todo recurso pecuniario, con el gravamen forzoso de sostener a mi familia y sin poder seguir pagando el arrendamiento de la casa que ocupo a consecuencia de encontrarme además bastante enfermo e imposibilitado por lo mismo para arbitrarme recursos con que seguir fomentando mi establecimiento [...]." The syllabus content was quite comprehensive and included logic, ideology, morality, sociology, phrenology, hygiene, physics, chemistry, natural history, pure mathematics, mechanics, humanities (Greek, Latin, universal history, rhetoric, mythology and classical biographies) and languages ("Mexican", Spanish, German, French, English, Italian). The price was two pesos, "payable in advance".

56 The record is in the D.F. Civil Register of 1881, fols 109, 111, MHMM, unclassified: "compareció el señor Plotino Rhodakanaty de Grecia, de 47 años, soltero, médico homeópata, vive en el callejón de la Santa Veracruz núm. 10 y presentó vivo al niño Plotino Nefi, que nació en el de López número 9 el día 8 ocho de julio de 1879 a las 2 de la mañana, hijo natural suyo."

57 "Carta de Plotino Rhodakanaty al presidente del ayuntamiento. (Letter from Plotino Rhodakanaty to the president of the city council, Mexico D.F. 5 December 1884), AHDF, Fondo Ayuntamiento, Terrenos, vol. 4088, exp. 1429; AHDF, Fondo Ayuntamiento, Terrenos, vol. 2491, exp. 1557, fols 3, 28, 30.

58 "Carta de Plotino Rhodakanaty al presidente del ayuntamiento, México D.F., 19 de enero de 1886" (Letter from Plotino Rhodakanaty to the president of the city council, Mexico D.F., 19 January 1886), AHDF, Fondo Ayuntamiento, Terrenos, vol. 2491, exp. 1557, fol. 33: "Mas no siéndome absolutamente posible exhibir cantidades fijas cada mes, en atención a la eventualidad de lo que voy generando en mi raquítica profesión de médico homeópata, y la cual además ejerzo con bastante inseguridad, por adolecer de un mal crónico que me imposibilita para salir libremente de mi domicilio y a razón a tan lamentable estado a usted suplico ciudadano presidente se digne concederme el que verifique yo el pago de la referida cantidad o valor del terreno, en el mismo plazo de un año que me ha sido señalado, pero mediante la exhibición de partidas libres, comprometiéndome solemnemente a integrar en dicho plazo, toda la cantidad ya mencionada a la vez, el que desde luego pueda hacer yo el uso que me convenga del citado terreno a fin de tener mayor facilidad de expeditar su pago."

59 AHDF, Fondo Ayuntamiento, Terrenos, vol. 2491, exp. 1557, fol. 35; "Carta de Plotino Rhodakanaty al presidente del ayuntamiento, México D.F., 2 de junio de 1886" (Letter from Plotino Rhodakanaty to the president of the city council, Mexico D.F., 2 June 1886), AHDF, Fondo Ayuntamiento, Terrenos, vol. 2491, exp. 1557, fol. 39. In July 1888, Virginia Alcalde had still not settled the payment for the piece of land. AHDF, Fondo Ayuntamiento, Terrenos, vol. 2491, exp. 1557, fol. 44; AHDF, Fondo Ayuntamiento, Terrenos, vol. 2491, exp. 1557, fol. 28; "Carta de Plotino Rhodakanaty a Porfirio Díaz, 20 de abril de 1886", (Letter from Plotino Rhodakanaty to Porfirio Díaz, 20 April 1886), Colección Porfirio Díaz, Universidad Iberoamericana [CPD], leg. 11, box 9, doc. 4145.

60 "Journal of Helaman Pratt", MHMM, unclassified.

61 "Instituto Monasterio", *El Tiempo*, 4 February 1886; "La Arcadia Mexicana", *El Siglo Diecinueve*, 28 January 1886; Valadés, *El socialismo libertario mexicano*, p. 144; John Mason Hart, *El anarquismo y la clase obrera mexicana, 1860–1931* (Mexico: Siglo XXI, 1980), p. 97.

9 The Cecilia Colony: Echoes of an Amorous Utopia in the Libertarian Press

Laura Fernández Cordero

The experiment known as the Cecilia Colony was founded in Brazil by Giovanni Rossi and a group of Italian anarchists at the end of the nineteenth century. The aim of this chapter is to explore the intense coverage given to the experiment in the anarchist press in Argentina and which has so far received little attention. A potted history of the brief life of the colony appears in a pamphlet entitled *Un episodio de amor en la Colonia Cecilia* (An Episode of Love in the Cecilia Colony), published in 1893 by Livorno in their Biblioteca del Sempre Avanti series. Two years later, the Buenos Aires publishing group La Questione Sociale selected it for a series aimed at women. The author was trying to narrate the most dramatic side of the experiment: the practice of "amor múltiplo" (plural, multiple love). The colony survived for only three years; nevertheless it proved to be very effective in terms of propaganda. In the early months of 1892, there was already news of Cecilia in the anarchist newspapers, *El Perseguido* and *La Révolte*. Shortly afterwards, other anarchist newspapers circulating in Argentina, such as *La Protesta Humana*, *La Revolución Social*, *El Oprimido*, and *La Fuerza de la Razón*, also reported on the experiment and contributed their own critical viewpoints. This chapter presents an account of these traces in the press and sets them in dialogue with the most general framework of anarchist ideology relating to the utopian tradition, free love and the emancipation of women.[1]

Anarchisms and utopias

The points of contact between anarchism and utopian thought are not as obvious as might appear. For many anarchists, the utopian imagination was too organized. For others, it was an exercise that distracted them from the fight against capitalism.[2] The experimental colony also aroused suspicion, above all when it had recourse to the money of the bourgeoisie or royalty to finance itself, or when it led to workers emigrating to unknown countries, so weakening the battle front in Europe. For these reasons, there are few utopian accounts that are anarchist as such, although several can be found presenting libertarian points of view or elements that anarchism shared with other currents concerned with sociality.

The resistance of anarchism towards the utopian tradition stems from its notion of freedom, which was so extreme that it prevented them, in many cases, from thinking ahead to what the organizational coordinates of the ideal city might look like. Determining the guidelines for communal living or government in advance must have seemed a complete contradiction to them. At the same time, the epithet of utopian was laden with certain negative connotations related to the projects being flights of fancy or pipe-dreams and so impossible to realize. When the publishers, La Protesta, launched a series called "Los utopistas" (The Utopians), the introduction made it clear that:

> Utopia is not, as is often believed, a pipe-dream or the result of a fanciful desire to offer social evolution a ready-made plan. [...] libertarian utopias are much less numerous, because the energy of the Libertarians in the fight against so many obstacles is very frequently absorbed by other kinds of militancy.[3]

The anarchist Matt Nettlau was pointing out there that the Utopian had to resolve one central question: forms of authority. In many cases, despite the good intentions of its creator, the perfect city eventually turned into an implacable order maintained with obvious authoritarianism,[4] a tendency that the convinced anarchist could not overlook. For some, though, the reasons for writing utopian accounts were overwhelming, partly because anarchism was a tributary of the same broader Enlightened tradition that revived the genre and linked it to scientific progress. At the same time, it offered them an excellent vehicle for propaganda since many readers were drawn to the attractive, intriguing narrations about utopias.[5] It also enabled the anarchists to offset their reputation for being "destroyers of order" and to demonstrate that they had thought of a viable organization for the day after the revolution.

Finally, another point of contact between anarchism and utopian thought is that they both coincide in not subordinating the organization of everyday life and affective relations to the alleged superiority of economic transformation; in this sense, both traditions show that they are highly sensitive and intuitive with regard to the impact of social orders on subjectivity. In the most classic accounts of the utopian genre, the system of government and the arrangements for eating, the local economy and dress, urban planning and forms of love are all described with equal care. Amorous relationships and sexuality (generally linked to reproduction) are never left out of the utopian plan. Furthermore, most of the authors of this tradition during the nineteenth century try to provide some response to the demands for the emancipation of women.

In the case of Joseph Déjacque (1821–1864) — whose work, *L'Humanisphère* (*The Humanisphere*), is recognized as a precursor of the libertarian utopias — the intention to rethink the role of women was not only vividly expressed in the account, but also in the author's explicit response to the misogyny of Proudhon.[6] The cross between utopia and anarchism was particularly fruitful in Argentina where the French anarchist, Joaquín Alejo Falconnet, under the pseudonym Pierre Quiroule, wrote a trilogy of utopias: *Sobre la ruta de la anarquía* (On the Road to Anarchy) (1912), *La ciudad anarquista americana* (The Anarchist City of America) (1914), *En la soñada tierra del ideal* (In the Promised Land of the Ideal) (1924).[7] In the second of the trilogy, which is the best known, the author spends twenty-four chapters describing the ideal city, the product of "a work of revolutionary construction" and adds a perfect, geometric plan laid out with houses and gardens, bathing and swimming establishments, and well-stocked storehouses, on streets called "Abundancia" (Abundance), "Actividad" (Activity) and "Humanidad" (Humanity). His proposal for love is not one of the most extreme, although he does imagine the family as being replaced by society, which is responsible for raising children in an efficient *pouponnière*, or crèche house. He plans, in turn, for women to have their own *home* where they can live alone.

Before going into detail about the repercussions of La Cecilia in the anarchist press, we shall briefly describe the experiment, before giving an account of the framework of the debates about free love.

An Italian-style amorous episode in Brazil[8]

One of the most famous utopians is undoubtedly Charles Fourier.[9] Ambitious in his writings and confident in his revolutionary plans, Fourier worked hard towards the building of phalansteries, the creation of small perfect societies based on rational combinations of types of passion. Their guiding principles were attractive labour and task rotation so that their inhabitants could exploit their personal tastes to the full, to the benefit of social order. He wrote various volumes explaining the global scope of his projects and, at the beginning of the nineteenth century, anticipated transformations of space and social relations, which, today, seem to us quite mundane.[10]

Marx and Engels were able to see the discerning critical sense that Fourier was applying to civilization, and his sensitivity to the historical subordination of women. Given his interest in the passions, he proposed a "new amorous world" in which desire was the mainspring of social consensus and certain "manias" would not be understood as aberrations or deviations, but as examples of the profound variability of humans. His closest disciples took pains to conceal the more audacious facets of it, especially those passages in which Fourier integrated into the phalanstery those who obtained pleasure from relations with persons of the same sex and who enjoyed group sex. "Sapphianism" or scenes between women lovers formed part of his own personal "mania" and the only problem he saw in those ways of experiencing pleasure was that he might not be able to find companions of both sexes willing and ready to share them. It was no coincidence that those old manuscripts, which had been gnawed by mice, saw the light of day again in the 1960s when the new winds of free love and sexual revolution began to blow, making it easier to read about that amorous utopia.

Despite the fact that the benefactor that Fourier apparently waited for at noon every day never appeared — which meant that he was unable to sow the world with phalansteries — his works were nevertheless widely read. Fifty years after his death, a restless Italian, Giovanni Rossi, was alternating his books on agronomy with writing inspiring passages about communities that lived in harmony based on principles that were neither authoritarian nor oppressive. Rossi was a committed anarchist and also a thoughtful scientist, convinced that an experiment was the best way to demonstrate that anarchy was viable. He used his newspaper, *Lo Sperimentale*, to encourage all types of utopian projects, as a result of which he ran into problems with those, like the well-known Errico Malatesta, who stated that setting up colonies in isolation was counterproductive to the revolutionary process, since those efforts needed to be concentrated in Europe or in the rapidly expanding cities of the American continent.[11]

Rossi planned and implemented a few utopian experiments in Italy until he found his own benefactor, Pedro II, the ruler of the Empire of Brazil. Shortly before the empire was replaced by the young republic, the ruler granted Rossi a piece of land in the area of Paraná. The future founder immediately assembled a small group of people and set sail, planning to work out the specific details of the long-awaited colony that was to be called Cecilia.[12] What met them was a country in political turmoil, a hot, very humid climate and a piece of land a long way away from the major cities. His plan was to make preparations for the arrival of a larger contingent that would bring families to join this advance party, in which there were two or three women, although this is not known for certain.

Within a few months, they had managed to clear the ground, build some dwellings and start the sowing that would make the small village of Anarquía self-sufficient. In those early stages, food was scarce and comforts non-existent. The work consumed their every effort and living without rules turned out not to be so simple. The environment was not easy either and those tempted by Rossi's propaganda joined and left the project whenever they were able to find less precarious opportunities elsewhere.

The colony began in 1890 and, despite the difficulties, remained viable for three years. Rossi was the community's internal leader and organizer, but was also responsible for letting the outside world know how the experiment was faring. He was convinced that the colony would make a major contribution by providing propaganda with a tried and tested experiment that showed the specific possibilities of organizing social life according to the principles of anarchism. He wanted to demonstrate that work for the common good was given spontaneously, as well as make the experimentation in love arrangements clear. His readings of Fourier had led him to think that one of the main problems in sociality was the way in which the passions were organized. Affective and amorous passions in the form in which they were known in the traditional family represented a dead weight, which Rossi invited new arrivals to cast off soon after they arrived in the colony. Inherited roles and marital jealousies, however, prevented new ties from being created and retained the selfishness typical of the family group. The founder, however, advocated the *finis familias* and exploring the promises of libertarian love. His hypothesis that "loving more than one person at the same time was a need of human nature" would not only be put to the test in remote *Anarquía,* but was also summarized in a pamphlet that would take the story to the world.[13]

Un episodio de amor en la Colonia Cecilia (An Episode of Love in the Cecilia Colony) was published as a pamphlet through the Biblioteca del Sempre Avanti in Livorno in 1893, that is, immediately following the dissolution of the village. Rossi signed it under his own name, but added the pseudonym of Cardias. Two years after that first edition, the Buenos Aires publishing group, La Questione Sociale, chose it to complete its own series aimed at women,[14] hence, following on from pamphlets called *A las Hijas del Pueblo* (To the Daughters of the People), *A las muchachas que estudian* (To the Girls Who Study) and *A las proletarias* (To the Proletarian Women), this short ideological essay appeared combining the characteristics of scientific treatise and romantic novel.[15]

The opening epigraph serves as a warning "Si la verdad te espanta, no leas; porque este librito está, para ti, lleno de espantos" (If the truth horrifies you, do not read it; because, for you, this little book will be full of horrors). The main characters are Cardias himself and a couple, Aníbal and Eléda, who had just arrived in the colony. This unusual triangle would revolutionize traditional theory about married couples and try to make it a reality. In this spirit, and while acknowledging a mild sensation of falling in love, Cardias communicates his reasoning to Eléda. Already convinced by the postulates of free love, she agrees to talk to Aníbal and to offer him the revolutionary deal: she will share her time, her love and her body with both men.

Aníbal is hurt but bravely accepts the agreement. Much of the language that runs through the pamphlet is of the sentimental novel, albeit restrained by rationality, the superiority of the libertarian cause and control of the passions. Aníbal suffers the undesirable bitter taste of jealousy, knows that Eléda is not his property, but prejudice and habit still eat away at him the first night that she sleeps with Cardias. Nonetheless, disinterested love, the joy of comforting another companion and his faith in freedom soon cure him and enable him to send fraternal greetings to the man who, in any other context, would be his rival. The remainder of the colonists look with respect upon the experiment that finally puts the theory of anarchist love into practice. The narrator has delicately sought to spare us any possible conflict. No passions are unleashed, there are no broken hearts or tragic overtones, and to test it, he subjects each of the participants to psychological questionnaires. The questions are diverse; in the case of Aníbal, he is interested in probing his faithfulness to the libertarian cause and his victory over doubt, fear, and selfishness; in relation to Eléda, he is concerned to demonstrate that she has not lost her health, moral integrity, nobility of character and, above all, that the experiment has not made her "more sensual". Over and above personal sentiments, what is valued is the outcome "con preferencia a todo, socialmente útil" (with a preference above all for the socially useful), as the undaunted Eléda asserts before answering the last question in the negative: "¿Te disgustaría no conocer la paternidad de un hijo que ahora generases?" (If you bore a child now, would you be upset not to know who the father was?)

The first part of the pamphlet closes, without outbursts of passion or major disagreements, with a reflection in which Cardias reviews his own feelings and confesses that the goodnight kiss that Aníbal sends when Eléda is with him has a calming effect. The second part is one long reasoned argument in which the theoretical assumptions of the thesis to be demonstrated are laid out. As the narrator has told us in advance, the central idea is *finis familias*, or the destruction of the family, the hard central core of selfishness, superstition, vices and slavery.

In the Italian edition, the *episode* was preceded by a kind of report about life in La Cecilia; it explained how a small number of pioneers set sail on February 20, 1890, with the idea of founding an experimental colony in the municipality of Palmeira, in Paraná state, Brazil. The initial work was arduous and they were constantly threatened by poverty, which, in the end, was what led to the dissolution of the village in only three years. Although the relationship with a hostile environment, their own lack of experience of agricultural work and the challenge posed by an organization without leadership all contributed to the downfall, Rossi identifies the survival of the family as a much larger problem. After giving an account of the economic and intellectual life of the village, he dwells on the moral aspect and the subtle commandments that persist in spite of the "morally hygienic environment" that they purported to construct. He asserts that "la destrucción progresiva o espontánea de la familia monogámica prepara el terreno al triunfo de nuestro ideal" (the progressive or spontaneous destruction of the monogamous family prepares the ground for the triumph of our ideal), although he warns time and again that family relationships remain firm and pose an obstacle to experimenting with other ways of organizing work and living together in harmony with others. As an example, he cites that "la casa social era confiada por pura formalidad a la única mujer del primer contingente" (the social house was entrusted as a pure formality to the only woman in the first contingent), but when families were added, the mothers spread themselves between the kitchen and the store.

The pamphlet in which the love episode appeared was advertised as being "full of horrors" because of the substantial transformation that it required of the emotions and the body, aspects that other emancipatory movements had paid least attention to. The fact is that Rossi was occupied with the critical problem of planting potatoes and cassava but he realized, at the same time, that the eventual fate of La Cecilia would hinge on the love life within the colony. The small village of Anarquía had to demonstrate, not only the viability of collectivist work, voluntary service and extreme democracy, but also the transformation of everyday life, love relationships and subjectivity itself.

We should not, however, be too quick to celebrate the "feminism" of Rossi, who seems to be giving women the benefits of polyandry, until we have checked the backdrop to the romance: the dearth of women in the colony. No lone female pioneers arrived in La Cecilia and those who came accompanied were reluctant to take part in the trios that could in theory be tried out in the poor village. Eléda's acceptance, therefore, is the answer to two problems with fatal consequences for the social order: celibacy and masturbation.

When, following the dissolution of La Cecilia, Rossi planned a new venture, he took the problem of the scarcity of women very seriously. In a letter to his friend Alfred Sanftleben, who was in Germany at the time, he confessed the following unsettling idea:

> It would be a question of setting up a distillery and [...] of buying young Indian girls from semi savage tribes with the aquavit! They would quickly become free comrades, but what a shameful/wicked way of founding one's freedom![16]

The exchange of letters with Sanftleben brings to light other details of the romance.[17] This is how we find out that Eléda is an anagram of Adele, a woman with the surname Serventi who had arrived in Paraná with her partner Annibale after being captivated by Rossi's talks in Italy.[18] In the letters sent by the leader to his German friend, the self-sacrificing Aníbal / Annibale seems to be more jealous of and less in thrall to the decisions of Rossi. It seems likely in fact that he intervened in the account of the love episode, for example, by obliging Rossi to leave out the participation of a third man, the young Breton, Jean Géléac, who had joined the romantic arrangement afflicted by a misfortune or sickness greater than hard labour or starvation: sexual abstinence. In one of his missives, Rossi comments that the young man would have died "por darse a la masturbación a causa de la preocupación de las mujeres de la colonia por preservar su honorabilidad" (of masturbation because of the concern of the women in the colony to preserve their honour).[19] Identifying masturbation with sickness, the dissipation of vital energies and death is a constant in local anarchist discourse.[20] Consistent with the prevailing ideas of their time, they rejected "masturbatory" practice and, in dramatic tone, listed it as one of the most repugnant vices.

Saved from such a fate, Géléac even turned out to be the father of Ebe, Adele's first daughter. Rossi confirms this detail, and it is curious that he should highlight it, because one of the pillars of the project that he himself strongly advocated was to blur the boundaries of certainty over paternity, and with it, its corollary of patrimony and inheritances. Nonetheless, before the disappearance of the colony and of Géléac, Adele and her two children left with Annibale, who was able to give them some support, in spite of his indolence and alleged alcoholism, two defects that any anarchist would deplore. By that time, Giovanni would have been living in Brazil on his own, pursuing his profession; nonetheless, years later, back in Italy, he set up a common family with Adele under the surname Rossi. According to her daughter, Ebe Rossi, Adele and Giovanni communicated very little of their life in Paraná to her and preferred not to recall that episode.[21]

Echoes in the press

Although the colony only survived three years, it nonetheless continued to be very much present in anarchist propaganda, since its founders and those who lived in it always made an effort to keep in touch with the libertarian newspapers. As was pointed out earlier, in 1892, there were allusions to La Cecilia in *El Perseguido* of Buenos Aires.[22] *La Révolte* — issues of which reached Argentina through an exchange that the newspapers organized themselves and were kept in the bookshop of Émile Piette[23] — published an informative letter by A. Capellaro and a discussion of the question some months later.[24]

In the letter, the informant describes the colony, giving an account of the surface area and number of inhabitants, the variety of foods and the organization of the work. It is also stated that they were going to demonstrate that anarchy was not a utopia, as their adversaries claimed, but that it could, in fact, be put into practice. In the following issue, Capellaro described the poor food, the meagre balance sheet (yet in spite of their poverty, they had set up a library) and the harshness of the work that was still carried out despite the absence of bosses and overseers. He announced that the arrival of more olonists, and especially, women, was expected. Some months later, under the heading "Colonisation anarchiste" (Anarchist Colonization), an unnamed staff writer complained that Capellaro's reference to the scarcity of women provided bourgeois journalists with the opportunity to speculate that the colonists would opt for sharing them, thus reverting to a state of "primitive promiscuity".[25] He, on the other hand, wanted to clarify that the ideal anarchist made sure that the total autonomy of women was recognized, especially in so far as "matters of love" were concerned.

Before the clarifications made in the French newspaper, the editors of *El Perseguido* in Buenos Aires published a very similar version of the letter, sent by C. A. (Amilcare Capellaro?) from the Cecilia Colony. To the comment that what tormented them most was the fact that "el amor libre todavía no ha penetrado en el corazón de nuestras compañeras" (free love has still not penetrated the hearts of our female companions), the editors add their unsigned opinion:

> As for free love, it is well understood that in a Colony, the single man must suffer because he cannot impose his love on his [female] companions in the Colony. It is our understanding that free love does not mean that the woman is obliged to give herself to all men, and one man to all women, but that, on the contrary, the woman is free to give herself to whomsoever she chooses and if she only wants to give herself to one man, she is free to do so. When an individual, whether man or woman, is acting of his or her own will, without subjecting himself or herself to any rule, he or she is acting freely, but everyone is free to make any rule they

> want for themselves, never for another person; so, the man who wants to go with many women, let him do so if he can find any who want to go with him, and the woman who wants to go with many men, let her do so if she can find any who want to go with her, but equally free is the man who only wants to love one woman, or the woman who only wants to love one man.[26]

Despite the fact that none of the letters reported amorous trios or quartets, both newspapers expressed misgivings about the radical nature of the experiment.[27] In other cases, its scientific side was highlighted; the editors of *La Revolución Social*, for example, announced the appearance of Rossi's pamphlet and commented that "it lends itself to an in-depth study of the practice of free love, this becoming a matter of utmost importance because of the psychological data that accompany it."[28] In the Bibliography section of the newspaper *El Oprimido*, there is praise for the publication and firm support for the thesis of plural, multiple love.[29] The author highlights the importance of destroying the family and points out:

> This, which scandalizes and makes the hair of the moralists stand on end. . . without morality, it is nonetheless, logical and reasonable; and, we would also say, indispensable for human well-being. [. . .] We recommend it to all women in particular, and to male companions in general.[30]

The group that published, *La Fuerza de la Razón* (The Force of Reason), however, included an unsigned opinion piece in which, not without praising the series of pamphlets, the author opposed Rossi's theses and the views expressed in *El Oprimido*:

> The ecstasy of the most intense pleasures cannot be multiform. One can move from ecstasy to ecstasy, but one cannot be in two different places at one and the same time [. . .] just as one cannot blow out and suck in at the same time. [. . .] As for the rest, the pamphlet is quite profoundly logical and scientific, especially in the part that anathematizes the bourgeois family.[31]

Months later, *El Oprimido* maintained its defence even when it published a note to the contrary in which R. Canto declared that some cases of polygamy could be accepted, but that there was no reason to think of it as the general state of affairs in amorous relationships. The editors went on to point out that his arguments for refuting Rossi's thesis were very poor.[32]

As can be seen, the controversy was never resolved, and the more extreme positions —in support of "plural, multiple love" — and the more moderate ones — restricted to "free union" — coexisted across the broad spectrum of the anarchist press.

Free union or absolutely free love?

Rossi's work and the fact that a group of editors should choose to publish it gives an idea of anarchism's contribution, typical of its time, to making the sexual question a topic of interest.[33] The particular feature of anarchism was that they debated those subjects in the press itself; for example, in 1897, it was possible to find a brief item praising a speaker who had given a talk to local village men and women and had addressed the topic of "sexual relations" without recourse to euphemisms.[34]

In spite of the fact that the idea of free love was readily identified with the anarchist manifesto, other political and cultural movements, such as socialists and freethinkers, also proposed it;[35] furthermore, not even those within the libertarian camp agreed on a single definition of the meaning of free love, which was constantly subject to debate.[36] By recuperating or reconstructing some of the debates and dialogues about free love that the militants — editors, correspondents and contributors, both male and female, rather than the major names or leading theorists — engaged in over many years, before and after the publication of the pamphlet, this multiplicity of opinions becomes even more obvious.

The series of pamphlets of *Propaganda entre las mujeres* (Propaganda Among Women) itself offers two quite dissimilar versions. The publishers inserted an unsigned appendix, entitled "La unión libre" (Free Union). By scouring the texts available at that time, one realizes that it is a paraphrase of a passage that Jean Grave devotes to love in *La société mourante et l'anarchie* (Moribund Society and Anarchy).[37] This way of defining relationships between couples was a synthesis of a minimal agreement: serial monogamous relationships with no legal or religious sanction. Only those participating in the "affinity" could mark the beginning and end of the bond founded on "free will". The text follows through by characterizing bourgeois marriage as a form of prostitution and predicting the eventual disappearance of adultery and hypocrisy, since the respect for "genetic necessity" would lead to a family based on true love and affection. They used this argument to defend themselves from the accusation of being "destroyers of the family", since they would only be fighting the juridical family. They consider it impossible to "subject sexual relations to rules and regulations" and assert that attempts to do so have only led to "new vices". They believe it necessary for human nature to develop in complete freedom in order for "evolution" to act and the "fittest" to reproduce and survive. In short, monogamy based on love, affection and sympathy, even if not governed by laws or other forms of regulation, will be enduring.

The addition of this passage indicates a tendency to believe that freedom leads to monogamy and not to the multiplicity of affections, as Rossi suggested in the following pamphlet. A year after the series appeared, the first locally produced pamphlet was published in Rosario devoted to the emancipation of women and affective and sexual relations. *La mujer y la familia* (Woman and the Family) was a lecture in which the well-known medical doctor Emilio Z. Arana set out an erudite analysis based on history, philosophy and anthropology.[38] After his daughter recited a poem, the doctor opened his lecture without moving too far away from the basic ideology: woman, "ese ser tan desgraciado" (that unfortunate being), had to be emancipated "con iguales derechos, sin más diferencia en sus derechos que los contingentes á su sexo" (with equal rights, with no more difference in her rights than those contingent upon her sex). According to the author, modern civilization had made the situation of the "fair sex" worse and condemned it to slavery, transforming the man "de compañero en amo" (from companion to master). Building his case on the well-worn thesis of original matriarchy, Arana tries to show the historicity of monogamy and concludes that "la institución matrimonial es la base del régimen económico vigente" (the institution of matrimony is the basis of the prevailing economic regime) and declares that it would, therefore, be necessary to abandon marriage as prostituted by bourgeois laws and give women all rights, including sexual ones. Nevertheless, those terms needed to be defined with utmost care and precision, since:

> There are those who, not being familiar with the intimate nature of love — of whose existence there can be no question — defend the multiple nature of this affection, the existence of more than one affection of this type, that "*loving more than one person at the same time is a necessity of human nature*"; this is quite simply a gratuitous statement without any basis to it, which reveals a lack of observation on the part of the person who makes such a claim, or [has] intentions, that, frankly, I neither understand nor wish to investigate further.[39]

Without naming the author, Arana is referring here to the thesis that Rossi sought to verify and publicize. Like the editors of *La Révolte*, he was concerned about relapsing into "primitive promiscuity", many memorable instances of which, according to him, could be found in the Bible. According to Arana, sexual love for an anarchist was "natural and noble" and the purpose of it was "the preservation of the species". Love could therefore be "elective and mutable" but not excessively so. In his version of free love, he explains that "no es posible que si la mujer tiene caprichos sexuales, haya el hombre de procurárselos" (it cannot be that, if a woman has sexual whims, it is the man who has to obtain them for her). For this speaker "well qualified in medical and social sciences", nature and morality were the limits within which "free union" should be contained and, as a synonym of free love, it should mean nothing more than

> the natural union of two individuals of a different sex for a more or less extended period of time, according to the duration of the effect they profess, whether of sympathy or tenderness that brought them together.[40]

He declares that in future society, the number of those who have short-term relationships "to satisfy their genetic needs" will be fewer and fewer since the majority will choose the "sweet delights of the home". The first part of Arana's pamphlet is a recital of the best-known arguments for explaining the historicity and evolution of the juridical family. The whole framework of the argument, without citing them, is based on two works widely circulated among anarchists, socialists, feminists and freethinkers at the time: *La mujer en el pasado, en el presente y en el porvenir* (Woman in the Past, Present and Future) by August Bebel (1879) and *El origen de la familia, la propiedad privada y el Estado* (The Origin of the Family, Private Property and the State) by Friedrich Engels (1884).[41] Both of these were based on the theses of the anthropologists Johan J. Bachofen (1861) and Lewis H. Morgan (1877) about the ancient world, especially those referring to a primitive state of promiscuity and the consequent demonstration of the historicity of patriarchy and the family, and to the hypothesis of an original matriarchy at a time when women would not have been subordinate to the patriarchal order.[42]

Dr Emilio Arana and the majority of the libertarians followed this line of thought because the works were considered as part of modern scientific knowledge about the social world and history. As has been stated, some of the central ideas of evolutionism, positivism and hygienism were shared by anarchism, socialism and liberalism.[43] Nonetheless, a single quotation from an author or a reference to a particular theoretical tradition might accompany quite opposite lines of argument. So, for example, there are echoes of those widely accepted anthropological notions in Rossi's pamphlet, although they are used in an argument structure that ends up justifying polygamy, not, as was customary, monogamy. This particular instance could have derived from his Fourierist readings; it is likely, therefore, that when Rossi recognized that he had gone beyond particular limits that anarchism itself dictated, he began his pamphlet by warning of the horrors that his study might cause. But, which of them would be the most shocking? Certainly not the idea of avoiding religious or legal sanctions, since not only was this part of the basic creed of anarchism but, as Arana himself said in passing, most workers were not legally married anyway; civil marriage was in fact a very recent phenomenon and, until the Civil Code was modified in 1889, religious marriage was enough to legalize the union.[44]

From the libertarian point of view, the corollary of this kind of union was adultery. Hence, for some anarchists it would actually be untenable for free love, in certain versions, to look like it (i.e. adultery). Far from adopting a tone of condemnation, Rossi explained that adultery was proof of the possibility of loving more than one person at the same time:

> It is free love minus the loyalty, or rather the lie, the enjoyable lie; that is the sophistication of free love; it is adultery. [. . .] People are well aware of these things and put them into practice every day. Except that right is exercised furtively, like theft; and what should be free trade takes on the pleasurable and provocative — but undignified — character of contraband.[45]

These are echoes of Fourier who thought of adultery as a seed that could be left to germinate because it would demonstrate "la posibilidad de compartir amigablemente en el amor" (the possibility of sharing love amicably).[46] As might be envisaged, multiple unions would entail not being able to determine paternity, for which reason, some anarchists preferred "free union" based on a pact of exclusivity.

This version was perfectly synthesized in the section that accompanied the series of pamphlets aimed at women. So, even though they denounced all laws that consecrated patrimony and matrimony, a limit seemed to be drawn by those anarchists at the point where blood lines became blurred, and as a result, extolling (serial) monogamy was very important in those groups that shouldered the reputation of being destroyers of the family and champions of the community of women. Adherence to evolutionist ideas and confidence in the unlimited progress of humanity seemed to have left some anarchist activists with the impression that multiple ties and not knowing who the biological father was entailed a return to "savage" states or, as most of the followers of this restricted version of free love repeated, one might lapse into "primitive promiscuity".

Whether primitive or in the present, promiscuity seemed to be the greatest horror that the pamphlet about La Cecilia could deliver, particularly if it spread to women who, in the case of a colony populated by many single males, would be the major figures in polyandrous love arrangements. The margins of female sexual freedom might be extended to unimaginable limits if the example of Eléda proliferated, as Rossi seemed to suggest in *Episodio de amor*:

> The heroine [...] was tied by eighteen years of married life and a crown of five children. Nonetheless, she too has felt a new affection arise alongside the old affection; and she has nobly expressed this to the father of her children [who] heroically drained the bitter chalice [...] We are all pleased with him for the strength of spirit with which he fulfilled his duty.[47]

Some writing in the press celebrated specific examples of women who were most free:

> The female companion of hero Paulino Pallás has united freely with one of those comrades who were prisoners in the warship "Navarra" during the persecutions. May the example spread and for black and white priests to remain idle is our desire, because, to enjoy sexual pleasures, no ceremonies are needed, but true affinities and the need for sexual pleasure.[48]

Others, such as Arana, preferred to restrict the number of love protagonists in the present or to postpone it to an anarchist future. For the most extreme cases, Rossi warned that "among many anarchists [there are those] who think they are the most fervent instigators of freedom, yet when it comes to love are still Muslims, or something worse."[49] He extends his call to women but warns that there is little hope for emancipation when the men who are interested disregard it themselves, and admits that even the greatest enemy of property will try and retain possession of women. When the women themselves talk about love, they do not share the most radical versions; on the contrary, because of their own *flesh and blood* experiences, they warn about the prejudices and violence that libertarian romances too may conceal.[50]

Some years later, the anarchists would hear of a new radical twist to free love from the virtuous lips of the respected Italian criminologist, Pietro Gori. Gori, a driving force behind the organizing sectors that eventually prevailed, travelled the country, enthusiastically spreading the word. In one of his most oft-quoted lectures, he addressed the women and attempted to demonstrate that women's and workers' causes were intimately linked.[51] The journalists of the time point out the hypnotic effect that Gori had on women, indicating to them, practically from the start of his talk, which model they should follow: the Spartan woman, the companion to the ancient warrior. After reciting two or three main points of doctrine, which included comparing bourgeois marriage to prostitution, he opened with a disturbing question: What happens when one is the companion of a socialist or an anarchist? He went on to develop a version of love that does not fall under the heading of free union and constituted one of the so-called most extreme versions of sexual freedom. At the beginning of the twentieth century, Gori told the women in his audience:

> If he sees you as a comrade, he will not let you succumb to the prejudices of a religious or bourgeois order, which relate that a woman cannot or should not feel love for more than one man at a time, and not experience the joys of sex with more than one man, always the same one; or which cause the pleasures of the flesh to be considered inferior to any other. If he thinks of you as a comrade, he will bring you up to date with liberating and subversive ideas and knowledge in the matter of sex, if you are unaware of them, even if only to let you make up your own mind. If he sees you as a friend, the voluptuous experiences that you can obtain for yourself outside the home will make him as happy as those that you can find within it, because nobody can call himself a friend of another if he does not share in his joys. And even if you desire to have a child with a man other than the one you cohabit with, is not your body your own?[52]

One might imagine that speeches of this type were infrequent or that, given the maelstrom of voices, they succumbed in favour of less extreme tones. However, Gori's speech was recuperated on various occasions, including at the end of the 1920s, when *La Protesta* did not seem to be receptive to such heterodox discourse on feminine sexuality. The publishing house La Protesta was also responsible for re-issuing Rossi's pamphlet in 1920 as part of a small series that also included Charles Albert's *El amor libre* (L'Amour Libre) first published in 1900. The theses of the two pamphlets could not have been more different. In the very first paragraph, Albert attacks those who understand free love as "a regression to the sexual promiscuity of the earliest primitive ages, or fortuitous pairing based on whim and giving free rein to the passions."[53] On the contrary, he argues:

> We know, in fact, that the love that has been the centre of so many rigged pedantic arguments is just the form of sexual attraction typical of beings who have reached a certain stage of development and which, for that very reason, forms the moral necessity of their reproduction.[54]

In 1921, another major publishing house belonging to Bautista Fueyo issued *Socialismo y Anarquía* (Socialism and Anarchy) by Errico Malatesta, in which the respected Italian author injected a certain note of realism against dalliances of a more romantic sort, warning that the anarchist solution did not prevent being hurt by love since the happy encounter of two lovers could never be guaranteed. The family, which could be relaunched, according to the moral criteria of anarchism, would still retain a certain raison d'être:

> Let us note first that, in spite of the regime of oppression and lies that has prevailed, and prevails still, within the family, it has been and continues to be the major factor in human development, since it is in the family that the normal man sacrifices himself for the sake of man, and does good for the sake of good, without wishing for any other reward than the love of his companion and his children.[55]

In short, over four decades, the libertarian camp oscillated between two poles: serial monogamy with no legal or religious sanction, on the one hand, and free, multiple, simultaneous experiences of love, on the other. The restricted version of "free union" was perhaps the most widespread; nonetheless, the most audacious strand of plural, multiple love at the end of the nineteenth century — embodied in the account of La Cecilia — was not lost and echoes of it resonated in two new, widely circulated, studies in the 1920s. In the first place, the most extreme elements reappeared in Julio Barcos' *Libertad sexual de las mujeres*, (Sexual Freedom of Women) a book that spread beyond libertarian spaces and was reissued many times. This work, whose introduction claimed that it was "modern, advanced and scientific", cited Bachofen, Morgan and Engels again,[56] while the second of these, Morgan, was even considered at length under the sub-heading of "Theory of the evolution of the family". Barcos pointed out that in multiple love affairs — whether simultaneous or serial — paternity would lose its importance and, conscious of the radical nature of his proposal, he elected to reinforce the maternal tie. So, implicitly recuperating the arguments of Bachofen, the "right of mothers" was adopted again in all its force, the mother-child relationship was enshrined as the only certain biological tie and this became the feature upon which the family relationship was built. His book would be transcribed and critically commented on by those responsible for the newspaper for women, *Nuestra Tribuna*.

In second place, there was the amorous camaraderie of Émile Armand, whose proposals circulated in Argentina via the columns of the *Ideas* newspaper in the city of La Plata in the early 1920s. Apart from offering a perspective from the standpoint of Individualist Anarchism, Armand recuperated the ideas of Fourier as a foundation for variation in sexual tastes, and opened up the possibility of wondering whether, in an absolutely free version of love, there would be a place for people who loved those of their own sex.

During these same years, the publishing house, La Protesta, re-issued the series of pamphlets from La Questione Sociale, and with them, the episode of love. Once again, the Colonia Cecilia reasserted its twofold utopian status. In principle, because of its very characteristics, it was an imagined city, utopian, a benefactor, an unexplored territory, a social reinvention, and so on. On the other, it was a vehicle for the dreams of loves that were vital, diverse, and mutable: the utopia of a new amorous world.

Notes

1 This chapter quotes sections of my book, *Amor y anarquismo: Experiencias pioneras que pensaron y ejercieron la libertad sexual* [Love and Anarchism: Pioneer Experiences that Thought and Exercised Sexual Freedom] (Buenos Aires: Siglo XXI, 2017).

2 Luis Gómez Tovar (comp.), *Utopías Libertarias Americanas. La ciudad anarquista americana de Pierre Quiroule* (Madrid: Tuero, 1991); Luis Gómez Tovar (comp.), *Utopías Libertarias. Esbozo de Historia de las Utopías de Max Nettlau* (Madrid: Tuero, 1991).

3 Max Nettlau, "Introducción", in *El humanisferio* (Buenos Aires: La Protesta, 1927), pp. 9–13: "La utopía no es, como se cree a menudo, el sueño ocioso o el resultado de un deseo quimérico para ofrecer a la evolución social un plan hecho de antemano. [...] las utopías libertarias son mucho menos numerosas, porque la energía de los libertarios en lucha contra tantos obstáculos es absorbida muy frecuentemente por otros géneros de militancia." Nettlau refers to the Cecilia Colony in glowing terms, as well as its utopian correlate *Paraná en el siglo XX*. Gómez Tovar, *Utopías libertarias*, p. 83.

4 Max Nettlau, *Esbozo de Historia de las Utopías* (Madrid: Tuero, 1991). Laura Fernández Cordero, "Buenos Aires de la utopía", in *El hilo rojo. Palabras y prácticas de la utopía en América Latina*, edited by Marisa González de Oleaga and Ernesto Bohoslavsky (Buenos Aires: Paidós, 2009), pp. 33–43.

5 Raymond Trousson, *Historia de la literatura utópica. Viajes a países inexistentes* (Barcelona: Península, 1995).

6 Déjacque, *El humanisferio*.

7 Adriana Petra, "¿Sueñan los anarquistas con mansiones eléctricas? Ciencia y utopía en las ciudades ideales de Pierre Quiroule", in *El hilo rojo,* González de Oleaga and Bohoslavsky (Buenos Aires: Paidós, 2009), pp. 55–70. Pierre Quiroule, the pseudonym of Joaquín Alejo Falconnet, Lyon, 1867–Buenos Aires 1938. He lived in Argentina as a child. He published the periodical *La Liberté* (1893–1894). He contributed to *El Perseguido* (1890–1897) and *Le Cyclone* (1895). He formed part of the editorial staff of *La Protesta* on various occasions. He was a prolific writer of utopias, dramas and essays. Horacio Tarcus, *Diccionario biográfico de la izquierda argentina. De los anarquistas a la "nueva izquierda" (1870–1976)* (Buenos Aires: Emecé, 2007), p. 541.

8 This section includes some passages from Laura Fernández Cordero, "Una utopía amorosa en Colonia Cecilia", *Políticas de la Memoria* 5 (2004): 57–62.

9 Pierre-Luc Abramson, *Las utopías sociales en América Latina en el siglo XIX* (Mexico: FCE, 1999), p. 304. Eduardo Subirats and Menene Gras, "Prólogo: La voluptuosidad subversiva", *La armonía pasional del nuevo mundo,* by Charles Fourier (Madrid: Taurus, 1973), pp. 9–31. Charles Fourier, *El nuevo mundo amoroso* (Buenos Aires: Siglo XXI, 1972). Roland Barthes, *Sade, Fourier, Loyola* (Madrid: Cátedra, 1997).

10 Charles Fourier, *El Falansterio* (Buenos Aires: Intermundo, 1946). Charles Fourier, *Teoría de los cuatro movimientos* (Barcelona: Barral, 1974).

11 Abramson, *Las utopías*, p. 320.

12 Cecilia was the name of an earlier tale written by Rossi. The author came from a family of musicians and Cecilia was the patron saint of music. It is also possible that the name was

Giovanni's tribute to La Cecilia, a mid-nineteenth century Italian revolutionary. See Abramson, *Las utopías.*

13 From the first letter that gave the news of its operation, the anarchist experiment has attracted the attention of the Academy, literature and the cinema. Classics are: Afonso Schmidt, *Colônia Cecília. Uma aventura anarquista na América, 1889 a 1893* (São Paulo: Anchieta, 1942). Newton Stadler de Sousa, *O anarquismo da Colônia Cecília*, (Rio de Janeiro: Civilização brasileira, 1970). See also the novel by Miguel Sanches Neto, *Un amor anarquista* (Rosario: Beatriz Viterbo, 2006). See also, Cândido de Mello Neto, *O anarquismo experimental de Giovanni Rossi de Poggio al mare à Colônia Cecília* (Ponta Grossa: UEPG, 1998); Helena Isabel Mueller, *Flores aos rebeldes que falharam. Giovanni Rossi e a utopia anarquista: colônia Cecília* (Curitiba: Aos quatro ventos, 1999); Rosellina Gosi, *Il socialismo utopistico. Giovanni Rossi e la colonia anarchica Cecilia* (Milan: Moizzi, 1977); Isabelle Felici, *La Cecilia. Histoire d'une communauté anarchiste et de son fondateur Giovanni Rossi* (Lyon: Atelier de Création Libertaire, 2001). See also, the film by Jean-Louis Comolli entitled "La Cecilia" filmed in 1975, and the cinema and anarchism database created by Santiago Juan-Navarro: http://www.cineyanarquismo.com/ For a critical evaluation of the historiography of the colony, see Isabelle Felici, "A verdadeira história da Colônia Cecília", *Cadernos AEL* 8/9 (1998): 9–61. Izis L. Felix Cararo and Hélio Sochodolak, "1890-Colônia Cecília, uma experiência anarquista no Paraná", *Revista Eletrônica Lato Sensu* 1 (2008): 1–25.

14 Héctor Recalde, "Sexo y amor en la propaganda anarquista", *Todo es Historia* 355 (1997): pp. 27–34. The pamphlets were recently re-published by Christian Ferrer and Martín Albornoz (comps.) as *Folletos anarquistas en Buenos Aires. Publicaciones de los grupos La Questione Sociale y La Expropiación* (Buenos Aires: Biblioteca Nacional, 2015).

15 Translated by José Prat, reissued in 1920 by La Protesta. Unlike the first Italian edition, its title in Buenos Aires was *Un episodio de amor en la Colonia Socialista Cecilia* [An Episode of Love in the Socialist Colony of Cecilia].

16 Felici, *La Cecilia*, p. 94: "Se trataría de instalar una destilería y [...] comprar con el agua de la vida jóvenes indias de tribus semi salvajes! Ellas devendrían rápidamente libres camaradas pero qué forma ignominiosa/infame de fundar su libertad!" (author's own translation).

17 Alfred Sanftleben compiled the data about the colony and this correspondence with Rossi in *Utopie und Experiment* (1897). The publication was welcomed almost immediately, as they announced it in *La Anarquía* 22, 8 August 1897.

18 Felici, *La Cecilia*, p. 57. Gosi, *Il socialismo.*

19 Felici, *La Cecilia*, p. 59.

20 N. d., "Notas y apuntes. Una contestación", *El Perseguido* 98, 7 March 1896. Félix Basterra, "El último desajusticiado", *La Protesta Humana* 82, 15 April 1900.

21 Felici, *La Cecilia*, p. 94.

22 *El Perseguido* 42, 22 May 1892.

23 Émile Piette was a bookseller, originally from Belgium, who arrived in Buenos Aires in 1885. His shop, which was well known for its "modern" and political literature and was a meeting point for anarchists, made the local and international press available to readers. Tarcus, *Diccionario*, p. 511.

24 *La Révolte*, 4 October 1892, and *La Révolte* 25, March 1893. I am grateful to Charles Fonlupt for sending me this newspaper.

25 "Colonisation anarchiste", *La Révolte* 25, 4–10 March 1893.

26 *El Perseguido* 55, 29 January 1893: "En cuanto al amor libre, se comprende perfectamente bien que en una Colonia el hombre soltero ha de sufrir porque él no puede imponer su amor á las compañeras de la Colonia. A nuestro entender el amor libre no quiere decir que la mujer estará obligada á darse á todos los hombres, y un hombre á todas las mujeres, sinó por lo contrario cada mujer es libre de darse al que ella quiera y si solo quiere darse á uno ella es libre. Cuando el individuo, sea hombre ó mujer, obra por su propia voluntad sin

sujetarse a regla alguna obra libremente, pero cada uno es libre de hacer la regla que quiera para sí mismo y nunca para otro, pues el hombre que quiere ir con muchas mujeres, que vaya si las encuentra que quieran ir con él, y la mujer que quiera ir con muchos hombres que vaya si los encuentra que quieran ir con ella, pero también es libre el que solo quiere amar á una, como la que solo quiera amar a uno [sic]."

27 Bearing in mind the period and the technical possibilities available, the speed at which news travelled is quite remarkable. The main figures would have arrived in the colony in November 1892. The text ends with the date April 1893. The Italian edition is at the end of that same year, and the Buenos Aires edition, less than three years later.

28 *La Revolución Social* 9, 15 August 1896: "se presta a profundo estudio de la práctica del amor libre, adquiriendo suma importancia éste, por los datos psicológicos que lo acompañan."

29 J. E., "Bibliografía", *El Oprimido* 15, 7 June 1896.

30 *Ibid.*: "Esto, que hace escandalizar y poner los pelos de punta a los moralistas . . . sin moral, es sin embargo, lógico, razonable; decimos más: indispensable para el bienestar humano. [...] Lo recomendamos a todas las mujeres en particular y á los compañeros en general."

31 N. d. "Apreciaciones y crítica", *La Fuerza de la Razón* 1, 23 July 1896 (Communist-anarchist publication of Chivilcoy): "La estasis de los más intensos placeres, no pueden ser multiformes. Se pasará de estasis á estasis, pero contemporáneamente no se puede estar en dos distintos sitios [...] como no se puede soplar y absorber al mismo tiempo. [...] En lo demás el folleto es muy profundamente lógico y científico, especialmente en la parte que anatemiza la familia burguesa [sic]."

32 R. Canto, "Tribuna Libre. Un episodio de amor en la colonia Cecilia", *El Oprimido* 25, 1 January 1897.

33 On this set of problems, we quote two pioneering studies: Dora Barrancos, *Anarquismo, educación y costumbres en la Argentina de principios de siglo* (Buenos Aires: Contrapunto, 1990). Mabel Bellucci, "Anarquismo, sexualidad y emancipación femenina. Argentina alrededor del 900", *Nueva Sociedad* 109 (1990): pp. 148–157.

34 "Reuniones, grupos, iniciativas", *La Protesta Humana* 3, 15 July 1897.

35 Saskia Poldervaart, "The Recurring Movements of 'Free Love'", workshop on 'Free Love and the Labour Movement', International Institute of Social History, Amsterdam, 6 October 2000.

36 A compilation by Osvaldo Baigorria, *El amor libre. Eros y Anarquía* (Buenos Aires: Libros de Anarres, 2006) offers us an overview of the different points of view.

37 Jean Grave (Breuil-sur-Couze, 1854–Vienne-en-Val, 1939) was an anarchist militant who led and took part in various undertakings in the press: *Révolté*, *La Révolte* and *Les Temps Nouveaux*. His best-known works are *La Société mourante et l'Anarchie* (1893), *La Société future* (1895), *L'Individu et la société* (1897), *L'Anarchie, son but, ses moyens* (1899). Jean Maitron, *Dictionnaire biographique du mouvement ouvrier français* (Paris: Les Éditions Ouvrières, 1981) was an author who was known and respected locally; in 1897, great efforts were being made to collect the money necessary to publish his long work, *La société future*. The author's warnings about the Utopian genre demonstrate the ambivalence of anarchism towards the tradition.

38 Emilio Z. Arana, *La mujer y la familia* (Rosario: Grupo de Propaganda Comunista Anárquica Ciencia y Progreso, 1897). Agustina Prieto, "Notas sobre la militancia anarquista. Rosario, 1890–1903", *Entrepasados. Revista de Historia* 32 (2007): 77–88.

39 *Ibid.*: "Hay quienes desconociendo la naturaleza íntima del amor, cuya existencia no puede ponerse en duda, sostienen la multiplicidad de este afecto, la existencia de más de una afección de esta índole, que "*amar á más de una persona contemporáneamente es una necesidad de la naturaleza humana*", lo que no pasa de ser una afirmación gratuita, sin base alguna, que revela una falta de observación en quien sostiene semejante tesis, ó fines que, francamente, no alcanzo a comprender ni quiero entrar á investigar." (The italics are the author's).

40 *Ibid.*: "la unión natural de dos individuos de diferente sexo por un tiempo más ó menos largo, según la duración del efecto que se profesen, de la simpatía ó cariño que les ha hecho buscar su acercamiento [sic]."

41 August Bebel, *La mujer en el pasado, en el presente y en el porvenir* (Barcelona: Fontamara, 1980). Friedrich Engels, *El origen de la familia, la propiedad privada y el Estado* (Buenos Aires: Claridad, 1941).

42 Stella Georgoudi, "Bachofen, el matriarcado y el mundo antiguo: reflexiones sobre la creación de un mito", in Georges Duby and Michelle Perrot, *Historia de las mujeres: La Antigüedad* (Madrid: Taurus, 2003), pp. 517–536.

43 Juan Suriano, *Anarquistas. Cultura y política libertaria en Buenos Aires. 1890–1910* (Buenos Aires: Manantial, 2001), p. 27. It should be pointed out that, among the anarchists, this did not mean an uncritical reading.

44 Héctor Recalde, *Matrimonio civil y divorcio* (Buenos Aires: Centro Editor de América Latina, 1986).

45 "Es el amor libre menos la lealtad, o más la mentira, la grata mentira; es la sofisticación del amor libre; es el adulterio. [...] Estas cosas las sabe muy bien la gente, y las pone en práctica cada día. Sólo que, el derecho, ejércese en el misterio, como el hurto; y aquello que debería ser el libre comercio, asume el carácter placentero y provocativo — pero poco digno — del contrabando."

46 Charles Fourier, *Jerarquía de cornudos* (Mexico: Coyoacán, 2000). Charles Fourier, *El nuevo mundo amoroso* (Buenos Aires: Siglo XXI, 1972), p. 237.

47 "la heroína [...] estaba ligada por diez y ocho años de vida matrimonial y por una corona de cinco hijos. Sin embargo, ella también ha sentido surgir un nuevo afecto al lado del afecto antiguo; y noblemente lo ha manifestado al padre de sus hijos [quien] apuró heroicamente el amargo cáliz [...] Todos nos hemos alegrado con él por la fuerza de ánimo con la cual ha sabido cumplir su deber."

48 N.d., *El Perseguido* 83, 16 June 1895: "La compañera del héroe Paulino Pallás, se ha unido libremente con uno de los compañeros que durante las persecuciones estaban presos en el buque de guerra "Navarra". Que el ejemplo se propague y que se queden olgando los curas negros y blancos es nuestro deseo; pues, para gozar de los placeres sexuales no hacen falta seremonias; sino verdaderas afinidades y necesidad de gozar [sic]."

49 Juan Rossi, *Un episodio de amor en la Colonia Cecilia*, translated by J. Prat (Buenos Aires: Biblioteca de La Questione Sociale, 1895). Reissue of *La Protesta* in 1920: "entre muchos anarquistas que creen ser los más férvidos fautores [sic] de libertad pero que en el caso del amor son aún musulmanes o algo peor."

50 Laura Fernandez Cordero, "The Anarchist Wager of Sexual Emancipation in Argentina, 1900–1930", in Geoffroy de Laforcade and Kirwin Shaffer (eds.), *In Defiance of Boundaries: Anarchism in Latin American History* (Gainesville: University Press of Florida), pp. 302–325.

51 Pietro Gori (Messina, Sicily 1865–Isle of Elba, 1911) was a lawyer, criminologist, journalist and anarchist orator. He lived in Buenos Aires between 1898 and 1902. Tarcus, *Diccionario*, p. 286.

52 Pietro Gori, "La Mujer y la Familia", *La Protesta* 5841, 3 January1928: "Si te mira como camarada no te dejará sucumbir en los prejuicios de orden religioso o burgués que cuentan que una mujer no puede o debe sentir amor más que hacia un hombre al mismo tiempo y no experimentar alegrías sexuales más que con un hombre, siempre el mismo; o que hacen considerar como inferiores a los otros los goces de la carne. Si te considera como camarada, te pondrá si lo ignoras al corriente de las ideas y los conocimientos liberadores o subversivos en materia sexual, aunque no sea más que para permitirte determinarte. Si ve en tí una amiga, las voluptuosidades que puedas procurarte fuera del hogar lo harán tan feliz como las que encuentres en la casa, porque nadie puede decirse amigo de otro si no se regocija

con sus alegrías. Y aun cuando deseases tener un hijo de otro hombre que aquél con quien cohabitas, tu cuerpo, ¿no es tuyo?"

53 Charles Albert, *El amor libre* (Buenos Aires: Librería Sociológica, 1900): "una regresión á la promiscuidad de sexos de las primeras edades, ó el aparejamiento fortuito de los caprichos y la rienda suelta a las pasiones."

54 *Ibid.*: "Nosotros sabemos, en efecto, que el amor, en torno del cual han sido amañadas tantas discusiones pedantescas, no es otra cosa que la forma de atracción sexual propia de los seres llegados á cierto grado de desenvolvimiento y que por este hecho mismo forma la necesidad moral de su reproducción."

55 Errico Malatesta, "El problema del amor", in *Socialismo y Anarquía* (Madrid: Ayuso, 1975), p. 65: "Hagamos observar antes que nada, que, a pesar del régimen de opresión y de mentira que ha prevalecido y prevalece aún en la familia, ésta ha sido y continua siendo el más grande factor de desarrollo humano, pues en la familia es donde el hombre normal se sacrifica por el hombre y cumple el bien por el bien, sin desear otra compensación que el amor de la compañera y de los hijos." The newspaper *Ideas* included the note "El problema del amor"in issue no. 144, 15 May 1925, extracted from that work by Malatesta.

56 Julio Barcos, *Libertad sexual de las mujeres* (Buenos Aires: n.d., 1925).

10 Technologies of the Afterlife: Spiritualism and Social Imagination in Nineteenth-Century Mexico

Ana Sabau

On March 26, 1874, the *Teatro Nacional* in Mexico City offered its stage to a group of mysterious performers who sought to unleash and invoke, on stage, the forces of the supernatural. Journalists for Mexican daily newspapers — from *El Monitor Republicano* to *La Voz de México* — chronicled the spectacle with scepticism and disbelief, uncertain if the performance they had witnessed had been merely a charade or perhaps the outcome of a true diabolical intervention. This commotion was prompted by two illusionists from the United States, William Fay and Harry Kellar. They were members of a troupe of magicians, The Davenport Brothers, who had begun their careers in show business in 1854, travelling around the world and amazing audiences with their ability to communicate with the powers beyond. An important local newspaper narrated the performance as follows:

> The *Teatro Nacional* had lowered the front curtain, and three rows of chairs were arranged in a semi-circle with their backs to it. In the centre was an oval table with four guitars and six small bells, a chair to one side, and at the edges of the stage, two tables with candelabra; the apparatus could not have been simpler—no one could have suspected some hidden gimmick. Kellar invited two people to tie his hands and feet as tightly and carefully as they could; the two appointed, Santiago Sierra and Baron Gostowski proceeded to do so. The lights went out and some moments later, guitars and bells went flying in a thunderous din. They moved with extraordinary force, towards the feet of those present, then over their heads, the guitar strings producing rare harmonies and the clappers of the bells chiming incessantly. When the lights came on again, guitars and bells were strewn over the stage and Professor Kellar, still tied, was sitting quietly with his bonds untouched. The phenomenon was inexplicable.[1]

Many interpreted the unusual spectacle through the lens of spiritualism, which after gestating in the United States, had spread throughout the far reaches of the Mexican Republic starting in the 1850's.[2] In one of the articles published in *La Ilustración Espírita*, Santiago Sierra, one of the most prominent members of Mexican spiritualist circles, investigated the possible connection between the spectacle at the *Teatro Nacional* and the emerging doctrine.[3] His willingness to embrace a supernatural interpretation of Fay and Kellar's show might seem incomprehensible to twenty-first century sensibilities. However, at the end of the nineteenth century, amid rapid yet unexpected technological development, the wilting power of the Catholic Church over the public sphere, and the numerous imperial and civil wars that devastated large parts of the population in different countries, spiritualism offered its followers a framework through which they could assimilate and interpret the new experiences that modernity was bringing with it.

Indeed the attempt to historicize spiritualism even today entails starting from an uncomfortable position that challenges the hegemonic narratives of secularization and nation building that have predominated in nineteenth century Latin American studies. Perhaps for this reason, spiritualism and its impact on nineteenth century culture and politics has remained a marginal topic in the historiography of this period. Followers of spiritualist doctrine have long been seen as occupying a no-man's-land between religion and science, falling outside the framework of secularism because they clung to beliefs about a tangible spiritual world, and excluded from institutionalized religion due to their unorthodox beliefs. Nonetheless, it seems that this ambivalent position is precisely what renders spiritualism an important topic for discussion today. It not only opens a window to analyse aspects of the past that have received little attention, but forces us to confront the internal contradictions that inhabit many current conceptualizations of modernity. Through the convergence of trends that were, in principle, in opposition to each other, spiritualism presented a case in which the conceptual cartography of nineteenth-century Mexican society — sustained by dichotomies such as "modernity" and "tradition", "liberalism" and "conservatism" and "science" and "religion" — became problematic. The phenomenon of spiritualism, which promoted direct communication with the dead, blurred the boundaries in each of these dichotomies, leaving a chaotic space where the concepts that supposedly ordered the experience of the world collapsed. To approach that historical moment while recognizing and reflecting on the porosity of those boundaries also forces us to subvert and reconsider the limits that constitute our own paradigms of interpretation.

In *Ghosts of Futures Past*, for example, Molly McGarry argues that reactions to the technological developments of the second half of the nineteenth century were not only channelled towards a tendency to secularization that would lead to the so-called "disenchantment of the modern world".[4] On the contrary, along with the furore caused by the telephone, the telegraph and photography, among other things, there was a great proliferation and variety of spiritual practices in which many men and women of different social classes sought to incorporate the new inventions into their transcendentalist visions of the world. To some extent, as McGarry states, the fact that "this broadly popular movement has been marginalized in standard histories of the nineteenth century tells us more about historiography and the history of secularism than it does about religious life in the nineteenth century United States".[5] Far from thinking of themselves as anti-modern or reactionaries, the followers of spiritualism saw themselves as members of an avant-garde who had been able to find the perfect balance between materialism and spirituality. The ease with which they inhabited both spaces invites us to reflect with more complexity upon the contradictions that defined their experience of "modernity" and also to reconsider the ways in which we currently approach the flows and counter-flows between the dominant perspectives of a moment and those occupying a more subaltern position.[6]

Spiritualism in Mexico not only provided its followers with techniques and methods to be used in the practice of communicating with spirits; it also offered a structure — a lens that could both encompass and make more manageable popular as well as learned ideas about contact between the world of the living and the dead. In a century of change and political instability, spiritualist practices promoted new forms of mourning, which, in turn, transformed ways of thinking and relating to historical time. Despite the positivist influences introduced into spiritualism via the works of Allan Kardec, the emerging doctrine exposed the contradictions of a linear, teleological temporality in which the past remained fixed and remote.[7] By claiming that it was possible to engage in dialogues with spirits from any time or place in the universe, spiritualist practices set in motion an experience of time based on anachronistic interruptions. The past ceased to be a uniform dimension and was fragmented into infinitely small pieces that could reappear at any given moment to engage in dialogue with the present without any chronological order or logic. That past, which was pulverized but still available, vital and within reach, opened the doors to an affective experience of history that traversed multiple temporalities. Each spiritualist séance offered the possibility to hold a conversation with a mother, a brother or a dead uncle, but also to talk with figures of the like of Sor Juana Inés de la Cruz, priest Miguel Hidalgo y Costilla, Bartolomé de las Casas or Swedenborg himself.

Spiritualist gatherings constituted the material expression of a new way of engaging with the relevance of the past. The process of conjuring a spirit and communicating with it conveyed a desire for establishing fluid channels of communication between various phases of history, challenging ideas related to the evanescence and irretrievability of past moments. In communing with the world of the beyond, historicity was negated in favour of a different conception of temporality where access to any moment of the past was granted via direct communications with either figures of historical importance or anonymous people who had something to communicate to the living:

> The spirits manifest themselves spontaneously or in response to an invocation. All spirits can be invoked: those that have animated the darkest mortals, as well as the most illustrious characters, irrespective of the time in which they lived. We are able to invoke our relations, friends, or enemies and we can receive oral or written communications, advice and general information about their state beyond the grave, their thoughts about us, and any revelation that they are permitted to share with us.[8]

As can be appreciated in the above quotation, spirits were also conjured based on the conviction that the time for social regeneration had arrived. According to the emerging spiritualist doctrine, spirits had overcome the obstacles of materiality in their transition to the afterlife, and from this vantage point were sending messages about how living humans could form a universal egalitarian community, promoted and attainable by means of the study and practice of spiritualism itself.

In spiritualist séances, not only was the past reconfigured as multidimensional, unexpectedly open and accessible, but imagination and the place of the future were also at stake. Through the messages that the spirits sent from the afterlife, followers of spiritualist teaching started to imagine an alternative society that could escape current social pressures and what they saw as "moral decline" driven by hegemonic materialistic philosophies. For the men and women who invoked the dead, the future oscillated between the promise of something better to come and the constant realization of the ephemeral, anachronistic promise announced every time one of the spirits incarnated in order to communicate something. The invocations that took place in such gatherings facilitated an experience in which immaterial beings who existed outside historical time were brought into the present. These beings inhabited a time where past and future folded into each other. As such, spiritualist séances were spaces and gatherings that dismantled linear views of time (albeit in a contradictory and momentary way).

As might be imagined, the intersections between spiritualism and utopian thought were many. The coincidences between these two strands of thought went beyond sharing a historical moment and several of their thinkers. In the second half of the nineteenth century, when men and women from different parts of Mexico gathered round tables in clubs or houses with the intention of conjuring spirits and communicating with the beyond, they also participated in reshaping the social order and the polis. In the face of new, unsettling technologies, spiritualism provided a thought and practice that made it possible to contain the fears and anxieties arising in response to the rapidly changing world. Spiritualism was a way to tame and redirect those concerns towards imagining a new society characterized by harmony, order and equity—one in which even the grief caused by the death of loved ones could be avoided by the fluid, open contact between the world of the living and the dead.

The emerging doctrine assimilated its historical moment by finding a crack in which to imagine a harmonious confluence between tradition and the latest transformations of the time. The social readjustment facilitated by spiritualism can be analysed on two main axes: the local and the cosmopolitan. At the heart of both sets of issues lay, among other things, a dispute about notions of political representation (or non-representation). Spiritualism — contradictory, fragmented and multi-temporal — encapsulates the shocks and tensions, but also the possibilities of other social configurations glimpsed from the multiple facets of that historical period. This period at the end of the nineteenth century was experienced as a turning point. In the remainder of this chapter, I shall try to expand that picture in order to trace the tensions and imaginations that spiritualism mobilized along both those axes.

Negotiating the local: spiritualism and "new subjectivities"

Many studies dedicated to the final decades of the nineteenth century in Latin America often mentioned the democratizing tendencies that presaged the consolidation of mass culture. Increasingly expanding education systems, the proliferation of newspapers and growth of the reading public, added to the migration of people to urban areas, were translated as new collective subjectivities emerging in the social sphere and imaginary. The irruption of the masses onto the social stage, documented by *fin de siècle* writers like José Martí, Rubén Darío and José María Ramos Mejía has been carefully and insightfully studied by Graciela Montaldo.[9] This phenomenon of new subjectivities became more and more tangible as the turn of the century drew closer and brought with it a social reorientation, especially at the local level. Spiritualism, a doctrine daughter of its time, was one of the spaces that documented and contended with this readjustment.

On January 1st 1876, a short article published in the Yucatan spiritualist paper *La ley del amor,* reported that its editors had received a group of experimental photographs from New York. The images were the fruit of the work of Dr. J. R. Simoni, who frequently sent the newspaper updates regarding his findings on the subject of spiritualist photography. The photographs included in his post revealed the blurred figures of seventeen spirits, six of which were alleged to be those of people who had lived in Merida. Apart from the obvious significance they had to the local community, Simoni sought to add the photographs to the growing list of evidence that proved the existence of a world after death that could be captured by emerging technologies. Subscribers to *La ley del amor* who might be interested in seeing the photographs for themselves could go to the newspaper's offices and contact the editors to make an appointment to view the images with their own eyes.[10]

Apart from this information, the article offered a detailed description of Simoni's photographs, providing the readers with all sorts of particularities in order to capture their attention. Although the identities of the six resident-spirits of Merida appearing in the images were unclear, one of them at least was beyond doubt: "the identity of one of them is absolute, with the notable circumstance that he died without ever having had his picture taken because he lived in poverty. Despite this, he was a person known to all by virtue of the fact that he always issued the subpoenas for the office of the magistrates' court of this capital."[11] Although it may seem a trivial detail, this quotation illustrates why the study of spiritualism can shed some light on the unforeseen upsurge of marginal subjectivities in the political sphere. If, as can be seen in the quotation, photography in the nineteenth century was normally reserved to the upper classes and "cartes de visite" were common practice in middle-class bourgeois families, the space for representation and visibility that spiritualism imagined beyond the grave seemed to be configured quite differently.[12] While that anonymous person who lived in poverty on the streets of Merida could not possibly have had a portrait or photograph of himself taken while he was alive because of his social condition, the rules for gaining access to the photographic medium, and so to the public sphere and visibility, operated differently once the threshold of death had been crossed.

In the "Little History of Photography", Walter Benjamin maintains that one of the greatest impacts of the invention of photography was, for the first time, the "revelation" or the "awareness-building" of the existence of an optical unconscious. The first photographic images enlarged the field of vision, enabling glimpses of aspects of the physical world that would otherwise have remained invisible to the naked eye.[13] Legitimizing spiritualist photography hinged precisely on this argument, which revealed an epistemology sustained by the intimate relationship between visuality and evidence. If the photographic paper of Dr Simoni (in this case) had succeeded in capturing the shadows and silhouettes of the invisible bodies of seventeen spirits, the pictures could then be produced as irrefutable proof of their existence.

After the invention of photography, the telephone and other technologies were also used to enter into contact with the beyond. Spiritualism can therefore be seen as a symptom of the expansion of sensorial experience taking place at that time. Attempts to communicate with the realm beyond the grave can be understood as another manifestation of an age that was beginning to consider the possibility that the senses and consciousness did not necessarily have the capacity to capture the totality of physical or objective reality that was developing external to the subject.

In another brief item published in 1887 in the daily newspaper, *El Tiempo*, spiritualist theory suggested that mediums who mediated between the material and the spiritual world were operating in an analogous way to Benjamin's description of the camera, whereby their sensorial apparatus enabled them to capture information that was inaccessible or imperceptible to the majority of people:

> Mediums who saw, felt and used all their bodily senses in a way entirely different from other men; in such a way that it might be said of them that they had a sixth sense that engulfed all the others and with a power superior by far to the five put together [. . .] I have met mediums who could see regions that were thousands of kilometres away; and could see them better than we could see them at a distance of less than a hundred metres. They could see the people and animals that inhabited them and moved about in them; they could listen to human voices and observe with precision whatever came under the power of their senses.[14]

In this way, spiritualism joined other phenomena of the time that favoured crystallizing ideas about certain spaces of human experience that lay beyond consciousness. In debates about the truthfulness of spiritualism and the manifestation of the presence of spirits by means of so-called "rotating tables" (tables that moved in order to pass on messages from some spirit that had been invoked), sceptics argued that supernatural experiences could be explained, not because beings from the other side had materialized for a moment, but because the participants in spiritualist séances were moving the furniture without being completely conscious of doing so. It does not seem to be a coincidence that just a few years after these changes were taking place, Freud should think of the unconscious as something internal to subjectivity. As the above quotation shows, the Mexican daily newspapers, which had been debating the meaning of spiritualist knowledge since the 1850s, were offering expressions of an experience that could not be completely absorbed or controlled through the conscious use of reason and the senses.

This idea of the optical unconscious proposed by Benjamin — what the world concealed from the eye of man but exposed to that of the camera — can also be thought of in socio-political terms. Reconsider, for instance, the news article cited in the Yucatan daily newspaper. The ghost of that anonymous person who had gone through life as if invisible, from the poverty of the streets to the magistrates' court in the city of Merida, was now making his ghostly appearance in the photographic prints that had been sent from New York to Yucatan, "revealing" on a social level what would otherwise have remained invisible. So, it becomes almost inevitable to read the article as if it were a document about the experience of the upper or educated classes in the face of the upsurge of the popular onto the political stage. The idea conveyed by the newspaper that if one lived in poverty, it was only possible to enter photographic space as a ghost, can be seen as an indicator that these "marginal subjectivities" already had a space — albeit an otherworldly one that was perhaps feared and unwanted — in the imaginary of the bourgeoisie. The ghosts in the photographs from the newspaper article inhabited a temporality flickering between announcement and arrival, lurking in the historical present without fully materializing in it.

The newspaper's mention of Dr J. R. Simoni's images of spirits presents to its readers the two sides, or the double-edged nature, of representation. Georges Didi-Huberman recently expressed the idea more fully in *Pueblos expuestos, pueblos figurantes* (*Peuples exposés, peuples figurants*). He argues that not only is it essential to continue thinking about the multiple ways in which aesthetic and political representation intersect each other, but that this must also take into consideration that the leap from the field of the invisible to that of material representation simultaneously, entails new forms of erasure and exclusion; it implies the conquering of space by new subjectivities (or peoples, in the case of Didi-Huberman) that, in the process, are exposed to their own extinction. These considerations further lead to questions concerning the ownership of the images and the loss of control over their circulation and dissemination which accompanies the entrance of the "representational" onto the social stage.[15]

Spiritualism, in this respect, was a field that was split in two. In its sustaining ideology coexist the two sides of representation described by Didi-Huberman in his book: the lurking of the peoples, but also the threat of their extinction, and the constant efforts made to successfully neutralize and control their emergence in the social sphere. To put it another way, through spiritualism's doctrine and practice, we can find traces of an emancipatory potential, on the one hand, while on the other, we also see attempts to tame, control or bring back into harmony the many social changes taking place at the time. Browsing through the spiritualist dailies exposes cultural historians to various reflections on how the new doctrine encouraged the imagining of possible ties and alliances through the multiple social strata, even if this seemed to satisfy the anxious desire of upper classes to control the changes in that new social mapping. By the end of the nineteenth century, the Spiritualist Society of Mexico was one of the largest in Latin America, with a network of interconnected clubs and groups scattered throughout the Republic, especially in the north of the country and in the Yucatan Peninsula, as outlined above. *La Ilustración Espírita*, one of the major publications dedicated to spreading spiritualism in Mexico, was founded by Refugio González in 1870.[16] The issues of this publication are of unparalleled value in documenting how spiritualism navigated and imagined new configurations of the increasingly close interaction between elite and popular classes.

One of the most marked examples on the subject, which captured the attention of many nineteenth-century Mexican readers, was that of Teresa Urrea, the Saint of Cabora. She was a woman who allegedly had powers as a medium and healer, and in whose name a number of popular uprisings took place on the northern border of the country. Through the various articles that appeared about her in several issues of the *Ilustración Espírita*, it is easy to appreciate how the platform of spiritualism helped shape the figure of Saint of Cabora into a symbol capable of organizing different social strata. Spiritualists recognized Teresa as a natural medium, "intuitive" and autodidact, who had learnt the teachings of the spirit-world through trances and prophecies communicated to her from the beyond, not through a bookish or formal education. In this way, it did not matter that her knowledge of the spirit-world was cloaked in a Christian, non-scientific discourse, because "true science" was said to be hiding behind such popular manifestations:

> Unfortunately, doctors in general despise anything that does not originate from them, and they do not want to understand that behind ways that frequently seem ridiculous, behind the ignorance of ordinary people who do not understand the truth of facts except by sorting out their own concerns and vitiated ideas, behind folk medicine without the humbug, healing without drugs or potions, and behind those ordinary men and women with no more knowledge than their hands, and no more wisdom than the phenomenon that they can see in front of them but cannot explain, there exists a type of true knowledge, a new truth, a surprising revelation and a wonderful world whose virtue is unknown. Among healers and doctors alike there are self-interested charlatans. No attention should be paid to one side or the other, neither deserves consideration; but consideration is demanded by those who devote their lives with noble disinterest to doing good to their brothers and sisters, who expose their lives, spend their resources, deprive themselves of their own comforts and endure all sorts of discomforts for the good and benefit of the sufferer. Teresa Urrea falls into this category.[17]

As the quotation shows, educated spiritualists used such speeches to justify their role in extracting the "scientific truth" from popular myths. In a world that was becoming increasingly democratized, folk wisdom could not remain ignored; it was better to assimilate it and find new ways of validating the pre-eminence of scientific knowledge and learning over it. After all, the fact that there were "intuitive" mediums who received their spiritualist knowledge without having been indoctrinated was even more conclusive proof of the real existence of those laws of the beyond that were starting to oppose materialism and positivism—two of the major currents of thought upheld by *Porfiriato* [the *Porfiriato* was the name given to Porfirio Díaz's thirty-five year presidency].[18] That someone like Teresa, without training in spiritualist ideology, should receive messages from beyond the grave, was another way of proving the veracity of the ideas that were giving shape to spiritualism. What is remarkable is that, by promoting this line of argument, spiritualism was not only deploying its efforts to re-validate formal learning over folk wisdom, but was also, paradoxically, opening itself up to the possibility of disseminating a current of thought and a set of practices that would be accessible to everybody. Mediums, to whom spiritual reality manifested itself in many ways, could be found in any social class. Dialogues with the dead were not then just an urban, upper-class phenomenon, but also occurred in the rural areas and among the poorest social groups.

Reflections on the changes implied by technological developments frequently went hand in hand with deliberations about the possibility of using them to overcome the barriers that separated social classes from each other. In one of the many articles that the renowned Spanish writer and spiritualist Amalia Domingo Soler contributed to the *Ilustración Espírita*, the author wrote at length about how the increasing reach of the press, in addition to opening the doors to communication with the afterlife, also helped spread knowledge of the new doctrine through the different strata of society: "The press is one, and perhaps the first of human advances because it revitalizes thought, because it binds all social classes together, because a newspaper is a universal letter, and because a book is a bouquet of sweet-smelling flowers that never withers."[19]

From this vantage point, spiritualism found a way to embrace the recent transformations by taking the latest inventions and making good use of them as educational tools. For Domingo Soler, spiritualism offered a framework that could be used for profit by involving the emerging mass media. The fluid relationship between mediums and spirits was repeatedly articulated as a relationship between readers and writers, which, with the expansion of the press, had now extended considerably. In another section of the article just cited, Domingo Soler wrote:

> For some time now the number of writers has increased because our friends from the other side are sending us their thoughts through recording, mechanical, intuitive and aural mediums, and philosophical and historical works for pleasure have come to enrich literature on both continents. The spirits said, Start writing! The spiritualists replied, Start reading! And study groups and centres have been created and a part of humanity maintains active correspondence with the souls of those who have departed. The universal family has ceased to be a myth; we are related to beings who have left their corporeal forms. They write. We read. Blessed are those who can write, and happy are those who make haste to read.[20]

Following Robert Darnton's suggestion — which is detailed in his study on the role of mesmerism in the early days of the French Revolution — we could well imagine that the flexible nature of spiritualism, its openness to assimilating a wide variety of traditions and discourses, as well as its ambivalent position between science and religion, captured the attention of many people at the same time that it eluded political censorship.[21] Cloaked in rhetoric with heavy overtones of morality and the spiritual world, spiritualism promoted ideas that were quite radical for the time and which went unnoticed to the censoring eyes of the *Porfiriato*. The doctrine did come up against harsh criticism as soon as it arrived in Mexico, beginning with the 1875 debate in the Hidalgo Lyceum where such prominent figures as José Martí, Ignacio Ramírez, the Sierra brothers and Gabino Barreda all sought to clarify the place of spiritualism in the realm of science. Nevertheless, the attacks directed at followers of spiritualism were, more often than not, based on their participation in a movement that had strong tendencies to charlatanism and that made it particularly difficult to find solid proof of their beliefs — not for the politics that spiritualism promoted.[22]

Perhaps because they were masked by moralistic excess and adorned in Christian rhetoric, the most radical egalitarian messages of spiritualist discourse filtered through without attracting too much attention or causing scandal. However, a single quotation from the numerous articles that were published on the subject reveals the political seed one could speculate spiritualism helped to sow:

> Spiritualism, this science founded on the fairest and most equitable principles of the most pure philosophy — this doctrine, based, sustained and demonstrated in the Gospel; debated, reasoned and accepted by many men distinguished for their wisdom, both ancient and modern; and finally, revealed and also demonstrated by mutual consent, by a multitude of intelligent beings from the afterlife — its throbbing heartbeat has already appeared among all the peoples of the Earth and all classes of society with neither distinction nor privilege.[23]

What makes this quotation politically interesting is not just the idea of equality being able to penetrate all layers of society, but also the contradictory power of incorporating, in a single body of text, diverse traditions of thought that would generally be at odds with each other.

One of the most important aspects of the local political reconfiguration of spiritualism in Mexico is the way in which the basic tenets of the doctrine made it possible for various women, in a century still marked by considerable gender inequality, to intervene in the public sphere and the political arena in subject areas almost exclusively reserved for men.[24] Spiritualism in Mexico enjoyed its greatest boost in popularity in the wake of the translation of the work of Allan Kardec, the famous French *spiritiste*, who codified the principles of the doctrine in works such as *El Libro de los Espíritus* (The Spirits' Book) and *El Libro de los Médiums* (The Book of Mediums); nevertheless, from the start of the movement in 1848, with the Fox sisters in the United States, spiritualism would give priority, or at least a position of equality, to women.

The fact that the vast majority of mediums were women did occasionally reproduce stereotypes linked to the female character by perpetuating ideas on how female sensibility favoured the reception of messages from the other side, or that women had a certain predisposition towards morality. Nevertheless, mediumistic practices opened up spaces for women to define themselves with greater autonomy, especially in the case of young women who came from more rural areas, as Molly McGarry suggests in her book *Ghosts of Futures Past*.[25]

At the core of spiritualism lay the assertion that human beings had individual souls that survived the body after death. Two tenets in particular favoured gender equality: the first linked to the world of the spirits and the second to the earthly world and the role of mediums in it. With regard to the first tenet, Kardec claimed that once the spirit was released from the body after death, gender distinctions were overcome. Sexual difference in the afterlife did not function as it did in the living world. In a sense, the world of the spirits took on the form of a utopian projection, reflecting the imagination and political desire which shaped the moment in which spiritualism arose. Hierarchies were not totally abolished, since some spirits were clearly more enlightened than others; however, their position was not determined in the same way as it was in the social hierarchy of nineteenth-century societies. Family ancestry was irrelevant, as were gender, physical attributes and economic wealth. The structure of the imagined community of spirits depended rather on noble and virtuous leadership, in which wisdom, morality and, above all, the common good were fundamental:

> Hierarchies among men can and should only be admitted through the power of virtue and knowledge, and the man who is truly a virtuous sage is only a humble adviser, a passive director of all those who are in need of his example and instruction, and he does good for the sake of good and never to see what interest he can derive.[26]

With respect to the second tenet, the role of mediums in the earthly world, it is important to refer to an article by Refugio González, which appeared in *La Ilustración Espírita* and condensed the most important principles of spiritualist doctrine. After participating in a séance, González stated that he had received the following message from a 'protective spirit' assuring him that: "Every human being is a medium".[27] By opening up spaces that were eroding the hegemonic stratification of nineteenth-century Mexican society, spiritualist clubs were expanding the political imagination of the time, which was conducive to the participation (but also control) of groups that were normally relegated to the margins, among them, women.

It is no coincidence that we are able to study the nexus between the early manifestations of nineteenth-century feminism in Mexico and spiritualism. Apart from Teresa Urrea, who was mentioned prior, there were other, perhaps more obvious connections between spiritualism and the struggle for women's rights in Mexico.

Laureana Wright de Kleinhans, who had been the editor of one of the first Mexican magazines written by and for women, *Violetas del Anáhuac*, was also one of the most important contributors to *Ilustración Espírita* and one of the few women who were considered part of the Hidalgo Lyceum. Little is known of her conversion to spiritualism, although it has been stated that in the first years after the doctrine was introduced to Mexico, the writer and intellectual did not take the emerging doctrine seriously. It is possible, as Lucrecia Infante Vargas suggests in an article on the subject, that Wright gradually became attracted to spiritualism because of its implicit openness to matters related to gender equality, rather than because it was promoting beliefs or 8precepts about what was happening in the afterlife.[28] Unlike *La Ilustración Espírita*, almost no other publication of the time printed articles written by women in its general sections — that is, in those sections not specifically devoted to topics traditionally related to women's interests. Hence, many of the issues of *La Ilustración Espírita* published between 1890 and 1893 feature reflections by Laureana Wright on the front page. Such articles help support the idea that Wright approached spiritualism because of its openness to women. Moreover, they frequently offer thoughts about the place of women and incisive criticisms of materialism, rather than addressing the topic of séances and spirit messages from beyond the grave.

The universal family: spiritualism and cosmopolitanism

This chapter has concentrated so far on analysing the ways in which spiritualism contended with social and political change on the local stage. However, spiritualism also supplied tools for assimilating the transformations that were taking place as a result of the growth and acceleration of transnational interactions. As indicated at the beginning, spiritualism occupied a blind spot in nineteenth-century Latin American studies for quite some time. One possible explanation for this was the tension between religion and science, tradition and modernity, which was not only apparent in spiritualist doctrine, but also challenged historiographical thinking that sought to study the period by making a clear-cut distinction between these areas from the outset. Another explanation was the emphasis given to the study of nation building narratives that started with the independence movements in Spanish America at the turn of the 18th century.[29]

The image of spiritualism in Mexico that emerges from articles on the subject published in newspapers and magazines is that it was a doctrine having little interest in the national scene. The networks emerging from the spiritualist movement and the imagining of that community of the future, where the material and spiritual worlds and the living and the dead would converge, would be organized like a "universal family" — to put it in the words of Amalia Domingo Soler in the article cited earlier. This is perhaps the reason why spiritualism remained incomprehensible for some time and why it seems to resonate more nowadays with an academic space that is increasingly open and interested in studying the transnational and the global.

The pages of magazines and periodicals like *La Ilustración Espírita* and *La Ley del Amor* allow us to read about the real and imaginary social networks sowed by spiritualism and the ways in which these went beyond the limits of the national gaze. Through numerous transcriptions of conversations with spirits — from Blaise Pascal, Ignacio de Loyola and Allan Kardec to Miguel Hidalgo and Sor Juana Inés de la Cruz — articles in spiritualist publications punctuated the linear understanding of historical time with their anachronism and also helped construct a discourse and experience of history that lay beyond a narrow national perspective. The messages that the spirits sent from the world beyond generally had little to do with strengthening the ties amongst national communities and instead facilitated dialogues between different places in the world, contexts and times, and also cut through or overturned the filters separating the private and public spheres.

The latter occurred particularly when historical figures were said to have transmitted knowledge that had little to do with the place they occupied in historical memory, but that was directed rather to calming the mortal anxieties of those who invoked them. A good example of this can be seen in the transcription of a conversation with Bartolomé de las Casas that appeared in *La Luz de México* in 1873. In the conversation, the spirit of Las Casas was answering the questions that those attending the spiritualist séance had anxiously asked him about the state of a friend by the name of Castera, who had recently died. Las Casas' replies contained no historical dates and were confined to describing the state of Castera, who was completing the process of transitioning from the body to the spirit and taking the last steps before he could be released from his earthly worries: "But this does not cause him great suffering as might be believed, but is a passing disturbance that slightly dulls the perceptions of the soul."[30] This conversation with the spirit of Las Casas allows us to see history through spiritualist experience, as something where the past was far from distant and bookish and instead acquired a refreshing vitality. The transcription of this supposed chat with the spirit of the renowned Dominican friar reveals how spiritualism activated the desire for an affective, personal experience of history, in which conversations with the past provided an intimate knowledge, making it possible to cope with the recent subjective crises.

In addition to the numerous references to spirits that present a historical panorama spanning time and geography, spiritualist magazines were full of articles originating in the foreign press, especially from France, Spain and the United States, which had been translated and republished. There were also, as we saw in the case of Amalia Domingo Soler, periodical contributions by international figures. Spiritualism, like Freemasonry, worked through societies and clubs that had international, as well as local reach. The comings and goings between different countries are documented in these publications as a fundamental element in the spiritualist imaginary from the middle to the end of the nineteenth century.

The transnational map traced by spiritualism corresponded to the increasingly frequent and accelerated availability of exchanges between different parts of the world. Once again, as was the case on the local stage, it is possible to detect in the ideas of spiritualist doctrine a certain ambivalence towards the changes that growing openness and contact between multiple economies and cultures around the world entailed. On the one hand, spiritualism manifested a utopian impulse — the desire for transformation and social regeneration, the scale of which had to be felt in global or "universal" terms, to use the words of the time. On the other hand, it looked for ways to neutralize the problematic disagreements to which transactions between different languages, cultures and perspectives gave rise. In view of the tensions that could arise from the new *fin-de-siècle* geopolitical order, spiritualism offered the harmonious, idealized image of cross-linguistic encounters between different regions. In articles dealing with the subject, very few make reference to problems or obstacles when it came to communicating with the spirits. Different languages or failure to understand the mechanisms and codes used by the spirits to send their messages are rarely mentioned, but when they are, the purpose is rather to stress the cosmopolitan nature of the spiritualist community than to point out any friction in the dialogues between the world of the living and the beyond. As early as 1853, a note in *El Universal* published under the heading "El diablo hospedado en una mesa" (The devil a guest at a table), republished and translated selections from a letter sent to the Catholic daily newspaper in Paris by the canon of Limoges and Tulle, M. Charles Gay. The passages in the letter told of the experience of the priest and a further group of people who, after invoking the spirits, had mistakenly managed to make contact with the Devil. The most striking feature of the note is the conversation recreated between the members of the group and the spirit invoked, since it is a scene that highlights the poliglot desire of spiritualism:

> I begged for it not to proceed and I asked the priest if he would be so kind as to ask the spirit in Latin. — 'Loqueris ne latina?' — Mr Bertrand said to him — There was no reply —'Do

> you speak Latin?' he then asked him in French. — The table knocked once — And once again in Latin 'quis es tu? Dic nobis nomen tuum?'.[31]

Although perhaps the French version might have been read in a different way, the translation and publication of passages from the letter in the Mexican newspaper reveal multiple tongues converging at the heart of spiritualism. This scene presented to the Spanish-speaking reader thus opened a way to envision the ideal spiritualist séance as one that involved members from different places, or at least with different knowledge and experiences. Those who could embrace the greatest number of languages and possible codes were better prepared to respond to and understand the messages that would be transmitted as soon as a spirit materialized.

It is highly likely that the communicative transparency frequently described in spiritualist séances, and which seemed to have been one of their constant desires, may have been a response to the notion that the afterlife was considered to be part of the search to overcome all obstacles to communication. Although new technologies, such as the telephone, the telegraph and, later, the radio, had managed to connect different social centres scattered across the world efficiently and rapidly, communication was far from uncomplicated and free of technical difficulties. Inspired by the potential of the new inventions, spiritualism promoted the fantasy of immediate and transparent communication that could expand beyond the global scale: "In short, M. Allan Kardec promises us a new telegraphic method that will exceed all others in speed and accuracy. Two people, mutually invoking each other, are able to transmit their thoughts to each other. 'This human telegraphy will one day be a universal medium of correspondence'."[32]

Accordingly, spiritualism responded to the social and political changes that were occurring along two axes. While locally spiritualism allowed us to trace the increasingly apparent emergence of the popular classes and women into the political arena, at a wider level, spiritualism imagined and gave form to a transnational polyglot community who felt limited by a national paradigm. By drawing upon universal history, multiple languages, and collaboration between spiritualist intellectuals residing in different countries, followers of spiritualism in Mexico prepared themselves to shape the social regeneration which they felt would eventually succeed in linking the spiritual and material realms in one egalitarian social fabric: the religious and the scientific; the local and the transnational; life and death.

Consequently, the doctrine that was initiated in the United States in 1848 and went round the world in the twinkling of an eye, in Mexico fluctuated between domesticating the changes proclaimed by the new century and breaking with social hierarchies. The fate of entering history as a pseudo-science and a religion not to be taken seriously has impeded thinking about the fundamental place of spiritualism in the politicization and inclusion of certain marginalized groups, as well as their possible links with the galvanization of the Mexican Revolution. In an era when the narrative of secularization has been naturalized, it is commonly difficult to imagine a belief with spiritual and religious overtones promoting revolutionary rather than conservative values.

As I have tried to show here, spiritualism invoked the past and the spirits to legitimate a rupture—a social regeneration articulated not only in material but also spiritual terms. While to our eyes, the show put on by Fay and Kellar at the National Theatre would not have had anything supernatural about it, in their own time, it paved the way for much broader questions. Spiritualism today can help us reflect not only on issues that were important at the time, but also make us think about how the past mobilizes and interacts with the present.

Acknowledgement

A shortened version of this chapter was published online in Spanish as "Ideas para el XIX: Espiritismo", *Gaceta Frontal* (7 July 2015).

Notes

1 Enrique Ovalaría, *Reseña Histórica del Teatro*, vol. 3 (Mexico: Imprenta, Encuadernación y Papelería La Europea, 1895), p. 166. "El foro del Teatro Nacional tenía bajado el telón de boca, y dando espalda a él, había tres órdenes de sillas formando un semi-círculo, en cuyo centro se veía una mesa ovalada con cuatro guitarras y seis campanas, a un lado una silla, y en los extremos de la escena dos mesas con candelabros: el aparato no podía ser más sencillo, sin que nadie pudiese ni aun sospechar oculta máquina alguna. A invitación de Kell[a]r, se nombraron dos personas que le ataran de pies y manos con cuanta fuerza y precauciones pudieran; así lo hicieron Santiago Sierra y el Barón Gostowski. Apagadas las luces, pocos momentos tardaron en volar guitarras y campanillas con atronador estrépito, moviéndose con extraordinaria violencia ya a los pies de los concurrentes, ya sobre sus cabezas, produciendo las cuerdas de las unas raras armonías, repicando los badajos de las otras con incesante repetición: al volver a encenderse las luces, guitarras y campanas estaban esparcidas por el suelo y el profesor Kell[a]r perfectamente atado y tranquilo; nada ni nadie había tocado sus ligaduras: el fenómeno era inexplicable."

2 As mentioned in José Ricardo Chaves, *México Heterodoxo: diversidad religiosa en las letras del siglo XIX y comienzos del XX* (Mexico: Bonilla Artigas, 2013), p. 57, recent studies about spiritualism in Mexico disagree about the origins of the spiritualist doctrine in Mexico. Some situate the epicentre in the capital, while others insist that the doctrine originated in the north of the country or in states of the Bajío, such as Jalisco. For references to spiritualism in Mexico, see José Mariano Leyva, *El ocaso de los espíritus: el espiritismo en México en el siglo XIX* (Mexico: Cal y Arena, 2005); Gonzalo Rojas Flores, *El movimiento espiritista en México (1858–1895)* (Master's thesis: UNAM Facultad de Filosofía y Letras, Mexico, 2000); Lia Theresa Schaefer, *The Spirits of the Times: The Mexican Spiritist Movement from Reform to Revolution* (PhD diss.: Davis, University of California, 2009); Antonio Saborit, *Pedro Castera* (Selection and Prologue) (Mexico: Cal y Arena, 2004). For a study of the development of spiritualism in the United States, see Molly McGarry, *Ghosts of Futures Past: Spiritualism and the Cultural Politics of Nineteenth-Century America* (Berkeley–Los Angeles: University of California Press, 2008). For studies of spiritualism in other parts of Latin America, see Manuel Vicuña, *Voces de ultratumba: historia del espiritismo en Chile* (Santiago de Chile: Taurus, 2006); Francisco Ferrándiz, *Escenarios del cuerpo: espiritismo y sociedad en Venezuela* (Bilbao: Universidad de Deusto, 2004); Michael Taussig, *The Magic of the State* (New York: Routledge, 1997).

3 See Leyva, *El Ocaso de los espíritus*, pp.100–130.

4 McGarry is thinking of the United States, although certain aspects of her analysis apply equally to the case of Mexico.

5 McGarry, *Ghosts of Futures Past*, p. 5.

6 *Ibid.*, p. 7.

7 Allan Kardec, born Hippolyte Léon Denizard Rivail, was a French intellectual, regarded as the figure who systematized spiritism. One of his most important and most frequently translated books (in Mexico it appeared in the 1860s) was, as Leyva points out in his work, *El libro de los espíritus*.

8 Quoted from one of Allan Kardec's texts, translated in *La Ilustración Espírita*: "Los espíritus se manifiestan espontáneamente o en respuesta a una invocación. Se puede invocar a todos

los espíritus: a aquellos que han animado a los más oscuros mortales, así como también a los personajes más ilustres, sin importar la época en que hayan vivido. Podemos invocar a nuestros parientes, amigos, o enemigos, y podemos obtener de ellos por vía oral o escrita comunicaciones, consejos e información general sobre su estado de ultratumba, sus pensamientos respecto a nosotros, y cualquier revelación que les sea permitida compartir con nosotros."

9 For her most recent study on the topic, see Graciela Montaldo, *Zonas ciegas: Populismos y Experimentos culturales en Argentina* (Buenos Aires: Fondo de Cultura Económica, 2010). For her study on modernism, and the figure of Rubén Darío in particular, see Graciela Montaldo, *La Sensibilidad amenazada: Modernismo y Fin de Siglo* (Rosario: Beatriz Viterbo, 1994).

10 For this brief item, see "Fotografías Espíritas", *La Ley del Amor,* Merida, Yucatan (1 January 1876). The daily can be consulted online via the Yucatan virtual library.

11 *Ibid.*: "la identidad de una de ellas es absoluta, con la notable circunstancia de haber fallecido sin haberse jamás retratado, porque vivía en la miseria, no obstante lo cual, en virtud de que se ocupaba constantemente de hacer las citaciones de los juzgados de paz de esta capital, era persona muy conocida de todos."

12 For a study of the history of photography in Mexico, see Olivier Debroise, *Mexican Suite: A History of Photography in Mexico*, trans. Stella de Sá Rego (Austin: University of Texas Press, 2001). For a study of different uses of photography in its infancy in the nineteenth century, see Allan Sekula, "The Body and the Archive", *October* 39 (Winter 1986): 3–64.

13 Walter Benjamin, "Little History of Photography", in *Walter Benjamin: Selected Writings (1931–1934)*, vol. 2 (Cambridge, MA: Belknap Press, 2005), pp. 510 and 512.

14 Brief item, "El Hipnotismo vuelto a la moda", *El Tiempo,* Mexico, 11 August 1887: "Los mediums que veían, sentían y usaban todos los sentidos de su propio cuerpo de un modo enteramente diferente de los demás hombres; de tal modo, que pudiera decirse de ellos que tenían un sexto sentido que sumía todos los otros y de una potencia con mucho superior a la de los cinco juntos [...] he conocido mediums que veían las regiones que se encontraban a millares de kilómetros de distancia; y las veían mejor que lo que nosotros las pudiéramos ver a una lejanía menor que cien metros. Veían las gentes y los animales que las habitaban y en ellas se movían; escuchaban las voces humanas y observaban con precisión cuanto podía caer bajo el dominio de sus sentidos."

15 See Georges Didi-Huberman, *Pueblos expuestos, pueblos figurantes* (Buenos Aires: Manantial, 2014), pp.14–16. In *Escenarios del cuerpo*, Francisco Ferrándiz develops a similar thesis to the one presented here, about the possibility of also tracing at the same time emancipatory aspects of spiritualism so that they could be used as attempts to control popular demonstrations.

16 For a detailed study of the publication, *La Ilustración Espírita*, see chapter 2 of Leyva, *El Ocaso de los espíritus.*

17 Short article published as "Un problema para la medicina", *La Ilustración Espírita*, Mexico, June 1891: "Desgraciadamente, los médicos en general desprecian todo lo que no parte de ellos, y no quieren comprender que, tras formas ridículas muchas veces, tras la ignorancia del pueblo que no se explica la verdad de los hechos sino arreglándose a sus propias preocupaciones y viciadas ideas, tras el curanderismo sin charlatanería, tras las curaciones sin drogas ni brebajes y tras esos hombres y mujeres del pueblo, sin más ciencia que sus manos y sin más sabiduría que el fenómeno que se ofrece a su vista sin explicárselo, existe una verdadera ciencia, hay una verdad nueva, una revelación sorprendente y un mundo maravilloso cuya virtud se desconoce. Entre los curanderos y entre los médicos hay charlatanes interesados. Ni a unos ni a otros debe atendérseles, ni unos ni otros merecen consideración; pero sí la exigen aquellos que con noble desinterés se dedican a hacer el bien a sus hermanos exponiendo su vida, gastando sus recursos, privándose de las comodidades propias y

sufriendo todas las molestias en bien y provecho del que sufre. En este concepto está Teresa Urrea."

18 It is important to point out that, in the years of its publication until 1893, *La Ilustración Espírita* did not have an explicit anti-Porfirist policy, although there are many articles that express this rhetoric opposed to materialism and progress without a spiritual dimension.

19 Quotation from a feature published by Amalia Domingo Soler, "Escribir", *La ilustración Espírita*, Mexico, September 1890: "La prensa es uno, quizá el primero de los adelantos humanos, porque vitaliza el pensamiento, porque enlaza a todas las clases sociales, porque un periódico es una carta universal, porque un libro es un ramillete de fragantes flores que no se marchita jamás."

20 *Ibid.*: "De algún tiempo a esta parte ha aumentado el número de los escritores, porque nuestros amigos de ultratumba nos envían sus pensamientos por conducto de médiums escribientes, mecánicos, intuitivos y auditivos, y obras filosóficas e históricas recreativas han venido a enriquecer la literatura de ambos continentes. A escribir! han dicho los espíritus. A leer! responden los espiritistas y se han formado centros y grupos de estudios y una parte de la humanidad sostiene activa correspondencia con las almas de los que fueron. La familia universal dejó de ser un mito, estamos relacionados con los seres que dejaron su envoltura material. Ellos escriben, nosotros leemos. Venturosos los que saben escribir, y felices los que se apresuran a leer!"

21 Robert Darnton, *Mesmerism and the End of Enlightenment in France* (Cambridge, MA: Harvard University Press, 1968).

22 For information on the Hidalgo Lyceum polemic, see Zenia Yébens Escardó, "Las vicisitudes del desencantamiento del mundo: espiritistas y positivistas en la polémica del Liceo Hidalgo de 1875", in *Polémicas intelectuales del México moderno*, edited by Carlos Illades and Georg Leidenberger (Mexico: UAM–CONACULTA, 2008).

23 "El espiritismo, esta ciencia fundada en los principios más justos y equitativos de la más pura filosofía, esta doctrina, basada, sostenida y demostrada por el Evangelio; discutida, razonada y aceptada por muchos hombres prominentes en sabiduría, tanto antiguos como modernos; y por fin, revelada y también demostrada de común acuerdo, por multitud de seres inteligentes de ultratumba, ya ha aparecido palpitante entre todos los pueblos de la Tierra y entre todas las clases de la sociedad sin distinción ni privilegio."

24 For a study on the links between spiritualism and feminism in Mexico, see Lucrecia Infante Vargas, "De espíritus, mujeres e igualdad: Laureana Wright y el espiritismo kardeciano en el México finisecular", in *Disidencia y disidentes en la historia de México*, compiled by Felipe Castro and Marcela Terrazas (Mexico: UNAM, Instituto de Investigaciones Estéticas, 2003), pp. 277–294. For a study on feminism and spiritualism in the United States, see McGarry, *Ghosts of Futures Past*.

25 McGarry, *Ghosts of Futures Past*, p. 29.

26 Brief item "Sin título", *La Luz en México*, Mexico, 4 April 1873, p. 3: "Las jerarquías entre los hombres solo pueden y deben ser admisibles por el poder de la virtud y del saber, y el hombre que verdaderamente es un sabio virtuoso, solo es un consejero humilde, un director pasivo de todos lo que necesitan de su ejemplo e instrucción, y practica el bien tan sólo por el bien y nunca por ver al interés."

27 See the published article, Refugio I. González, "Materia elástica manual de espiritismo y magnetismo práctico, escrito por Refugio I. González. Para ayudar a todos los adeptos que quieran practicar en círculos de evocación, conforme a la doctrina enseñada por el maestro Allan Kardec", *La Ilustración Espírita*, Mexico, May 1889.

28 See Infante Vargas, "De espíritus, mujeres e igualdad".

29 For two of the most important studies on the narrative of nation building in Latin America, see Benedict Anderson, *Imagined Communities* (New York: Verso, 1983) and Doris Sommer, *Foundational Fictions: The National Romances of Latin America* (Berkeley: University of California Press, 1993).

30 See Bartolomé de las Casas, "He venido a responder vuestras preguntas", *La Luz en México,* Mexico, 1 August 1873: "Pero no le produce esto gran sufrimiento, como sería de creerse, sino que es una perturbación pasajera que tiene embotadas levemente las percepciones del alma."

31 "El diablo hospedado en una mesa", *El Universal,* Mexico, 27 December 1853: "Supliqué que no se siguiese adelante, y rogué al señor cura que tuviese la bondad de preguntar al espíritu en latín. — 'Loqueris ne latina?' — le dijo Mr. Bertrand — No hubo respuesta — '¿Hablas latín?' preguntóle entonces en francés. — La mesa dio un golpe — Y otra vez en latín 'quis es tu? Dic nobis nomen tuum?'"

32 Brief item "Doctrina espiritista", *La Unidad Católica,* Mexico, 13 November 1861: "En fin, M. Allan Kardec nos promete un nuevo método de telegrafía que excederá a todos los otros en velocidad y exactitud. Dos personas, evocándose recíprocamente, pueden transmitirse sus pensamientos. 'Esta telegrafía humana será un día un medio universal de correspondencia'."

11 Universopolis: The Universal in a Place and Time

Andrew Ginger

In his essay *La raza cosmica* (The Cosmic Race) (1925), the Mexican politician and thinker José Vasconcelos issued one of the most resonant prophecies in Latin American history. "By the great river" — the Amazon — he foresaw, "Universopolis will arise, and from there will issue forth the preaching, squadrons, and planes propagating Good News."[1] Universopolis would be home to the cosmic race of the essay's title, a future humanity characterized by its "universal human sentiment."[2] In turn, the formation of this new people would find its precedent in the mixing of what Vasconcelos calls *races* in the Spanish and Portuguese possessions in the Americas, and in their post-independence successor states. "A history marked by war and atrocity paradoxically produces utopia", Marilyn Grace Miller observes of this vein of thought, paraphrasing in turn the Mexican intellectual Leopoldo Zea.[3] In his philosophical essay "Mestizaje and Hispanic Identity" (2010), Gregory Velazco y Trianosky remarks that this notion of the cosmic race is "certainly utopian in a literal sense, since it is clearly envisioned by many of its advocates, past and present, both in North and in Latin America, as defining a valued future state."[4]

John A. Ochoa has explained how Vasconcelos wrote his essay at a time when intellectuals and writers both in the United States — the Young Americans movement — and in Mexico — among those associated with the cultural institution, the Ateneo — shared an obsession with the history of visions of utopia, and especially of its relationship to the Americas. This is as much the case in Lewis Mumford's *The Story of Utopias* (1922) as in Alfonso Reyes's "The Presage of America" (El presagio de América) (1920). Both groups were interested primarily in the "usable past": that is, the ways in which things found in history might be relevant to the present day in imagining a future. More still, the rootedness of such utopias in the historical past "was meant to give America a sense of validation. Utopias provided a sense of belonging to a continuous universal culture, a fulfilled destiny". Yet, for all this common interest, "only Vasconcelos offered an actual utopia".[5] He did so on the back of his role as Minister of Education in the revolutionary Mexican government of the early 1920s (1921–24), where he coined his famous motto for the National University of Mexico (UNAM), "Through my race the spirit will speak" (Por mi raza hablará el espíritu).[6] His vision of a cosmic race — as much in its broad-brush effect as in its details — was to prove suggestive for later movements, from student protests in the Mexico of the 1960s to Chicano intellectuals and artists, that is, those of broadly Latin descent living in the United States. In particular, it has been taken up by the influential writer Gloria Anzaldúa.[7]

I am centrally concerned here with the twin character of Vasconcelos's prophecy. On the one hand, it is intensely local and specific: Universopolis will rise up in a given place, on the banks of the Amazon at some future juncture, and good news will be carried outwards from it by its legions. On the other hand, this is a universalizing prophecy, destined ultimately for the redemption of all humanity. The compelling force of *The Cosmic Race* derives not least from its powerful, vivid affirmation that the universal will emerge through a concrete and particular time and place without ceasing either to be universal or specific. This is at the heart of its utopian resonance. I want to explore what might be involved in Vasconcelos's envisioning that such a thing is possible. In so doing, I will address some of the key objections to his prophecy, which amount, in varying ways, to the preoccupation that his alleged universalism is ultimately either bogus or necessarily a failure, or (often) both. At the heart of these is the suspicion that particularity and its limitations must triumph, either because Vasconcelos's vision effectively promotes the goals of particular interest groups, or because it will always be held back by historical realities. This is the view that — as Nancy Stepan puts it in her book *The Triumph of Eugenics* (1991) — "the 'cosmic' race was not to be so cosmic or universal after all."[8]

My interest is not simply in the detail of this debate, but also in what assumptions about universality and particularity are involved on either side, and in whether and in what way the discussion might be resolved, if at all. In so doing, I will consider how Vasconcelos might understand transfiguration — the transformation of something specific into something transcendent —, and its relationship to physical intimacy. I will attend also to two challenging dimensions of the essay *The Cosmic Race.* The first is its relationship to chronological time. Many accounts, understandably, seek to interpret Vasconcelos's words by contextualizing them within Mexican intellectual life of the 1920s. Conversely, the essay itself ostensibly conjures up an unspecified future time free of the limitations of Vasconcelos's own context. The second is the style and language of Vasconcelos's prophecy — memorably described by Illan Stavans in his introduction to an English-language edition of 2011 as "maddeningly obtuse" and "a bumpy read".[9] My own interest here is not in trying to locate the essay within Vasconcelos' own extremely extensive oeuvre or in a deep contextualisation within the Mexico of the 1920s. I am not even especially concerned with the relationship between the essay on cosmic race itself and the lengthy travelogue that accompanied it when it was published.[10] I make no claim whatsoever to rival other scholars in those respects. I am primarily preoccupied with elucidating the notions of universality and particularity as expressed within Vasconcelos' resonant and influential essay.

Vasconcelos's account of the path to universality is, in some measure, a sort of history of civilization and a projection of it into the future. Early on, he claims that there was an ancient civilization in the lost society of Atlantis, and that among its descendents are the native peoples of the Americas (pp.17–19). While this may sound — and is — an extravagant claim, it had a distinguished lineage in the historical culture of Mexico since the Spanish conquest. The great Mexican thinker of the late seventeenth century, Carlos Sigüenza y Góngora, inspired by the German Jesuit Athanasius Kricher, had proclaimed the indigenous inhabitants of Mexico to descendents of peoples who had occupied Atlantis. In alluding to such assertions, Vasconcelos, like Góngora y Siguenza before him, is preparing the way for an affirmation of the dignity of Ibero-America in the face of outsiders who might look down upon it.[11] The past link to Atlantis is something preserved for a glorious future: the land has been kept to one side from the march of civilization for 5000 years so it may serve as the cradle of a new people (p. 33). For Vasconcelos, the Inca and Aztec empires were mere decaying remnants of Atlantis, for by then the impulse of civilization had passed elsewhere: first to Egypt, then to India, and on to Greece which provided the basis for Europe (p. 20). The point of Vasconcelos's remarks, though, is not so much to exalt European culture and society, as rather to see these as a route back to the primacy of the Americas. The key event in history that prepares the way to the future is the violent arrival of what he terms the Anglo-Saxons (the British) and the Latins (the Spanish, Portuguese, and to a degree the French) on that continent. This development sets up a new historical dynamic: a struggle for supremacy between Latins and Anglo-Saxons, which must be resolved outside Europe (p. 21). Vasconcelos singles out the French Emperor Napoleon in the early nineteenth century for failing to see this, and for assuming that the Americas were a sideshow to European conflict. His casual renunciation of Louisiana tipped the balance of power in favour of the Anglo-Saxons by opening their route to the Mississippi (p. 25).

What Vasconcelos is doing is to put the Americas at the heart of world history, displacing accounts centred around Europe, shifting manifest destiny from the United States southwards to Ibero-America. The manoeuvre is not dissimilar to that of the influential twentieth-century artist Joaquín Torres García who famously depicted the Americas with the south at the top, inverting the usual form of maps.[12] For the time being, and especially since the United States and the successor states of the Spanish and Portuguese possessions gained independence, the Anglo-Saxon is dominant, according to Vasconcelos. Anglophone success is based on an obsessive pursuit of material advances such as railways, bridges, and businesses (p. 43). It is also rooted in a defence of the power of one homogenous race and the destruction of others (p. 32). To these two elements of the Anglo-Saxon triumph correspond two flaws that point to an expiry date. First, the very exclusivism of the Anglo-Saxon triumph marks its limits. All prior civilizations built on the hegemony of one group of human beings have fallen, and therefore "it is clear that the dominance of the white will also be temporary" (es claro que el predominio del blanco será también temporal) (p. 20). Ultimately, such hegemony "does not resolve the problem of humanity" (no resuelve el problema humano) (p. 33). By this, Vasconcelos means that any civilization that is based on the exclusion of the vast swathes of human beings in the world will, sooner or later, fall. Presumably there are two reasons for this: that the opposition will, in the end, be too great; and that the hegemonic civilization itself will have denied itself the insights and opportunities that the rest of the globe's population might provide.

Second, the scientific and technological knowledge of the Anglo-Saxons can easily be learnt and taken on by other peoples (p. 41). We might gloss this remark: it is inherent to the claim that something is scientific that its truth should not be specific to a given culture or society. At the same time, the very science promoted by the Anglo-Saxons undermines any claim they might have to exclusive superiority: Vasconcelos cites how Mendel's work on genetic variations in inheritance has this effect (pp. 58–59). The two flaws in present-day hegemony come together in the problem of social injustice, which Vasconcelos predicts will split the Anglo-Saxons themselves. There is not just a concentration of power in one race, but of wealth and authority in an elite, as well as neglect of things other than such material pursuits: "The whites themselves, discontented with the materialism and social injustice into which their race has fallen [...] will come to us to assist in the conquest of freedom" (Los mismos blancos, descontentos con el materialismo y de la injusticia social en que ha caído su raza [...] vendrán a nosotros para ayudar en la conquista de la libertad) (p. 43). This observation connects to Vasconcelos's view that the future will require "socialism in government" (socialismo en el gobierno) (p. 62). He is looking to a thoroughgoing social and economic transformation.

Crucial here is the notion that the present hegemony will "serve as a bridge" (servir de Puente) (p. 20) and that whites "by fulfilling their destiny of mechanizing the world, have themselves put in place, without knowing it, the bases of a new era" (al cumplir su destino de mecanizar el mundo, ellos mismos han puesto, sin saberlo, las bases de un período Nuevo) (p. 30). This is in part for the reasons given above, but it is also — as the image of a bridge suggests — because of the effect of technology and large-scale enterprise on the degree of integration of the world. The examples Vasconcelos gives of what Latin America owes to the United States seem not coincidental: railways, bridges, and businesses, that is, communications infrastructure combined with international capital and commerce.

Conversely, Vasconcelos sees the post-independence Latin American states as possessing both a flaw, which underwrites their present weakness, and an advantage which points to their future success. Unlike the United States to their north, the successor states of the Iberian monarchies created a series of separate countries, which Vasconcelos disdainfully describes as "wee little nations and sovereign principalities" (nacioncitas y soberanías de principado) (p. 28). This leaves them in a situation of basic strategic weakness: "We jealously guard our independence in relation to one another; but one or other way we are subject or allied to the Saxon Union" (Nos mantenemos celosamente independientes respecto de nosotros mismos; pero de una o de otra manera nos sometemos o nos aliamos con la Unión sajona) (p. 23). The notion of a larger federation spanning the Atlantic and Pacific territories, and proposed in the Spanish revolutionary constitution of 1812, was rejected, as was the plan of the rebel Mina to defeat the old monarchy first in the Americas and then to take the battle back to metropolitan Spain (p. 31). Despite the best efforts of the leading light of the independence uprisings, Simón Bolívar, this has left the successor states insufficient larger vision for the world and for humanity as a whole: their concerns are notable for their "provincialism" and "lack of important/transcendent plans" (provincialismo; la ausencia de planes trascendentales) (p. 28). More still, in splitting in such a hostile way from Spain, the successor states have ignored relevant aspects of their own cultural inheritance, and have become obsessed with the predominant values of the hegemonic Anglo-Saxons. Vasconcelos refers to "de-Spanishizing sermons and corresponding Englishment" (las prédicas despañolizantes y el inglesamiento correlativo) (p. 28).

On the other hand, the Latin American states inherit from the earlier monarchies and from the independence period two latent advantages. These are occluded at present. In both cases, Vasconcelos is looking to earlier history less as a series of accomplished and realised events and outcomes, than as potentiality for the future: the seeds of a *might be* more than a *was*. The first is the as-yet-unfulfilled legacy of those earlier proponents that Latin America should be something more than a collection of independent countries. Vasconcelos notes that, by and large, even those who opposed a greater federation "felt driven by a feeling of universal humanity" (se sintieron animados de un sentimiento humano universal) (p. 33). This is seen, for example, in their declared — if not always realized — aim of abolishing slavery, and their statements about human equality (p. 34).

The second — and the one which has most occupied readers' attention — is the willingness among Iberian colonists and their descendants to have children with other peoples. Such behaviour contrasts markedly with the predominant culture of the United States, and more broadly the Anglo-Saxons. "It is", he says, "in this fusion of different stock that we must find the fundamental character of Latin-American distinctiveness" (Es en esta fusión de estirpes donde debemos buscar el rasgo fundamental de la idiosincrasia latinoamericana) (p. 34). Such attitudes and behaviour provide a precedent — relatively absent in the Anglo-Saxon world — for an inclusiveness towards all humanity that will characterize the future Cosmic Race. It is a route towards "the fusion of the various present-day races in a new one that completes and surpasses them all" (la fusión de las distintas razas contemporáneas en una nueva que complete y supere a todas) (p. 44). In significant measure here, Vasconcelos appears to be evoking a biological and specifically genetic process. In conjuring up the "marriage of a black Apollo to a white Venus", he refers to the effects of "the crossing of opposites, in line with the Mendelian law of inheritance" (enlace de un Apolo negro con una Venus rubia; el cruce de contrarios, conforme a la ley mendeliana de la herencia) (p. 51).

Before proceeding, I want to bring to the fore two characteristics of this account of human civilization and its future. First, running through *The Cosmic Race* there is a contrast between what Vasconcelos calls *limitation* and what he terms *universality* and *cosmic sentiment* (limitación; universalidad; sentimiento cosmico) (pp. 33, 57). Given that no formal definitions are offered for these or related terms, I will follow Wittgenstein's dictum that often "the meaning of a word is its use".[13] *Limitation* is employed by way of summary of a series of things that Vasconcelos finds ultimately defective, such as the "provincialism" described above. It refers to ways of being, feeling, and thinking that are primarily confined to a specific society, culture, civilization, or people, and, therefore, to a particular place and time. It likewise involves efforts to restrict (genetic) variation among human beings. Often in human history, it is associated with "violent domination" (violento predominio) (p. 43), that is, oppression. *Universality* and *cosmic sentiment* are, in this usage, a release from *limitation*. The contrast is not, however, quite so straightforward, for all these terms are presented as value judgements in Vasconcelos's usage. Not every form of constraint is a *limitation*, and not every surpassing of a boundary is per se *universality*. He is, for example, concerned to stave off a potential accusation of foretelling "a repugnant process of anarchic hybridism" (un proceso repugnante de anárquico hibridismo) (p. 44). *Universality* and *cosmic sentiment* involve not simply being open to all humanity and human possibility, but doing so in a specific spirit of freedom which characterizes a new ethos and aesthetics. *Universality* and *cosmic sentiment*, then, are used to describe a good life for humanity, and *limitation* is what has not attained that state of appropriate openness.

Second, all these things — *limitation*, *universality*, *cosmic sentiment* and related notions — are presented as historical realities, whether actual or potential; that is, whether things that have happened and do happen in human history up to the present, or, alternatively, things that — it may plausibly be judged on the face of history thus far — could happen. This means that — at least so far as Vasconcelos sees it — *universality* is not an abstraction from the specificities of human existence in place and time: quite the opposite, in fact. By extension, to seek to identify the sources of *universality* is not in contradiction with locating things in a time and place. Universopolis is, after all, to rise up in a specific geographical situation, by the Amazon, at a given (if unspecified) juncture in time. Even *limitation* itself is not a simple opposite of *universality*. It is, after all, a history thus far overwhelmingly characterized by such *limitation* that ultimately provides the conditions for *Cosmic sentiment*. This is the case, for example, in the bridging role undertaken by whites. Moreover, things that happen historically are important, in Vasconcelos, not simply for their having done so, or for their use in explaining present-day circumstances, but rather as indicators and sources of what might be, even if in the past or present such an outcome was defeated or oppressed. History, we might say, contains unrealised potential, in the particular shape of Latin America.

Vasconcelos's presentation of *limitation*, *universality*, and *cosmic sentiment* has the potential to play havoc with some understandings of *context* — perhaps most obviously exemplified in the Marxist Fredric Jameson's cry, "Always historicize."[14] Things which are, of their very nature, resistant to — indeed fundamentally opposed to — being located in a particular place and time (*universality*), are, simultaneously, things that happen in specific places and times (and vice-versa). Moreover, (many) things that occur at a given point chronologically and geographically cannot be understood fully as things proper to that point in place and time, because what matters about them is a potentiality that resonates through and may be activated in their future. The role of the historian, at least as Vasconcelos undertakes the task, is to appreciate the ways in which all this may be so. I use the word *appreciate* not simply to mean *notice*, but in its fuller aesthetic sense of experiencing with discernment and valuing (whether positively or otherwise). This is one of the main senses in which Vasconcelos rejects "empirical history, sick with short-sightedness", and embraces "intuition based on the data of history and science" (la historia empírica, enferma de miopía; intuición que se apoya en los datos de la historia y la ciencia) (pp. 19–20). Again, no formal definition is offered for *intuition*, but its usage within *The Cosmic Race* is clearly not reducible to something like *instinctive* or *irrational* but rather expresses the more refined modes of observation, feeling, and thought described here. This is consistent with the departures from positivism, which Guillermo Hurtado has shown to be fashionable within the Mexican Ateneo from the end of the first decade of the twentieth century. For example, the French philosopher Henri Bergson — serious interest in whom has been revived by the theorist Giles Deleuze — used the term *intuition* to describe a way of knowing things in the world intimately in their particularity.[15] An outlook such as that of Vasconcelos — if accepted — necessarily involves, not just assessments of past events, but predictive judgements concerning which futures, arising from them, are and are not possible and plausible, as well as desirable. That is a test at once of verisimilitude and of ethics. It supposes not simply that such judgements must be made, but that they can be viable. It is in this way that the historian of civilization is also the prophet of utopia, and the writing of history is, at the same time, the uttering of prophecy.

The principal objection often raised to *The Cosmic Race* is that its prophecy of universality is, in fact, a perpetuation of limitation. In its broadest terms, this is the affirmation that the vision of a cosmic race is either not what it purports to be, or, at the very least, that, carefully considered, it is not nearly as ethically enticing as it might appear. Such claims may bleed into the wider view that things simply cannot be other than what they are within the constraints of a particular time and place; imagining otherwise is utopian in the most pejorative sense of the word. Either way, the recommended approach is to situate *The Cosmic Race* firmly back in its place and time so as to show what constrains its outlook: in short, to contextualize it.

In her balanced account, Marilyn Grace Miller notes that even if a majority of Mexicans were by Vasconcelos's time *mestizos* — of mixed "race" —, his prophecy of *mestizaje* casts to the margins the large number of indigenous peoples — in the Chiapas region for example — who had preserved more of a separate identity.[16] Moreover, there is, she says, a "disconnect in the text between lived conditions in the Americas and Vasconcelos' philosophical prophesies." Miller observes that he enthuses about some regions of the continent and noticeably less about others, and that in the accompanying travelogue, Vasconcelos is overwhelmingly enthused by one of the most "white" of Latin American locations of the 1920s, supposedly cosmopolitan Buenos Aires in Argentina, a place characterized by large-scale European immigration.[17] She sees strong parallels, in fact, between *The Cosmic Race* and the positive consideration of eugenics and hopes for so-called "whitening" throughout Latin America.[18] A key point of the prophecy, then, is that it underwrites the hegemony of the already dominant groups in Mexican society, and that it contemplates mixed sexual relationships and assimilation primarily in order to rid Latin America of a strong indigenous and Afro-American presence.

In an essay on the *mestizaje* published in 2000, Catherine Poupenay-Hart puts this viewpoint bluntly: "Vasconcelos' program for the subcontinent is explicitly eugenistic", she says.[19] In *"The Hour of Eugenics"* (1991), Nancy Stepan looked in detail at Mexican interest in eugenics, including the role of Vasconcelos in particular. Stepan underlines four distinctive factors. First there was a rejection of views promulgated elsewhere in powerful countries that the mixing of "races" led to degeneration. Such a view was self-evidently damaging to the standing of the Mexican ruling groups, and great swathes of the country's population. In response, there was an assertion of the benefits of such mixing, with opinions ranging from the desirability of assimilation into a dominant white group, to eulogies for a new racial type born of mixing.[20] Second, there was Catholic-inspired opposition to direct interference with reproductive processes, which was shared even in many of the highly anti-clerical circles of Mexico's post-revolutionary leadership. Sterilization was widely opposed in Mexico, and introduced only briefly, and probably more in law than in deed, in the state of Veracruz during the year 1932.[21] Third, none of this meant that a form of eugenics was not widely endorsed, but rather that it took the form of what Stepan calls "preventive eugenics" and "matrimonial eugenics". That is to say, there were attempts to use marriage laws and practices, as well as to promote immigration controls, so as to defend and secure an improved Mexican stock. Notably, in 1928, three years after the publication of *The Cosmic Race*, the government introduced controls on the issuing of marriage licences, which were supposed to be refused to those with specific illnesses or vices.[22] Fourth, there was a marked preference for the account of heredity provided by the late-eighteenth- and early-nineteenth-century French biologist Jean-Baptiste Lamarck, in particular his Second Law. On this view — "soft inheritance" — characteristics acquired in a lifetime and in a given environment would then be transmitted to offspring. (While the notion that such traits would be overwhelmingly inherited has long since been discredited, some recent scientists hold that, by a diversity of means, some of them may be transmitted across generations.[23]) The Lamarckian view meant that as Stepan puts it, "no sharp boundaries between nature and nurture were drawn." In consequence, Mexican intellectuals concluded that refining their "race" entailed great attention to cultural and societal conditions and to individual and collective behaviour.[24]

Stepan points out the close — and all too apparent — parallels between *The Cosmic Race* and these dominant views about eugenics in Mexico. There is a marginalization of indigenous peoples in their own right, an explicit but modified invocation of eugenics — what Vasconcelos called "spiritual eugenics" —, a desire to create a purified and beautified race through selective biological reproduction, and advocacy of immigration controls against the Chinese. It is striking that the Mexican scientist, Alfonso L. Herrera had predicted four years before the publication of *The Cosmic Race*, in 1921, that laboratory work would eventually generate an earthly paradise inhabited by humans of outstanding beauty and merit.[25] Stepan summarises the real purpose as follows: "to eliminate heterogeneity in favor of a new homogeneity, the Europeanized mestizo."[26] In John A. Ochoa's view, Vasconcelos, through his writing, expressed his response to the earlier years of life in the border zone of Mexico and the United States, preoccupied with the menace both of the "Yankee" and indigenous peoples. In the face of this, Ochoa suggests, Vasconcelos comes up with a "self-aggrandizing attitude" in which he sees himself as "forced to become the personal representative of his entire race". From here derive the celebrated motto "Through *my* race the spirit will speak" (my italics), and what Ochoa describes as the "utopian fantasy" of Universopolis. Through such things, Vasconcelos ostensibly "joins the adamantly individual to an indefinite, universal entity", but is basically concerned in so doing with creating a projection of his own biography and nationality.[27] More still, he presents himself as the embodiment of Mexican national identity, and vice-versa. Discussing Vasconcelos in her study of Latin American autobiography, *At Face Value* (1991), Sylvia Molloy evokes the Mexican muralist movement of artists such as Diego Rivera. Molloy comments that "Vasconcelos "muralizes" himself, presenting a gigantic self-image that overshadows that of Mexico while feeding off it [...] assigning a national identity to his mythical persona".[28]

Equally, it is possible to find ways of writing about *The Cosmic Race* that appear, if not to contradict, at least to qualify the kinds of objection set out above. Among the issues at stake are: whether *The Cosmic Race* is really concerned with Mexican identity (a society in a given place); the degree to which, or way in which, it is about the 1920s (a society at a given time); whether it advocates homogeneity; and how it treats a range of historical causalities and material realities or alleged realities, biological and otherwise. In her study of Latin American intellectuals, *In the Shadow of the State* (1999), Nicola Miller bluntly observes that "neither *La raza cósmica* nor the subsequent *Indología* (1927) was written to address *mexicanidad* (that is, Mexicanness)", and that "Vasconcelos regarded national identity as a spurious issue for Spanish American countries". What is more, Miller notes, Vasconcelos himself certainly did not identify the existing "mixed race" society of Mexico with the Cosmic Race. His primary and overwhelming concern was with "universalism" and with opposition to "chauvinistic nationalism". If the Mexican establishment subsequently re-invented his vision of a Cosmic Race to their own nationalistic ends, this was a distortion of Vasconcelos's essay.[29] Years earlier, in his introduction to *The Cosmic Race* (1979), Didier T. Jaén alerted readers to what he found to be misguided tendencies in readings of the famous essay. He remarks that "the idea of a superior Latin American or Hispanic race was not in Vasconcelos's thesis." Meantime, Jaén notes with concern, "his prediction of a new age for humanity goes, not only unchallenged, but suspiciously ignored" in favour of debates and interpretations centred on identity politics and "race" as presently and historically understood. Readings of Vasconcelos are often, he says, "'sociological' and 'from the point of view of genetics'".[30]

There is some evidence in *The Cosmic Race* to support such qualifications, not least the utter contempt that Vasconcelos voices for the obsession with nation-building in Latin America since the independence wars, which he refers to as "idiocy" (imbecilidad) (p.31). More subtly, while he does affirm that Universopolis will be built in the southern American tropics, Vasconcelos does not describe it as a Latin American, less still a Mexican, polity. He speaks instead of "the One world" (el mundo Uno) (p. 32). He even considers potential scenarios in which Latin Americans have little role in the founding of utopia. "If our race turns out to be unworthy of this sacred soil [...] it will find itself supplanted by peoples who are more able to realize the fateful mission of those lands" (si nuestra raza se muestra indigna de este suelo consagrado, si llega a / faltarle el amor, se verá suplantada por pueblos más capaces de realizar la misión fatal de aquellas tierras) (pp. 56–57). In the event that the Anglo-Saxons were to succeed in conquering the Amazon, Vasconcelos contemplates how, they themselves might ultimately change and look to the formation of the Cosmic Race (p. 42).

The point of Latin American involvement, then, seems to be twofold in Vasconcelos's account. First, the openness to mixture in Latin America provides an opportunity for its peoples to take up the leading position on the world stage. In short, he is pointing to a strategic opening, a way out of second-class status, and is urging his continental compatriots to seize it: "our civilization, with all its faults, *may* be the chosen one", he says (my italics; nuestra civilización, con todos sus defectos, *puede* ser la elegida) (p. 32). Even there, Vasconcelos expects that the Latin Americans would operate in alliance with many Anglo-Saxon whites rebelling against global injustice (p. 43). Second, there is an apparent distinction in Vasconcelos's account between the causes and agents that might bring about Universopolis and, on the other hand, the latter's own identity. Put another way, for Vasconcelos, identity is not reducible to causes. It does not follow from (a) the conditions and peoples of Latin America will lead the way to the foundation of Universopolis, that (b) Universopolis will be a Latin American state.

Moreover, so far as the creation of Universopolis goes, *The Cosmic Race* is clearly projecting far, far into the future, presenting the circumstances and attitudes of the 1920s, in Mexico or anywhere else, as utterly inadequate to the task. "We are as yet nothing," Vasconcelos categorically remarks (nada somos aún) (p. 56). To the extent that Latin Americans might have the opportunity to build utopia, they will clearly have to change substantially in order to ready themselves for such an eventuality, not least by giving up on their "purely national patriotism" (patriotismo puramente nacional) (p. 22). What we have of value, he says, is "in potential" (en potencia) (p. 56). Mexico and Latin America, then, are more of a tiny embryo, a hint of what might be, than they are a blueprint for Universopolis. This is not to say, though, that the context of the 1920s is irrelevant to the vision of Universopolis, any more than it is to affirm that specifically Latin American peoples would not found the One world. Rather, *The*

Cosmic Race is suggesting the way or ways in which the 1920s might, over a long time period, extrapolate to the moment when Universopolis rises by the banks of the Amazon. It is an expression of possibilities latent, however remotely, in those years. These are presented in a very broadbrush fashion — perhaps because by the time the extrapolation is complete, and given that Vasconcelos envisages a variety of different possible time frames and agents, the details of 1920s Mexico will hardly be relevant. The specifics of the context of 1920s Mexico only matter insofar as they might — and the operative word is *might* — contribute to (or inhibit) these larger patterns.

The Cosmic Race is not, then, set simply either in the utopian future or in 1920s Mexico or Latin America, but, rather, is presented as a perception of how the former may extend out of the latter, as if seeing an opening or openings in present-time out onto a larger horizon, a way in which the specifics, and even the *limitations* of the chronological present, may transfigure into future *universality*. (I use italics here to designate Vasconcelos's specific usage of these terms.) Time thus expressed has both chronological and non-chronological characteristics. These are not in simple conflict with one another.[31] Time proceeds through a series of ages, one after the other (it is stadial, as people say), each led by a given "race" until the Cosmic Race arises in Universopolis. That progression, however, involves the invocation of and connection to "usable pasts", and indeed usable presents. As we have seen, a series of moments across chronological time, whether the 1920s or the conquest of the Americas, are experienced and perceived as redolent with possibilities that open up routes to the future Universopolis. Moments in time are, as it were, elastic or capacious: the specific form they take in their own period can be stretched and opened into a new shape at and for a later juncture. This means that to pin the words of *The Cosmic Race* onto specific meanings and deeds precisely situated in 1920s Mexico is to work — rightly or wrongly — against the grain of how Vasconcelos employs contextualization, and thus, of how — putatively — he deploys the language and terminology of his own time.

At the heart of debate about the contextualization of *The Cosmic Race* is the vexed question of "race". The nature of causal relationships is, once more, crucial here, in asking both whether the essay's account of "race" arises specifically from and is shaped by hegemonic elite views in 1920s Mexico, and, thence, whether *The Cosmic Race* thereby supposes — in line with contemporaneous eugenics — that supposed racial characteristics directly cause specific cultural, societal, and psychological outcomes. At stake — once again — is the nature of the objectives that Vasconcelos envisages for Universopolis, and whether its creation is to be an exercise in so-called "whitening" or homogenisation of humanity or both.

Vasconcelos's obsession with cross-breeding and his invocation of (Mendel's) genetics as routes to utopia support the view that his work is, in the end, racist and eugenicist. This is true even of his apparently most generous sentiments, because of the genetic, reproductive process that underpins the attainment of the good life, and his insistence on the merits of so-called miscegenation in the former Spanish and Portuguese possessions: "the crossing of blood" (el cruce de sangre) (p. 45). He appears to identify a specific genetic make-up (which he supposes to create a given "race") with the causes of given phases of civilization: "We have then", he says, "the four stages and the four trunks [of humanity]" (Tenemos entonces las cuatro etapas y los cuatro troncos) (p. 20). As Alan Knight puts it in his influential essay on "Racism, Revolution, and *Indigenismo*" (1990), "it is the inescapable ascription that counts".[32]

Moreover, the ideology of "whitening" finds strong echoes in Vasconcelos's extremely disturbing remarks that "in a few decades of aesthetic eugenesia, the black could disappear" (en unas décadas de eugenesia estética podría desaparecer el negro) and that "the low types of the species will be absorbed by the superior type" (los tipos bajos de la especie serán absorbidos por el tipo superior) (pp. 52–53).

In other passages, and whatever one's final judgement, *The Cosmic Race* proves a slippery text. There is the question of what Vasconcelos actually means by the very term *raza* (*race*). Joseba Gabilondo has observed how "he simultaneously deploys different historical theories of race, which are incompatible from a modern point of view."[33] In *The Rise and Fall of the Cosmic Race*, Miller likewise notes how the "notion of a 'mestizo soul,'" is "indicative of the elision between the physical and spiritual definitions of the condition".[34] This is in part a matter of Vasconcelos's own personal usage, in part a consequence of neo-Lamarckian genetic thinking, and in part a semantic feature of the Spanish language. *Raza* certainly encompasses common present-day English-usage of *Race* as "a group [...] with particular similar physical characteristics" (Cambridge English Dictionary), and it was used that way in relation to classifications of human beings into *castas* in the colonial period, for example. But it is also a looser term, defined historically and now by the Royal Spanish Academy as a "quality of origin or lineage" (calidad del origen o linaje). This means that it could historically refer to groups with a shared cultural inheritance. That interpretation might well be detected in statements such as: "The advantage of our tradition is that it possesses a greater facility for sympathy with others" (La ventaja de nuestra tradición es que posee mayor facilidad de simpatía con los extraño) (p. 32). Loose usage of terms is apparent in places in Vasconcelos's phrasing where he describes causality. We are told that the Cosmic Race will be made "with the genius and with the blood of all peoples" (con el genio y con la sangre de todos los pueblos) (p. 36). The repetition of *with* (*con*) — unforced in Spanish as in English — implies that *genius* and *blood* are complementary elements, but does little to elucidate their relative role or importance or degree of causal connection to one another. On one occasion — when offering an explanation for the course of Latin American independence —, Vasconcelos uses the rather offhand wording "whether by dint of blood or culture", as if it was of no great importance which (por la sangre o por la cultura) (pp. 27–28). In another place, we are informed that the world is heading to a state of affairs where "all types and all cultures will be able to fuse" (todos los tipos y todas las culturas puedan fundarse) (p. 21). The term *type* (*tipo*), often employed by Vasconcelos, was used both in relation to "racial" categories, and more broadly to sociological groups (or for that matter, physiognomic ones), having, as it does, one of its origins in nineteenth-century attempts to find ways to describe social and semiotic tendencies and groups. In turn, the nineteenth-century precedents often explicitly recognized the contingent or partial nature of such agglomerations.[35] At one point, Vasconcelos speaks of "a conglomerate of types and races" (un conglomerado de tipos y razas), as if the two were not always identical (p. 34).

To complicate matters further, it is not always clear whether, when Vasconcelos speaks of any given "race", he is referring to a group with homogenous biological characteristics, or whether he is simply evoking a commonly used descriptor or a contingent predominance without making any strong intellectual commitments in so doing. He certainly speaks of the whites as being the latest of the strong homogenous races (p. 33). Yet, equally, he describes the white civilization as originating in Hellenic

Greece through a mixture of Arians, Dravidians, Hindustanis, and unspecified others (p. 29), and he says that all the great moments in history arise when races mix. In other places, Vasconcelos even seems to draw attention to the way in which racial categories are constructed by cultures and societies contingently. He talks of the "so-called Latins" who are "not really Latins properly speaking" (llamados latinos; no son propiamente tales latinos) (p. 21). He likewise speaks of "the race we have come to call [or, agreed to call] Atlantean" (La raza que hemos convenido en llamar Atlántida) (p. 20).

Alongside passages where Vasconcelos waxes obsessively on genetics, there are parts where the historical stages and the "races" or "peoples" in *The Cosmic Race* could be understood as contingent phenomena and in a non-essentialist way. I mean by this that the major characteristics of these things might be taken to come about due to a myriad of historical causes. Likewise, on such a reading, the fact that a given time period or people can be described as exhibiting specific characteristics does not mean that this is a direct causal result of it being a particular moment in chronology or of there being a given set of genetic characteristics. This would give us an understanding of the term *race* akin to that proposed by Gregory Velazco as a non-essentialist re-use of the notion of *mestizaje*: that, in practice, people who have been treated as having certain similar surface physical characteristics have often — for example, through abuse or privilege — had certain shared political, cultural, societal, and economic experiences that are relevant to their present situation. (Velazco holds that Vasconcelos himself was an essentialist.)[36] For example, parts of the essay might be glossed as follows. The English (or more accurately, British), a major seaboard people, happened to be those who became one of the dominant set of colonists in the Americas; the struggles in the Americas opened up a new chapter of the economic, social, cultural, and political life of the globe, and a singularly important one; a series of factors in Anglo-Saxon cultures and societies meant that they took a lead globally; this involved an emphasis on capital and infrastructure which enhanced global integration, even if it did so in a way that exacerbated many racial and social injustices. It is in this way that the "Anglo-Saxons" set the global stage for the rise of Universopolis, and that their culture is, as Vasconcelos puts it, "pre-universal" (p. 21). That historical reality has nothing causally to do with their being "white": it is nonsense, Vasconcelos says, to affirm that the whites are at the upper end of evolution. More than nonsense, it is a false ideology produced by their own hegemony: "All imperialisms need a philosophy that justifies them" (Todo imperialismo necesita de una filosofía que lo justifique) (pp. 54–55). Conversely, the peoples now habitually grouped together as "Latins", notably the Spanish and Portuguese, while also playing a major role in the colonization of the Americas, did not happen to develop cultural and societal conditions that generated the advantages claimed by their "Anglo-Saxon" rivals; however, they did have a greater propensity for mixing sexually with other peoples, even as they were often cruel and oppressive in their conduct; the break-up of these large polities happened in a way that exacerbated fragmentation, and left them exposed to "Anglo-Saxon" hegemony; but the openness to other peoples, with all its historical qualifications, presents a potential long-term advantage. The broad outline of such an account would remain in many respects familiar today.[37] This would mean that the stages of civilization that Vasconcelos describes are understood as such only in retrospect once events have conspired to bring about overarching patterns. This would be consistent with some recent interpretations of nineteenth-century historiographical interest in periodization, notably Žižek's version of Hegel.[38] In that case, when

Vasconcelos speaks of historical periods and peoples having a "mission" (mission) (p. 57), or of "predestination" (predestinación) (p. 33), he would mean something to the effect that historical factors load the dice in favour of particular outcomes: some eventualities are simply far more probable than others. That would be consistent with his exploration of the different scenarios that could lead to Universopolis: it is, in his judgement, overwhelmingly likely that the peoples of the world will come together and mix, but rather less certain that the "Latins" will be those who bring this about. In turn, the *universality* embodied in the Cosmic Race would itself be a contingent outcome of a complex series of specific historical events.

Didier T. Jaén proposed that the operative word in the phrase *Cosmic Race* was not, in fact, the noun but the adjective, sardonically noting that "biological mixture would not fulfill what Vasconcelos expresses with the idea of the Cosmic race, any more than a mixture of different breeds of chicken would produce a Cosmic chicken". Jaén's point is that Vasconcelos only sees genetic — as opposed to other kinds of — mixing as one aspect of the formation of the Cosmic Race, and not necessarily the most crucial one.[39] That it is only part of the package is suggested in Vasconcelos' description of the path to utopia as "Mendelism in biology, Socialism in government, growing sympathy in souls" (mendelismo en biología, socialismo en el gobierno, simpatía creciente en las almas) (p. 62). Many of the guiding principles that Vasconcelos holds will direct both the formation of Universopolis and human reproductive activity are realizations of human freedom and of refined mental, emotional, and sensual experience. "Norms will be provided by the supreme faculty: fantasia", he affirms (Las normas las dará la facultad suprema, la fantasía). What will truly matter is "creative sentiment and compelling beauty" (el sentimiento creador y la belleza que convence) (p. 48). The point will be "to live in *pathos*" (vivir en *pathos*) (p. 49). On that account, genetics may be very much the servant and not the master in causing the rise of Universopolis. In neo-Lamarckian terms — which Vasconcelos curiously attributes to Mendel —, it will be culture and environment, nurture, that shape genetic outcomes, not the other way around: "the various faculties of the spirit take part in the processes of destiny" (las distintas facultades del espíritu toman parte en los procesos del destino) (p. 59). More still, Vasconcelos speaks of replacing "scientific eugenics" with "the mysterious eugenics of aesthetic taste" (la eugénica científica; la eugénica misteriosa del gusto estético) (p. 50). That phrasing may be glossed as *a form of genetic selection based on aesthetic criteria*, but also with a sentence like: "the key place occupied in many people's minds in the 1920s by the term *scientific eugenics* and its practices as a way to bring about utopia will be taken over by aesthetics and its mysteries." Vasconcelos, after all, rebukes the Anglo-Saxons for their application of zoology to sociology (pp. 54–55). On that gloss, this would not be *eugenics* as understood in the 1920s at all: Vasconcelos is playing with and changing substantively the meaning of words. Doing so would be consistent with his efforts to find the elasticity in and open out the resonance of any given moment in time so as to trace a route to Universopolis.

Such is Vasconcelos's emphasis in places upon a spiritual dimension to the Cosmic Race that some observers have seen in the essay an effort to occlude or depart from the real limitations of the physical and material reality of history. For example, John A. Ochoa argues in *The Uses of Failure* that only in such an ideal domain could the tensions and limitations of the Americas in the 1920s be ostensibly overcome. Particularly telling here is Ochoa's remark that "the only realistic way of transubstantiating the ideal into practice is by definition an incomplete version".[40] Ochoa's stance here is a variation upon the ongoing theme of objections to *The Cosmic Race*: that, in one sense or another, its prophecy of Universopolis and of *universality* is ultimately constrained by contextual conditions. In counterpoint to Ochoa's observation, it might be said that the fusion of ideal and practice is precisely and by definition what *transubstantiation* is: God's incarnate body is literally there in the bread and wine of communion, just as He is literally incarnate in Jesus Christ. Vasconcelos repeatedly and explicitly invokes the legacy of (Catholic) Christianity as a major source of his vision of *universality*: "Christianity liberates and engenders life", he says, "because it contains universal revelation" (el cristianismo liberta y engendra vida porque contiene revelación universal) (p. 56). Vasconcelos's assumption — unlike Ochoa's — would be that human history has the capacity to, and will, transubstantiate into Universopolis where *universality* will be incarnate: the particular can indeed become universal.

Such a presupposition might have some explanatory power in elucidating *The Cosmic Race*. It is resonant with echoes of influential Spanish-language thought about the conquest of the Americas and human and international rights, notably that of the sixteenth-century theologian Francisco de Vitoria. The latter speaks of "Christ's universal kingdom" whose "purpose" is "redemption".[41] Evidently, the incarnation is such that this "universal kingdom" has an inception at a specific place and time: in Palestine during the Roman Empire, just as it will have its critical moment in the crucifixion on Golgotha, and as transubstantiation occurs when and where Mass is conducted. Put another way, the universal has to have entry points where it can become embodied in human history, and these are locatable in geography and chronology. Vitoria remarks that "*positive divine law is effective for all and binding on all before it is promulgated to all.* The mere fact of promulgation is sufficient. Thus the law of the Gospel began to be binding on all men from the moment it was first promulgated". Incarnation — and even transubstantiation in the Mass — therefore entails a necessary problem: it cannot reach everyone at the same time, because the universal is of a particular place and time. The universal, on this account, needs to begin to travel through place and time, geography and chronology, to reach all humanity. So, Vitoria goes on: "those whom news of it had not reached were pardoned. And it is still true today that everyone in the world is bound by the law of Christ."[42] Following a like turn of thought, if there is to be a Universopolis, if human beings are to embody *universality*, this must happen at a place and at a time which incarnates those qualities in a transubstantiation. Thence, Universopolis will arise in a location (by the Amazon), at a particular juncture (an unspecified future time), "and from there will issue forth the preaching, squadrons, and planes propagating Good News."

It is at least conceivable that Vasconcelos is making use of the terminology of genetics and of breeding, not just for their own sake, but because they serve to evoke an embodied, incarnate redemption: a transfiguration not just of some abstract mental domain but of the whole human being. It is striking that just after alluding to Mendel, Vasconcelos proceeds to affirm: "it might be said that it is Christianity itself that is going to be consummated, but not now just in the souls but in the root of beings" (Se diría que es el cristianismo el que va a consumarse, pero ya no sólo en las almas, sino en la raíz de los seres) (p. 59). The verb *consummate* (consumarse) is redolent with Christ's last words on the cross: *consummatum est*, "it is accomplished" (John, 19.30). It is as if Universpolis will offer a more fully realized incarnation and transfiguration than even the Catholic Church has thus far proclaimed. To read *The Cosmic Race* this way is to go some considerable distance towards explaining the heady mixture, muddle even, of allusions to cultural, aesthetic, and biological factors. These are ultimately inseparable, not per se or not just because of the insights of Lamarck's science, but because a truly incarnational, transfiguring *universality* must involve them all together to the ends of redemption.

The physical and emotion come together in Universopolis because it is as much as anything a utopia of sexual intimacy. If we pay too much attention to matters of lineage and to the nature of the offspring in *The Cosmic Race*, we can become distracted from the importance of the sex act, of coupling, in and of itself to *universality*. Early in his account of why "Latins" might be well prepared to found Universopolis, Vasconcelos observes that they "persist in not taking much account of the ethnic factor in their sexual relationships" (persisten en no tomar muy en cuenta el factor étnico para sus relaciones sexuales) (p. 34). His sexual use of the word *love* is apparent in his attributing this to the "abundance of love" (abundancia de amor) exhibited by the Spanish conquerors (p. 32). Following through on this usage, Vasconcelos' concern that the Latins might not love enough to found Universopolis, presumably implies that they might fall short in not having sufficient sex with enough different kinds of people (p. 57). Vasconcelos' clarification of his use of the term *spiritual* is especially striking here. It is, he says, "the spiritual factor that will direct and consummate the extraordinary enterprise" (el factor espiritual que ha de dirigir y consumar la extraordinaria empresa) (p. 44). Not only does the use of the verb *consummate* potentially blur redemption with sexual intercourse, but, if there were any doubt, Vasconcelos explains: "By spiritual motive should be understood, rather than [or more than] reflection, taste directed by [or, that directs] the mystery of choosing a person amid a multitude" (Por motivo espiritual ha de entenderse, más bien que la reflexión, el gusto que dirige el misterio de la elección de una persona entre una multitud) (p. 45). Sexual preference is the supreme spiritual guide of Universopolis, and it is subject "in the end, to curiosity" (en último caso, a la curiosidad) (p. 45).[43] These desires and activities will take a highly refined form, consistent with the aesthetic sophistication of a society centred on human creativity; they are something beyond "low appetites" (apetitos bajos) (p. 49). At the same time, the choice of partner and pleasure will not be reducible to some set of criteria: that is the sense in which it is a "mystery", the ultimately unfathomable depths — as Vasconcelos sees it — of why we prefer to have sex with some people and not others. While evidently preoccupied in many passages with the propagation of the Cosmic Race, the essay seems not to reduce the sexual relations of Universopolis to reproductive acts. Vasconcelos in fact complains that one of the problems with reproduction until now has been that it is "without any limit to quantity" (sin límite de cantidad) (p. 52).

It is certainly possible to read the text in such a way that the babies who are born in Universopolis form a homogeneous ethnic, "racial" group that marginalizes and supersedes all others. If they are indeed to form the final "raza", that may be thought to be so. Such an interpretation depends on the supposition that a *raza* is, by definition, such a group, and that the repetition of the noun, used to describe prior hegemonic peoples such as the "whites", implies the Cosmic Race is in that respect no different from what preceded it. Equally, there is a potential elision here between (a) there being a single overarching term for a group, and (b) that group being homogeneous, as more broadly between (a) the *one* of "One world" and (b) *monolithic*. Conversely, and invoking Mendelian thinking, *The Cosmic Race* describes the pattern of development of the population of Universopolis as consisting of an unending number of "discontinuous and enormously complex variations" (variaciones discontinuas y sumamente complejas) (p. 51). Moreover, given that the basis of association between its people is non-rational — and ultimately unfathomable — attraction between particular pairs of individuals (and not others), similar discontinuous variations must characterize the patterns of social interaction as a whole.

In some significant measure, the issue hinges on whether the Cosmic Race is meant to effect a "whitening" of all other races. As we have seen, at key junctures, Vasconcelos's language seems to point brutally and unpleasantly in that direction. In adjoining passages, the emphasis sometimes shifts. Vasconcelos stresses that the achievements of whites are often surpassed by those of other "races", and he states that the Cosmic Race will not be white but rather describes it as "that new race to which the whites themselves will have to aspire" (sino esa nueva raza a la que el mismo blanco tendrá que aspirar). This could mean that the whites will also disappear. Vasconcelos, after all, seems to go out of his way to provoke white supramacists by conjuring up their nightmares, as in his remark about a black Apollo with a white Venus, and when he laments the lack of sexual interest that white women in the United States feel for Asian officers (pp. 35, 51). His phrasing is, at times, ambiguous, as in the words: "Perhaps among the characteristics of the fifth race the characteristics of the white may predominate, but such supremacy must be the fruit of taste's free choice" (Quizás entre todos los caracteres de la quinta raza predominen los caracteres del blanco, pero tal supremacía debe ser fruto de elección libre del gusto) (p. 43). The use of the subjunctive (predominen) in combination with *perhaps* (quizás) means that this could be taken merely as a concession to acknowledge others' predictions, whatever Vasconcelos' own scepticism.

In some passages, it seems possible, as Didier Jaén suggests, that the "cosmic race" is, in fact, supposed to mark the end of the notion of *race* as understood in prior history.[44] In that case, the racial language and terminology of the 1920s, like the word *eugenics*, are themselves fundamentally transfigured as they are stretched into the new forms of Universopolis. *Raza* will mean something utterly different in the Good News. To borrow Gabilondo's words, the term *race* then becomes "simultaneously mythic, historical, present, and utopian" effecting a "utopian realism".[45] Perhaps, in turn, the fantasia of the rise of Universopolis, like the allusive use of the word *cosmic*, has something of the quality Paul Gilroy finds in "oracular [...] complex futuristic figures like Sun Ra". These explored a "racial" identity "blackness" in ways that "strove to be both nationalist and internationalist" combining "a universalistic appeal to spirituality" with "shared human characteristics".[46] The language of the *Cosmic Race*, calqued from a myriad of sources lying about in 1920s Mexico, so apparently slippery upon analysis, seemingly contradictory or ambiguous, actually traces a path from the present into the aesthetics of Universopolis. Like the city's endlessly spiralling curves (p. 41), existing language takes on recognizeable forms, but ones that are resistant to sharp-edged definition and that open up successive possibilities. Words attain the luminosity forecast for the arts of Universopolis (p. 41), in that their meaning is not constrained in its context, but opened up like a brilliant window out onto the future. "Reality will be like fantasy" (La realidad sera como la fantasia) (p. 41).

Yet, it is not at all clear to me that any of or all of these myriad nuances decisively refute the objections to *The Cosmic Race*: that it is, in the end, a prejudiced product of 1920s Mexico. This is in part a matter of evidence also stacking up on the side of the objectors. Vasconcelos is absolutely capable of appalling outbursts, as when, in line with contemporaneous scaremongering in Mexico, he describes the Chinese as multiplying like rats (p. 35).[47] His account of sexuality excludes, among many other things, same-sex relationships. He declares that in the future, really ugly people will not procreate (p. 50). But the problem goes beyond a question of detailed evidence. It comes down to the degree to which *The Cosmic Race* is constrained by its originating, immediate historical context, and perhaps even to the degree to which things in general are so constrained.

Knight comments that the essay "could not [...] break out of that [contemporaneous racist] paradigm, but chose rather to criticize and and invert several of its basic tenets."[48] Stepan concludes that "Vasconcelos's idealization was fundamentally structured by the racism of the period."[49] On such an account, no modification or qualification that Vasconcelos brings to the terminology of *eugenics*, *race*, *genetics* (and so forth) could overcome the fact that these are intimately linked to the defects of 1920s Mexico. To interpret *The Cosmic Race* is to trace back such links and vocabulary to their originating time and place: this is what a good, critically minded historian does. The terms *paradigm* and *fundamentally structured* suppose that, once tainted thus by a defective origin, the essay is pervasively, inescapably so, however much it tries or pretends otherwise. Those practicing such historicizing un-masking and concerned with contextual particularity are understandably suspicious of the essay's claims to universality. We have seen Ochoa's assertion that transubstantiation of the ideal is simply impossible and a distraction from historical realities. In an essay on hybridity in cultural theory, Susan Mabardi states that Vasconcelos, "under the pretense of creating unity and equality, in effect eliminates differences and ignores institutionalized racism."[50]

Conversely, the more sympathetic account of the vision of Universopolis entails two things. First, it involves a predisposition to search out and bring to the fore sources of positive human potential within Vasconcelos's 1920s account of the universal. The tendency here is to see the flaws of the essay as falling short of the overarching premise of a realizable universality, and thus as spurs to future revision. Such a reading parallels the influential sociologist Jeffrey C. Alexander's way of reading historical assertions of universal rights in his book *The Civil Sphere* (2006). The point of such formulations of "universalizing solidarity", Alexander argues, is not that, at any junction, they end or are free from abuses. Rather, precisely because they are part of history, the attempts to bring them about — to translate them — necessarily entail complex struggles, within which "claimants can make the demand for their 'civil rights'" and in which their implications are contested.[51] Similarly the more sympathetic reading of *The Cosmic Race* parallels James Chandler's observation of nineteenth-century historiography, that many of the later criticisms of it are made precisely in terms that it itself had set out but (arguably) failed to meet.[52] Vasconcelos himself evinces awareness that some forms of universalism serve hegemonic interests and may lead to the occlusion of real and important conflict and divisions. In turn, assertions of differential identity (such as nationhood) have a key historical role to play as a way of resisting the unity sought by hegemonic oppressors. "In the present circumstances of the world," Vasconcelos observes, "internationalism would only serve to finalize the consummation of the triumph of the strongest nations" (en las actuales circunstancias del mundo, el internacionalismo sólo serviría para acabar de consumar el triunfo de las naciones más fuertes) (p. 24).

This approach to reading the account of universality in *The Cosmic Race* is concerned, then, with its inherent potential to generate future possibilities for universal human self-realization beyond the essay's origins in 1920s Mexico. That outlook is tied to the second aspect of the sympathetic reading: a fundamental willingness to suppose that particular, concrete circumstances might embody positive universal values and experiences. Such a disposition is evident, for example, in some recent responses to Vasconcelos's South-Asian contemporaries. In the essay "Different Universalisms" (2010), Sugata Bosa remarks that the "colonized world" was a "fount of universalism" which challenged imperialism with its own "universalist patriotism". Such trends are at odds, Bosa notes, not just with many strains of analytical philosophy but with "the premises of French and North American intellectual currents that are deeply suspicious of all meta-narratives".[53] This, after all, is the gauntlet thrown down by *The Cosmic Race*: that Universopolis could arise in a specific place and at a given time in actual human bodies and minds through a series of historical events. Put at its strongest, the reader of *The Cosmic Race* would need to be prepared to accept the possibility of transfiguration. Alexander comments more mildly that "universalism anchors itself [...] in the everyday lifeworlds" and that "difference can be positively recognized only if the particular is viewed, [again] to paraphrase Hegel, as a concrete manifestation of the universal."[54]

We can believe that *The Cosmic Race* is fundamentally shaped by its contextual origins in 1920s Mexico, and that putative universal ideals are only ever manifestations and occlusions of particular interests. Or we can hold that *The Cosmic Race* is an outpouring of potentiality beyond its originating location, and that particular bodies, places, and times may indeed become universals incarnate. There is an incommensurability between the two positions, as there is between those who say that Universopolis will be built by the Amazon, and those that state it will never rise anywhere. The split arises from within *The Cosmic Race* itself: at times, Vasconcelos' feet seem firmly in the prejudices of the 1920s, at others, his arms appear to open out into new formations. In the end, the question is: to what life possibilities are we open, and, correspondingly, what potentiality either viewpoint opens up or closes for us? No system of reasoning can give us the answer to that question.

Notes

1 José Vasconcelos, *La raza cósmica: Misión de la raza iberoamericana* (Madrid: Aguilar, 1967), p. 42: "Cerca del gran río se levantará Universópolis, y de allí saldrán las predicaciones, las escuadras y los aviones de propaganda de buenas nuevas."

2 Vasconcelos, *La raza cosmica*, p. 33: "sentimiento universal humano."

3 Marilyn Grace Miller, *The Rise and Fall of the Cosmic Race: The Cult of Mestizaje in Latin America* (Austin: University of Texas Press, 2004), p. 14.

4 Gregory Velazco y Trianosky, "Mestizaje and Hispanic Identity", in *A Companion to Latin American Philosophy*, edited by Susana Nuccetelli, Ofelia Schutte and Otávio Bueno (Oxford: Wiley & Blackwell, 2010), pp. 283–96, esp. p. 286.

5 John A. Ochoa, *The Uses of Failure in Mexican Literature and Identity* (Austin: University of Texas Press, 2004), pp. 119–20.

6 Ochoa, *The Uses of Failure*, p. 122.

7 Miller, *The Rise and Fall*, pp. 37–38; Ilan Stavans, *José Vasconcelos: The Prophet of Race* (New Brunswick: Rutgers University Press, 2011), p. xiii. On the outlines of the Chicano interest, see, for example, Monika Kaup, "Crossing Borders: An Aesthetic Practice in Writings by Gloria Anzaldúa", in *Cultural Difference and the Literary Text: Pluralism & the*

Limits of Authenticity in North American Literatures, edited by Winfried Siemerling and Katrin Schwenk (Iowa City: University of Iowa Press, 1996), pp. 100–11.

8 Nancy Stepan, *"The Hour of Eugenics": Race, Gender, and Nation in Latin America* (Ithaca and London: Cornell University Press, 1991), p. 153.

9 Stavans, *José Vasconcelos: The Prophet of a Race*, p. 14.

10 The point is particularly pertinent to some accounts of Vasconcelos's utopianism, notably Ochoa's *The Uses of Failure* (see, for example, page 127).

11 On Sigüenza y Góngora's views, see: D. A. Brady, *Mexican Phoenix. Our Lady of Guadalupe: Image and Tradition across Five Centuries* (Cambridge: Cambridge University Press, 2001), p. 115.

12 On Torres García's map, see María Amalia García, "Cities of Abstract Art: Urban Journeys through South America", in *Radical Geometry: Modern Art of South America from the Patricia Phelps de Cisneros Collection* (London: Royal Academy of Arts, 2014), pp. 32–55, esp. pp. 36–37.

13 Ludwig Wittgenstein, *Philosophical Investigations*, edited by P. M. S. Hacker and Joachim Schulte (Oxford: Blackwell, 2015), p. 43.

14 Fredric Jameson, *The Political Unconscious: Narrative as a Socially Symbolic Act* (New York: Cornell University Press, 1981), p. 9.

15 Guillermo Hurtado, "The Anti-Positivist Movement in Mexico", in *A Companion to Latin American Philosophy*, pp. 82–94. For Bergson on intuition, see David Krepps, *Bergson, Complexity and Creative Emergence* (Houndmills: Palgrave Macmillan e-edition, 2015), chapter 2.

16 Miller, *The Rise and Fall*, pp. 28–29.

17 *Ibid.*, pp. 31–33.

18 *Ibid.*, pp. 34–35.

19 Catherine Poupenay-Hart, "*Mestizaje*: 'I understand the reality, I just do not like the word:' Perspectives on an Option", in *Unforseeable Americas: Questioning Cultural Hybridity in the Americas*, edited by Rita A. Grandis and Zila Bernd (Amsterdam: Rodopi, 2000), pp. 34–55, esp. p. 41.

20 Stepan, *The Hour of Eugenics*, pp. 137–39.

21 *Ibid.*, pp. 128–32.

22 *Ibid.*, pp. 17, 125, 152.

23 See Gabriel Motzkin, "Lamarck, Darwin, and the Contemporary Debate about Levels of Selection", p. 3–8, esp. p. 7, and Eva Jablonka, "Introduction: Lamarckian Problematics in Biology", pp. 145–55, in *Transformations of Lamarckism: From Subtle Fluids to Molecular Biology*, edited by Snait B. Gissis and Eva Jablonka (Cambridge, MA: MIT Press, 2011).

24 Stepan, *The Hour of Eugenics*, pp. 17, 111.

25 *Ibid.*, pp. 134, 148–52.

26 *Ibid.*, p. 130.

27 Ochoa, *The Uses of Failure*, pp. 119, 122, 124–25,

28 Sylvia Molloy, *At Face Value: Autobiographical Writing in Spanish America* (Cambridge: Cambridge University Press, 1991), pp. 189–90.

29 Nicola Miller, *In the Shadow of the State: Intellectuals and the Quest for National Identity in Twentieth-Century Spanish America* (London: Verso, 1999), pp. 142–43.

30 "Introduction", in José Vasconcelos, *The Cosmic Race: A Bilingual Edition*, edited by Didier T. Jaén (Los Angeles: Johns Hopkins Press, 1997), pp. ix–xxxiii (pp. xi, xvi xix).

31 On the complex nuances of stadial time see Geraldine Lawless, "How To Tell Time," in *Spain in the Nineteenth Century: New Essays on Experiences of Culture and Society*, edited by Andrew Ginger and Geraldine Lawless (Manchester: Manchester University Press, 2018) (forthcoming).

32 Alan Knight, "Racism, Revolution, and *Indigenismo*: Mexico, 1910–1940", in *The Idea of*

Race in Latin America, 1870–1940, edited by Richard Graham, Thomas E. Skidmore et al. (Austin: University of Texas Press, c.1990), pp. 71–101, esp. p. 93.

33 Joseba Gabilondo, "Afterword to the 1997 Edition", in Vasconcelos, *The Cosmic Race: A Bilingual Edition*, pp. 99–117, esp .p. 106.

34 Miller, *The Rise and Fall*, p. 3.

35 Martina Lauster, *Sketches of the Nineteenth Century: European Journalism and its Physiologies, 1830–50* (Houndmills: Palgrave Macmillan, 2007), pp. 85–128 and 309–27.

36 Velazco y Trianosky, "*Mestizaje* and Hispanic Identity", pp. 284, 286–95.

37 See, for example, John Elliot, *Empires of the Atlantic World: Britain and Spain in America 1492–1830* (Yale: Yale University Press, 2007) [kindle edition], locs. 3974–76, 6579, 6692, 6703, 580, 6669–70, 6981, 6984, 7009, 7042.

38 See, for example, *For They Know Not What They Do: Enjoyment as a Political Factor* (London: Verso, 2008).

39 Jaén, "Introduction", pp. xvi, xxix.

40 Ochoa, *The Uses of Failure*, p. 120.

41 Francisco de Vitoria, *Political Writings*, edited by Anthony Pagden and Jeremy Lawrance (Cambridge: Cambridge University Press, 1991), p. 98.

42 *Ibid.*, p. 161.

43 See also Jaén, "Introduction", pp. ix–x.

44 *Ibid.*, p. xvi.

45 Gabilondo, "Afterword", pp. 99, 106.

46 Paul Gilroy, *Against Race: Imagining Political Culture Beyond the Color Line* (Cambridge, MA: Harvard University Press, 2000), pp. 341–42.

47 Knight, "Racism, Revolution, and *Indigenismo*", p. 96.

48 *Ibid.*, p. 87.

49 Stepan, *The Hour of Eugenics*, p. 149.

50 Susan Mabardi, "Encounters of a Heterogeneous Kind", in *Unforseeable Americas*, pp. 1–20, esp. p. 13.

51 Jeffrey C. Alexander, *The Civil Sphere* (Oxford: Oxford University Press, 2006), pp. 65, 153–54.

52 James Chandler, *England in 1819: The Politics of Literary Culture and the Case of Literary Historicism* (Chicago: University of Chicago Press, 1998).

53 Sugata Bosa, "Different Universalisms; Colorful Cosmopolitanisms: The Global Imagination of the Colonized", in *Cosmopolitan Thought Zones: South Asia and the Global Circulation of Ideas*, edited by Sugata Bose and Kris Manjapra (Houndmills: Palgrave Macmillan, 2010), pp. 97–111, esp. pp. 98–99.

54 Alexander, *The Civil Sphere*, p. 259.

12 The Commune in Venezuela: A Utopian Prefiguration

Dario Azzellini

My objective in this chapter is to analyse the Venezuelan mechanism of local self-government, the commune, as a utopian prefiguration. Since 2005, the Venezuelan people has organized itself into communal councils, a non-representative form of local self-government based on assemblies and direct democracy. Each communal council comprises between 150 and 400 families. The commune is a higher level of local self-organization, made up of various communal councils and other organizations present in the territory. At the end of 2017, there were more than 47,000 communal councils and more than 1,850 communes in Venezuela. Local self-government is not merely a question of administering one's own local affairs, but a repository of hope; it is seen as an instrument for developing and trying out new ways of living. Communes can be thought of as concrete, or real utopias, rooted in Latin American social and political traditions — and linked to a specific form of Latin American utopianism whose temporality is clearly not the linear temporality of modernity — connected to revolutionary and socialist experiences and the history of indigenous, black, and popular resistance.

It is of critical importance to research and spread concrete utopias as constructions that offer an alternative to the capitalist system, which is destroying the pillars on which human life is based. Since the neoliberal offensive of the 1980s onwards, capitalism has largely been justified on the grounds that there is supposedly no alternative and, indeed, the slogan, "There is no alternative" (TINA), coined by the former British Prime Minister, Margaret Thatcher, has been the main line of argument of the neoliberals. Until a few decades ago, socialism was regarded by its followers and enemies alike as a possible alternative to the capitalist system, although, as Erik Olin Wright asserts:

> Most people in the world today, especially in its economically developed regions, no longer believe in this possibility. Capitalism seems to them part of the natural order of things, and pessimism has replaced the optimism of the will Gramsci once said would be essential if the world was to be transformed.[1]

Wright concludes that "envisioning real utopias [Wright's term for what Bloch or

Gutierrez call "concrete utopias", author's note] is a central component of a broader intellectual enterprise, which can be called *emancipatory social science.*"[2]

Immanuel Wallerstein and other researchers at the Fernand Braudel Center called their studies of historically practicable social alternatives to capitalism that opened up spaces in which social relations based on equality, democracy and autonomy could be experienced, the "*scientific study of utopistics*".[3] These alternatives, or *utopistics*, are relational social practices, based principally on what Wallerstein calls the *anti-systemic movements*, and are spread by these movements, rather than through instruments of the State, the economy, or even academia.[4]

Real, or concrete utopias have a long history in Latin America. They can be traced back at least as far as the resistance settlements founded by escaped slaves in remote areas, frequently with assistance from the indigenous peoples, which were based on alternative norms of social life to those of the colony. They gained strength in the independence movements, especially in their dimension as social revolutions, and reappeared later in the revolutionary movements of the twentieth century, notably in the explicitly Latin American idea of *poder popular* ('popular' or 'people' power). Liberation theology, in its opposition to an oppressive Church that sided with the powerful, articulated the struggle for earthly paradise through concrete utopias. The best theorist of this is Gustavo Gutiérrez, whose approach is based on Ernst Bloch.[5] During the last two decades, the influence of the thought of Bloch and Walter Benjamin — both of whom stressed utopian anticipation — has regained momentum among the Latin American Left, where it carries more weight than in the Global North. Unlike the modernizing, homogenizing logic of real socialism, these concepts are compatible with decolonizing thoughts, practices and projects.

In 1988, Anibal Quijano stated very clearly, "Toda utopía es después de todo, un proyecto de reconstitución del sentido histórico de una sociedad" (Every utopia is, after all, a project for the reconstitution of the historical meaning of a society).[6] In Latin America, for at least the last two or three decades, this project has included increasingly obvious decolonizing features.

The Venezuelan communes, as concrete or real utopias, are "propuesta[s] de racionalidad alternativa" (proposals of an alternative rationality) that go further than the postmodernist and antimodernist neo-conservatism that lauds the seductions of capital, and further than the project of State socialism.[7]

> Latin America, on the other hand, is beginning to constitute itself through new social practices of reciprocity, solidarity, equity, democracy, in institutions formed outside or against the State — in other words, as a private antagonist of private capital, of the State that belongs to this private capital, and of its bureaucracy — as the possible site of an alternative proposal of rationality to instrumental reason and to the same historical reason linked to the disenchantment of the world.[8]

Twenty years ago Latin American utopias may have appeared to be "isolated cells . . . with minimal contact with the rest of the social tissue . . . conceived of as 'islets of the future in the present'".[9] Today, however, it can be unequivocally stated that the communes in Venezuela are inscribed within a continental framework of proposals for an alternative rationality, alongside the experiences of the Zapatista autonomous communities in Chiapas, Mexico, the autonomous, self-governing communities in Oaxaca and Guerrero, Mexico, the MST settlements in Brazil, or the semi-urban settlements of popular organizations in Argentina, as well as many other examples of community self-organization. As we shall see in the case of the Venezuelan communes, these communities are constructing autonomies that are far from being projects of self-isolation and autarky, but are instead interconnected with the rest of society and make significant contributions to the construction of a new Left and a new society.

I shall begin by briefly outlining the way local self-government is structured and how it works. I shall then examine various Marxist and emancipatory concepts of utopia and link them to Latin American thought, in order to show why utopian prefiguration is regarded as a necessary part of the processes of revolutionary struggle. There then follows an analysis of how a utopian prefiguration is practised in the communes, which I demonstrate first, with the example of the names that the communards chose for their communes. The names express a temporality that is different from that of modernity, since past, present and future overlap and are interwoven with each other, thus forming a new meaning with respect to the future. Another element that shows the character of the communes as concrete or real utopias is their communal economy, the effort made by the communes to be self-sufficient and to build an economy by and for the local communities. Lastly, I shall analyse the communal state, which is the result of the coordination of the communes at the highest level and will, in the popular imaginary, replace the bourgeois state. This chapter is based principally on observation and dozens of interviews carried out in Venezuela between 2006 and 2013.

Communal councils and communes

From the year 2000 onwards, popular organizations, communities and even some institutions in Venezuela started to develop various local self-government initiatives. The communal council, a form of self-government at the level of the neighbourhood, arose from these experiences in 2005, followed by the commune, in 2007, as the tier of self-government above that. For a large part of the Venezuelan population, especially among the popular strata, the community (urban or rural) as life space constitutes the most powerful form of self-identification, and the place where their social relations tend to be constructed. It can be argued that class becomes community through stories of shared struggles and existing social relations.

In 2005, after the failure of the Local Public Planning Councils (the first initiative of local participation at national level, which had remained locked into and dependent on the liberal representative institutions),[10] the president of Venezuela at the time, Hugo Chávez, once more took up the idea of the communal councils, which had been created from below, spread awareness of them and gave them a significant boost. In many places, communal councils were created in the wake of the bill being made public.[11] Communes evolved later as a consequence of the need for communities to come together at a higher level than the communal council in order to undertake more wide-ranging projects. A commune is made up of various communal councils within the same territory and is able to develop longer-term projects and measures over a wider geographical area, while the decisions continue to be taken in the communal council assemblies. The commune coordinates the communal councils, social missions and grassroots organization so that projects are planned, implemented and assessed jointly.[12]

The search for a framework in which participatory democracy had a leading role, as contemplated in the 1999 constitution, led to popular power and socialism, both of which are connected with the historical line of the commune, not that of the State. The idea of local self-administration is also associated with the historical experiences of the indigenous peoples and the Afro-Venezuelans, as well as with some currents of Latin American Marxist thought, that of the Peruvian, José Carlos Mariátegui, for example.

Other council initiatives emerged (workers', students', fishermen's, artisans' councils, and so on) and, in 2007, Chávez launched the idea of communal cities as a level of self-government above the communes. "Councilist structures" in different sectors of society are understood as the basis of Venezuelan socialism-in-progress and must cooperate and coordinate with each other at a higher level so that the communal state can replace the bourgeois state. Chávez's proposal took up a debate that went back to the anti-systemic currents and applied it generally.[13] A key theoretical reference for the construction of the "communal system" is Istvan Mészáros' book, *Beyond Capital: Towards a Theory of Transition*, in which he outlines strategies for the transition to socialism.[14]

Forms of local self-government have far-reaching implications for the model of the State. The providentialist function of the State is no longer the responsibility of a specialized bureaucracy, but is accomplished by means of transfers of public financial and technical resources to the communities.[15] Nevertheless, local autonomy is neither isolation from State power nor a counterweight to it, but rather a networked self-administration that supersedes the separation into political, social and economic spheres and renders the State in its present form partly superfluous. Chávez defined the communal councils as constituent power[16] and the place where they take shape — whether rural, urban or metropolitan — as the *community* (defined in terms of social relations, rather than as an administrative entity). It is principally in the marginalized zones of the urban and metropolitan areas that popular participation has been seen to be greatest.[17] The percentage of the population organized into communal councils tends to be higher in rural areas than in urban areas.[18] The first communes were created in rural areas.

Participation helps break down "socio-spatial segregation".[19] The population living in shanty towns reconquers public space at three levels: collective space, living space and institutional space. Participation enables communities and their inhabitants to expand their horizons and plan their lives, which are now more self-determined, rather than reduced to a mere struggle to survive. It also makes it possible to develop a utopia that is not located on the far side of what is imaginable, but is connected to reality. It is a "concrete utopia", as Ernst Bloch defined any kind of utopia — from the field of art, architecture or technology to social utopias — that expresses "dreams of living together in a better way".[20] In the following sections, I shall first of all discuss various concepts associated with concrete utopias and then examine the communes of present-day Venezuela as concrete social utopias.

Concrete utopias and prefigurative politics

In contrast to the way the term *utopia* is commonly used and to the attitudes of the traditional Left, many authors have considered utopia and utopian prefiguration as fundamental to revolutionary movements, so taking up Marx's contention that utopian thinking is necessary, as he did in a letter to Arnold Ruge in 1843:

> It will then become evident that the world has long dreamed of possessing something of which it has only to be conscious in order to possess it in reality. It will become evident that it is not a question of drawing a great mental dividing line between past and future, but of realizing the thoughts of the past. Lastly, it will become evident that mankind is not beginning a new work, but is consciously carrying into effect its old work.[21]

Many authors, using a variety of terms and approaches, have asserted the need for utopia and utopian prefiguration in the struggle for a classless society. Walter Benjamin talked of *Now-time*,[22] John Holloway of the "cracks" in capitalism that provide a glimpse of a possible different future.[23] In the context of communist praxis, Bloch considered that the anticipation of social utopia was essential.[24] Apart from engaging with the unfinished past, Bloch's concept of the *not yet* also involves hope and the imagining of a possible different future and a critical praxis of freedom is necessarily founded on the hope of an achievable future. The importance of experiencing the new social relations is fundamental, since as Bloch stresses,[25] Marx always underlined, for example in the "Theses on Feuerbach", that "the essence of man is no abstraction inherent in each single individual. In its reality it is the ensemble of the social relations".[26]

Ernst Bloch developed the concept of *concrete utopia*,[27] and made the distinction between this and *abstract utopia*:[28]

> The point of contact between dreams and life, without which dreams only yield abstract utopia, life only triviality, is given in the utopian capacity which is set on its feet and connected to the Real-Possible. And which in fact tendentially transcends what exists in each respective case, not only in our nature, but in that of the entire external world of process. Thus the only seemingly paradoxical concept of a concrete utopia would be appropriate here, that is, of an anticipatory kind which by no means coincides with abstract utopian dreaminess, nor is directed by the immaturity of merely abstract utopian socialism.[29]

A very similar concept of utopia can be found in liberation theology, which developed, not uncoincidentally, in the Latin American context, where it had, and continues to have, more of a following. For one of the main theorists of liberation theology, Gustavo Gutiérrez, who makes direct reference to Bloch, utopia has to be a concrete utopia:

> Utopia, contrary to what current usage suggests, is characterized by its relation to present historical reality . . . Utopia necessarily means a denunciation of the existing order . . . But utopia is also an annunciation, an annunciation of what is not yet, but will be; it is the forecast of a different order of things, a new society . . . [Utopia is] the driving force of history and subversive of the existing order. If utopia does not lead to action in the present, it is an evasion of reality.[30]

Concrete utopia radicalizes the commitment of those men and women engaged in the struggle and prevents them from betraying their purpose, which is "to achieve a real encounter among persons in the midst of a free society without social inequalities".[31] In addition, according to Gutiérrez, utopia introduces a communal dimension — which is where concrete utopias are principally developed, practised and experienced — into the question of emancipation and liberation. The example of local self-government confirms this.

For Herbert Marcuse, utopia was a necessary exercise in political imagination in order to be able to develop alternative projects (although he was somewhat pessimistic that these alternatives could represent a challenge to capitalist ideology as it then existed in industrialized societies). The supposed "end of utopia", as Marcuse would say in his 1967 essay of the same title, was nothing other than "the refutation of those ideas and theories that use the concept of utopia to denounce certain socio-historical possibilities".[32]

Since the 1960s, a feature of the movements has been the creation of spaces of possibility for utopian creation. Wini Breines calls them *prefigurative movements*:

> The term *prefigurative politics* . . . may be recognized in counter institutions, demonstrations and the attempt to embody personal and anti-hierarchical values in politics. Participatory democracy was central to prefigurative politics. . . . The crux of prefigurative politics imposed substantial tasks, the central one being to create and sustain within the live practice of the movement, relationships and political forms that 'prefigured' and embodied the desired society.[33]

Marina Sitrin adopted the same term to describe the movements in Argentina after 2000.[34] *Prefigurative movements* are those that "are creating the future in their present social relationships."[35] The means are not subordinate to the ends. It is not a question of implementing a programme, but an ongoing process with no predefined outcome, an "inherently experimental and experiential practice".[36] The Zapatista slogan, "walking we ask questions" reflects the same idea. Taking into account that prefigurative practices are characteristic of many of the new global movements that have arisen since 2011, many authors have also called them *prefigurative movements*.[37] Teivanen suggests talking about *democratic prefiguration*, since, historically speaking, there have also been religious, esoteric and fascist prefigurative movements.[38]

Communes and local self-government in Venezuela

Shortly after the first communal councils were formed in 2005, forms of cooperation between various communal councils also appeared, known as *mancomunidades* (commonwealths) or *confederaciones* (confederations), or simply emerged in the form of networks. The discussion and search for a form of self-government on a wider scale began from below. Chávez took the initiative once more in 2007, when he started to talk in public about communes as the level above the communal councils. He paid special attention to the communes, which have been under construction autonomously and with no "official script" since 2007.

Communes started to spread across the national territory. The communities have taken over the concept and are now building communes, making it a priority in the process that the self-governing communal structure is appropriate for the community; to this end, the form and content of the communes are adapted to their needs and abilities. Self-government works on the basis of democracy and participation and is opposed to the logic of institutional representation. In order to develop a structural framework for the communes, several experiences of existing communes in the country were studied and discussion was encouraged in the communities. In February 2012, Atenea Jiménez of the *Red Nacional de Comuneros y Comuneras* (RNC) (National Network of Communards)[39] described the process as follows:

> In answer to a call from Chávez in 2007, debates started in each community as to what a commune should be like. A commune is made up of several communal councils, but there were historic popular movements that were not linked to the communal council, but then they could not remain outside the commune [either]. This debate started in almost all the experiences and it was agreed that all these movements as well as the communal councils had to start being organically integrated into the commune.
>
> There was also a risk that the commune would simply replace the municipality or the parish in terms of politico-administrative organization, and what we are proposing is not that kind of space, because otherwise, it would be, as in many other countries, just another entity in the bourgeois liberal state. It would mean changing the name but continuing to operate as before.
>
> We started to build with that in mind, and we also started studying other historical experiences of communes. We created a space for debate, and invited international guests who had given the matter some thought. We began the task of visualizing the whole country in terms of communes. If it is a construction process, it means the people being able to take back the constituent power that is in the constitution, which is creative, allowing you to open up spaces and create a whole number of things collectively. So then the people started to say, let's build the communes.
>
> We started by mapping and identifying twenty-one pilot experiences. But as everything at the level of bureaucracy is complicated, that process took a year. It made for a very interesting project because the communes made substantial progress. The Ministry of Communes was created and we were all thrown out and we decided then and there to start the process of organization using popular power. There were sixteen communes at that time. [We asked ourselves] how we could organize the work so as not to work on the state's initiative but on our own, with all the progress that we had made, how we could organize ourselves, by training, teaching ourselves and teaching each other, supporting the idea of endogenous development. We worked on popular education, the exchange of knowledge . . . We started out in 2008 with sixteen, and right now there are more than eighty experiences at national level in the Network, and with every activity that we do, more experiences are joining in. Driving forward the process of building the commune without being subordinated to any kind of power that is not the community itself is what unites us.[40]

The social bases see the communes as the most important instrument of self-organization for moving beyond representative democracy, the bourgeois state, and the prevailing capitalist model. The number of communes registered by the Ministry of Communes reached 1,862 by December 2017.[41] About half of them were registered in a single year, the majority since the Ministry of Communes appointed a new Minister in April 2013. Although popular initiatives to set up communes increased massively from 2010 onwards, the Ministry did not register a single commune until 2012, when it was obliged to do so because of protests and popular pressure from communes under construction.

The communes are not decreed by the government, but constructed through a collective process by the communities, communal councils and popular organizations. It was initially simpler to build the communes in rural and semi-rural suburban areas on the outskirts of the cities. In rural areas, there are usually fewer communal councils (between five and twenty) in each commune and the common needs are more obvious than in urban areas (where communes generally consist of between 20 and 40 communal councils). One of the first communes appeared in a suburban area south-east of Barquisimeto, the Ataroa Socialist Commune, which has a high density of community organizations and capacity for self-government. It is made up of some thirty communal councils and a large number of structures of popular organization.[42]

In the local self-government structures, consensual debate is the order of the day.[43] At a workshop for forming communal cooperatives in a commune in the Metropolitan Region of Caracas, Rafael Falcón explains that what is practised is not "the type of democracy that smothers, where there is a loser and a winner, but a consensual democracy that enables us all to see ourselves in the decisions that are made."[44] Decisions are put to the vote, although rarely with a simple majority. Spokespersons are selected. They cannot take decisions (or only very limited ones) and they can be removed from their posts at any time by the community.[45]

In the context of forming communes and communal cities, it is important to bear in mind that a distinction is made between (absolute) politico-administrative space and (relational) socio-cultural economic space.[46] Communes reflect the latter type of space. They do not have to correspond to existing politico-administrative spaces and they can cross municipal or even state borders, since the point is for the population to define and model its own socio-cultural and economic spaces. The mechanism for building socialist communes and communal cities is a flexible one and they themselves can define which tasks they are going to undertake. As a result, the construction of self-government can begin with what the population itself considers most important, necessary or appropriate. This flexibility makes it possible for them to define their own path to self-government. Non-representative self-government based on councils creates a "new geometry of power". According to Doreen Massey, the concept of power in human or social geography has been "put to positive political use . . . recognizing the existence and significance, within Venezuela, of highly unequal, and thus undemocratic, power-geometries".[47]

There is an inherent contradiction between representative democracy and its institutions on the one hand, and the structures of self-government on the other. Despite all the government declarations in favour of communal councils and communes and the valuable support that constituted power provides for their construction and consolidation, it is the mutual antagonism between constituted power and constituent power that lies at the root of the conflicts that arise during the process of construction. While representatives of constituted power tend for the most part to view local self-government structures as mere appendages of constituted power and try to reduce them to executive bodies responsible for implementing institutional decisions, those who take part in communal councils and communes regard them as embryonic forms of a structure that must, in the long term, replace the State and its old institutions:

> The communal councils are the right way to go for what the socialism of the future will be like. . . . Communities discussing budgets . . . , peasants marking out streets, producing their own electricity . . . the other thing is the workers' movement, the part that is self-management,

co-management, endogenous development, all these are key things for that future that we are beginning to see.[48]

True popular power is that which is framed within the communal councils, that the councils should take control of the State. . . . eliminate the parish councils and the bureaucracy of the councillors and all those things, and in the future, the town halls too. Let the people themselves deal with the public policies and have real control of this. OK, so there's quite a way to go, but it is being constructed.[49]

The process of building local self-government structures was marked at a very early stage by a simultaneous cooperation and conflict between constituent power and constituted power. Juan Carlos Pinto of the *Frente Nacional Comunal Simón Bolívar* (Simon Bolivar National Communal Front) (FNCSB),[50] in the State of Barinas, stated during a communal workshop for building a commune:

Usually the institutions of the state come together and say: this is the project, this is the *Comuna*, this is ready, and they present everything. And then you ask yourself: When was this project ever discussed at all? This destroys the essence, because the essence is the participation of the people and the people are writing their own history.[51]

As Melisa Orellana of the FNCSB explains, building a commune includes taking over various functions generally performed by the State:

The establishment of the Comuna has to do with the whole community. We would like to share this experience and to strengthen it from the grassroots level on, that's how we understand people's power. Concerning the political, the economic, the cultural and the military level.[52]

Contradictions and clashes arise especially where the structures of representative democracy and the new forms of local self-government compete directly for influence and political control:

Confrontations with the town halls . . . we've had them. They've called us anti-revolutionaries, troublemakers etc., etc. Initially, when we started to do the work, we wondered, could we be wrong about this? But no, we aren't wrong because we believe that this is the right way to go. . . . Not the way of the institutions, because the institutions are more of the same, whoever they put there, whether he's a mayor of the revolution or whatever type of mayor they set up there. It's always going to be a bureaucratic institution, and that's what we need to break with, because the system that the institutions have doesn't work. Many people say, 'It's not that the staff there are no good.' No. It's the system that's there. The system is no good.[53]

Those civil servants in the institutions of constituted power who fully support the process of local self-government building, such as Carmelo González of the Autonomous Municipal Institute for the Communes of Barinas, are in a minority. As bureaucratization takes hold, they are increasingly being sidelined by the institutions:

Water, electricity, telephone and the establishment of the EPS (social production companies). These are problems which are supposed to be managed by the assembly. This is your power

> and not ours as administrative officials. You have the possibility to acquire the power. This is something new. This is the creation of a new kind of socialism in which there is a real participation, which doesn't exist anywhere so far. If the Comuna becomes a reality in the whole country, in Barinas, in Venezuela, then as a result of the establishment of socialism — the new geometry of power — we can attempt to construct a communal government.[54]

While the grassroots have a very clear idea of what their goals are, the relationship between constituted and constituent power is defined by constituted power:

> We have always said that we must make progress in building the new society with the State, without the State and against the State. In other words, our relationship with the State is not defined by us, but by the willingness of the State to subordinate itself to the interests of the people.[55]

The Messianic power of the past in the construction of concrete utopia

In order to build a concrete utopia, the past is strategically invoked and constantly updated so that it remains relevant in the present and able to project itself towards the future. This idea is brought out, for example, by Adys Figuera of the *7 Pilares Socialistas* (The 7 Pillars of Socialism) Commune in Anaco (Anzoátegui State), who refers to what Marx described as "*realizing* the thoughts of the past . . . mankind is not beginning a *new* work, but is consciously carrying into effect its old work",[56] while simultaneously remaining optimistic and confident that the titanic work being undertaken will bear fruit in the future:

> You can create a new person with those values and principles that we do have but that have been forgotten; you are not sowing them, and suddenly you see that these people are starting to change. Your way of life changes too. And how do I feel? Happy, because I know we are going to reach our goal. Maybe we won't see the Communal State as constructed socialism, but we know that our children are going to see it: a better Venezuela, better people, better men.[57]

As Benjamin points out, an awareness of past generations is fundamental:

> Not man or men but the struggling, oppressed class itself is the depository of historical knowledge. In Marx it appears as the last enslaved class, as the avenger that completes the task of liberation in the name of generations of the downtrodden. . . . Social Democracy thought fit to assign to the working class the role of the redeemer of future generations, in this way cutting the sinews of its greatest strength. This training made the working class forget both its hatred and its spirit of sacrifice, for both are nourished by the image of enslaved ancestors rather than that of liberated grandchildren.[58]

The commune is not only linked to libertarian socialist concepts, but also to the experiences of the indigenous and Afro-Venezuelan peoples. In the Barlovento area, where many former slaves took refuge after they managed to escape and where many Afro-Venezuelans were also confined after the abolition of slavery, Afro-Venezuelan communes call themselves *cumbes*, a cultural reference to the communities that the runaway slaves formed during the time of slavery.

Chávez established a further connection with the early socialist, Simón Rodríguez, Bolívar's tutor and mentor and a central part of the ideological and cultural imaginary of Bolivarianism:

> Look at what Simón Rodríguez said; he was talking about toparchy in 1847. In a document addressed to Anselmo Pineda on 2 February 1847, Simón Rodríguez said: *The true utility of the creation of a republic is to make the inhabitants take an interest in the prosperity of their land. This is how provincial privileges are destroyed* (Bolívar used to say that in the towns, there was a caste, and he called it [the caste of] the doctors, the military and the priests, it is the caste in every place). If only every parish could set itself up as a toparchy. You know *topos* is the place and *archy* is the authority or the government, like in monarchy or oligarchy, in this case, it is toparchy, which is the government of the place, of the inhabitants of the place; it is popular government, it is communal government.[59]

The names of various communes reflect their optimism and utopian ideals. So, three communes include the term *renacer* (rebirth, revival) in their names, two make reference to *esperanza* (hope) or *futuro* (future), and there are yet others bearing names such as *Patria Nueva* (New Homeland),[60] *Hacia un Nuevo Mundo* (Towards a New World), *Una Nueva Independencia* (A New Independence), *Abriendo Caminos* (Blazing a Trail) or *Derribando Barreras* (Breaking Down Barriers). Even communes set up and registered during the worst days of the crisis in June 2016 bear names such as the *Comuna Socialista Agroalimentaria 'Sembrando Futuro'* (the 'Sowing the Future' Socialist Agri-Food Commune), *La Gran Esperanza del Sur* (The Great Hope of the South) or the *Guaicamacuto Bicentenario Limón Nuevo Día* (the 'Guaicamacuto Bicentenario' Commune in the *barrio* of El Limón, New Day sector).[61] Utopian names with religious and mystic overtones can also be found, as in the *Comuna Socialista San Francisco de Asís* (St. Francis of Assisi Socialist Commune), *Comuna Socialista Cristo Viene* (The Christ is Coming Socialist Commune), *Comuna Agroturística El Jinete del Caballo Blanco* (The Rider of the White Horse Agritourism Commune), the *Comuna Socialista Santa Rita de Casia* (Saint Rita of Cascia Socialist Commune) — Saint Rita was called the Patron Saint of Impossible Causes — and the *Comuna Socialista Estrellas del Futuro* (Stars of the Future Socialist Commune).

The most common name, found in seventy-seven cases, makes some reference to Chávez, either by including his name or the word *comandante* (commander), which is hardly surprising given Chávez's pivotal role in Venezuela and his support of local self-government. All the major figures in the revolutionary Bolivarian imaginary can be found in the names of communes. The Bolivarian imaginary[62] connects the popular imaginary to a revolutionary interpretation of republican heroes such as Simón Bolívar (commemorated in the names of 30 communes), the peasant general of the Federal War, Ezequiel Zamora (1817–1860) (19 communes),[63] the combatant and independence heroine, Manuela Sáenz (11),[64] the philosopher, Simón Rodríguez (1769–1854), Bolívar's tutor, who frequented utopian socialist circles in France at the beginning of the nineteenth century (4 communes), Antonio José de Sucre (4),[65] Francisco de Miranda (3),[66] Josefa Camejo (2),[67] José Félix Ribas (2),[68] Juan Jacinto Lara (1778–1859) (2), Eulalia Ramos Sánchez, also known as Eulalia Buroz (1 commune),[69] Josefa Joaquina Sánchez (1 commune),[70] and nearly two dozen other independence heroes. Alongside these major figures, revolutionary Bolivarianism situates those revolutionary heroes (the heroines are still few in number) and anti-system fighters who emerged from among the indigenous and African-descended Venezuelans, the peasants and the poor, the marginalized, the excluded and persecuted, the guerrillas of the 1960s and 1970s, as well as international revolutionaries and recently assassinated Bolivarian militants. All of these prominent figures are reflected in the names of communes.

Some twenty communes bear names associated with the indigenous resistance, including Chief Coromoto (2) and the chieftains Guaicaipuro (1),[71] Arichacan (1), Nigale (1) and Tamanaco (1). These communes are not registered as indigenous communes (and only three or four are in fact indigenous communities). Names of communes referring to the Afro-Venezuelan heritage include Negro Miguel (1),[72] José Leonardo Chirino(s) (3),[73] the Negra Hipólita (1),[74] and the fighters for independence, Pedro Camejo,[75] alias Negro Primero (4) and Juana Ramírez, "la Avanzadora" (1).[76] Five communes adopted the name of the anti-Gómez guerrilla leader, Maisanta (5).[77]

About fifteen communes have taken the names of guerrilla fighters of the 1960s and 1970s. Three are associated with Argimiro Gabaldón, alias Commander Chimiro (1919–1964), and one with Fabricio Ojeda (1929–1966).[78] Five communes are named after Ali Primera, known as El Cantor del Pueblo Venezolano (Singer of the Venezuelan People).[79] The names of international revolutionaries can also be found, such as Che Guevara (5), José Martí (1), Mao Zedong (1), Rosa Luxemburg (1) and Camilo Cienfuegos (1). In addition to a dozen communes bearing the names of local figures, such as rebels, musicians or historians, there is one named after Jesús Antonio Guerrero, the peasant who was murdered by landowners in 2008, and another after the Chavista member of parliament, Robert Serra, assassinated by paramilitaries in 2014.

As was mentioned earlier, the names of the communes provide evidence that Latin America inhabits a temporality in which past, present and future coexist simultaneously, which is therefore, different from the (false) linear temporality of modernity.[80] This same Latin American temporality is also a feature of Latin American utopianism (so contrasting with North American and European forms of utopianism), which takes "a dual perspective that looks simultaneously towards the past and towards the future [and] has been interpreted as a variant of the recovery of origins".[81]

Tradition and myth have permanent roles in social struggles in Latin America,[82] exerting what Walter Benjamin called "a *weak* Messianic power, a power to which the past has a claim".[83] For centuries, aspects of past movements and struggles against the established social order have been turned into banner issues in contemporary struggles.[84] This is a process that involves constant updating and should not be confused with nostalgia or folklore. Benjamin said that history and memory are "secret rendezvous between past generations and our own",[85] and Mazzeo, that "each update adds to the collective, historical composition of utopia, which thus displays superimposed folds and layers".[86] This, along with their optimistic view of the future, is particularly manifest in the names that the communards choose for their communes.[87]

The communal economy

The organized popular bases regard access to the financial resources necessary to build the new communal economy as a right, albeit with an autonomist perspective. Ramón Virigay, from a rural commune in the Llanos (the Plains) in the state of Barinas, commented in a meeting of spokesmen and spokeswomen of eleven communes convened to build the communal town, José Antonio de Sucre: "Even if we definitely need the government agencies at the moment, we have to be independent tomorrow due to our development. We cannot depend solely on the state forever."[88]

Generally speaking, the communes have attached great importance to social production projects under collective communal management. They are seen as providing a guarantee that each commune can develop in accordance with its own needs. Adys Figuera of the "Comuna 7 Pilares Socialistas" in Anaco, explains:

> If we are not the owners of our own system of production, how are we going to be a commune? It is just more of the same. We continue to be dependent on the same institutions, and that is not the idea. The idea is to break free from daddy . . . [and] mummy, mummy town hall and mummy regional government, and for us to own the means of production ourselves. In fact, in the communities where we are, none of us are developing projects to lay pavements, because we know that we can lay pavements later with the social production projects. So, the approach has been to concentrate on social production projects.[89]

Nevertheless, the function of such projects is not only to improve the conditions and quality of life in the communities and to lay the foundations for the financial autonomy of the communes; they are also seen within the wider perspective of the transformation of the relations of production and of the capitalist economic model. The central element of the utopia that has taken shape in the communes is to construct a model of society that does not involve exploitation or domination:

> No commune can be autonomous if it does not produce wealth that can be distributed first of all among its members. If we have a commune that depends on a third party — a governor, a mayor, whoever — and not on itself to generate its goods and services and wealth, then it is not a commune. How do we conceive these new socio-economic relationships that arise in the commune? And surpluses, how are they distributed? What are relationships like in these socially- or communally-owned enterprises, or whatever, inside the commune? What are social relationships like there? . . . In a commune, there has to be workers' control of the enterprises that are already there and of the ones that are going to be started up. And not only the workers, but the commune itself decides how it is going to function and how it is going to produce and what to do with surpluses. Management is socialist because it is the commune that decides.[90]

From 2008 onwards, the Enterprises of Communal Social Property (EPSC) model emerged. Institutions and State-owned enterprises began to adopt and promote this model of communal enterprise. The *Ley Orgánica del Sistema Económico Comunal* [Organic Law of the Communal Economic System] of 14 December 2010 provided the legal framework for Enterprises of Communal Social Property and there are now thousands of such enterprises at the level of the commune. In the communities today, EPSCs of all kinds can be found, although mainly in those sectors covering the most pressing needs in the shanty towns and rural communities: production of food and construction materials and the provision of transport services. As a general rule, productive enterprises such as textile plants, shoe factories and bakeries are common, with some involved in agricultural production, although there are others. As Pablo Arteaga from the "Eje de MACA" commune in Petare, Caracas, states:

> In over 40 years of democracy here, it has been proven that private companies have failed; even more so in means of service supplies: water, waste, electricity, energy, gas and other types of services. The people from the communal councils who know the functions of these services theoretically and practically, have taken up the task of finding a solution. That is what we are doing here; it isn't easy but it's not impossible either.[91]

Some state-owned enterprises are promoting the creation of direct distribution networks whose products are controlled by the community. In most cases, this arises from the initiative of the workers themselves, as in the case of several of the state's cement companies. The nationalized company, Cemento Andino, started out as the first cement company to encourage the setting up of community distributors of construction materials, as well as of cement block plants (firms that produce cement blocks for house building). Zoraida Benítez of Cemento Andino's Community and Environment Department noted that they managed to reduce speculation and lower prices by eliminating the middlemen.[92] Other State-owned cement companies followed their example. In 2013, the paper and paper products factory, Invepal, began promoting community shops to sell school supplies directly and, by the end of 2014, had set up thirty community shops.

Generally speaking, the communities are supported by state institutions — the Ministry of Popular Power for the Communes in particular — with workshops for designing the preferred form of organization for the community enterprise. So, after a long process of training and debate, the communities themselves decide on the structure and mission of their enterprises.[93] During August and September 2010, in Petare, Gran Caracas, at the Eje de MACA commune (comprising some thirty communal councils), the author of this chapter was able to attend some excellent workshops delivered by an employee from the Ministry of Popular Power for the Communes, to design, in conjunction with the community itself, the management structures for some community enterprises.[94] The debates and discussions concluded that the basic lines of the enterprises would be decided by the Commune assemblies, as well as who would work in them, and how any earnings over and above the costs needed to keep the enterprises operational would be managed.[95]

As Rafael Falcón, the Ministry of Communes promoter for the construction of EPSCs explains:

> The present models of leadership, organization and administration are made for individual interests. They are those of the capitalist companies. There is someone who decides, who enriches himself, and others that are being exploited and have no influence or control over the activities. We want to finish with that! How will we finish with it? The means for this, we must construct here, there is nothing like it existing yet. What should our model company look like?[96]

Apart from collective decision-making with respect to the structure and aims of the

EPSC, the issue that emerged as central for the communities was to avoid creating a hierarchy of activities (so that differentiation according to task or ability did not lead to a status or pay hierarchy). Other important questions were permanent training and mutual learning, job rotation (according to ability), and being of social benefit to the community, and beyond, if possible.[97]

The Eje de MACA's liquid gas distributor began operations in April 2011 and immediately began generating enough revenue to cover its operating costs and the wages of its four workers.[98] In June 2011, the Commune received six off-road minivans suitable for transporting passengers and began to run its own communal transport line in the shanty towns in the upper part of the Commune, which previously had no regular transport.[99]

As the communes become consolidated, the proposals for productive projects are increasing in number and size, and also becoming more sophisticated. Adys Figuera León and Delbia Rosa Avilés of the "7 Pilares Socialistas" Commune, which is part of the National Network of Communards (RNC), described the main productive project of their commune as follows:

> We have already brought some projects to fruition. We have the resources and they are in the process of being implemented. The most important is the tile factory, which came about precisely as the result of the meeting of the National Network of Communards in Carora, Lara. It was there that we visited their production site for artisanal tiles and brought the idea back to Anaco because we have the raw material, the clay. We were developing a project for a factory to build dwellings with prefabricated panels, and the roofs of the houses have tiles. . . .
>
> We contacted specialists who make tiles, and we went to different communities taking courses. The first funding we looked for was to build the kilns. We designed gas kilns . . . the kilns are built in the communities. We are going to bring the raw material to build them and what we are going to need is a space to put the kilns and space to store what we produce. For the house-building factory project, we were working with the Ministry of Science and Technology, and we worked jointly with the people in our communities: engineers, lathe operators, builders, and so on, and we drew up the plans to produce prefabricated panels.
>
> After firing clay tiles, the waste material is turned into a light material that can be processed for use in panel production. The panels we are making right now, we are making by hand, not with the moulds that we need, because the financing we need to make the moulds is very high, we are talking about almost fifteen million bolivars [at the time about 3.5 million US dollars].[100]

> The house-building factory project comprises the tile factory, the mould factory, production of the self-assembly kits of metal frames and . . . Right now, we're building the six kilns, distributed among groups of communal councils. With the kilns we are going to generate 1,326 jobs.[101]

> The idea is to keep expanding the kiln manufacture and to take them to other communities. It is not only the commune; our vision is the economic transformation of the whole municipality. In this same housing factory, there is also communal carpentry and a factory for sinks; so that we are the ones who build the houses, and not private companies, so that it will be managed by the community through the communal councils or the commune.[102]

More than a year later, Adys Figuera relates:

> We created the housing factory as the 'Revolutionary Forces for Everyone' Communal Direct Social Property Enterprise. We are installing an extrusion plant for tiles. The project has three phases and a fish-farming project attached to it. We have a project for controlled-environment greenhouses, the 'Cultivating Dreams' Communal Direct Social Property Enterprise, where we are going to produce vegetables, chilli peppers, onions, paprika . . . at the moment, because of the lack of financing, we have only installed four of the twelve greenhouses that we have planned. . . . We created the 'Hugo Chávez' school for productive socio-political training, where we support the other state communities and municipalities with their organization, planning and economic development. The facilitators and work team come from the 7 Pilares Socialistas commune.[103]

At the RNC meetings, the communes also agreed to exchange products with each other on a regular basis, for example, fish from a commune on the coast of Vargas and goat meat from another in the mountains of Lara. Associated with the RNC, there is also a barter network and a network of thirteen existing communal currencies, which can be regarded as an indirect form of barter. The use of local currencies is restricted to a very specific area, so that, apart from strengthening the local economy, they supposedly generate a different kind of logic from capitalist logic, because they are unsuitable for accumulation and their function is limited to exchange values. The government has promoted and supported the use of local currencies, which was also included in the Organic Law of the Communal Economic System (LOSEC) of December 2010.[104]

Communal State: State or Non-State?

The form of the *Communal State* is a "work in progress" that is realized through the construction of councils in different environments and territories, which then coordinate with each other. So far, the territorial council system has three levels: communal councils, communes and communal cities. In the Organic Law of Communes (LOC 2010), the *Communal State* is defined as a

> form of socio-political organization, founded on the Social State of Law and Justice established in the Constitution of the Bolivarian Republic of Venezuela, in which power is exercised directly by the *pueblo*, by means of self-governing communities with an economic model of social property and sustainable, endogenous development that makes it possible for the supreme social happiness of Venezuelan men and women in a socialist society to be achieved. The basic structural cell of the Communal State is the Commune.[105]

This implies a profound transformation of constituted power and the re-signification of the State. By this definition, the *Communal State* would be a non-State rather than a State.

According to the debate on the Communal State, the new structure would tend to gradually replace the old institutional structures. Nonetheless, there have also been statements by high-level government representatives to the effect that the old institutions and territorial divisions remain intact and that the new structures act as a parallel power. The policy guidelines provided by Chávez were clear: "a communal city, a city where there is no need for parish councils, where there is no need for the office of mayor or municipal councils, but Communal Power."[106] Chávez was also clear about the need to destroy the State and that this task could only be accomplished by popular power:

> In order to advance towards socialism, we need a popular power capable of dismantling the networks of oppression, exploitation and domination that linger in Venezuelan society, capable of shaping a new sociality, starting with everyday life, where fraternity and solidarity go hand in hand with the constant emergence of new ways of planning and producing the material life of our *pueblo*. This involves the complete pulverization of the bourgeois form of the State that we inherited — which is still reproducing itself through its pernicious old practices — and following through with the invention of new forms of political management.[107]

The proposals for a Communal (Communard) State and "communard democracy" go back to Kléber Ramírez (Ramírez Rojas 1991; 1998), who was one of the founders of the guerrilla group, FLN, then one of the leaders of the PRV-FALN guerrillas and PRV Ruptura (in 1978–1979, Chávez was also part of the clandestine leadership of PRV Ruptura, as he mentioned in his TV programme *Aló Presidente* 288 on 27 July 2007). Ramírez became one of the main ideologues of the clandestine civil-military organization founded by Chávez, MBR-200. Their strategy is consistent with the "from below" approach and the Communal State has become the political project of the movements:

> The question is how to begin to visualize the way to build socialism. At some point, we as a *pueblo* found ourselves trapped; we gained power, we have a revolutionary government, we have a revolutionary president, the banner of the left was raised . . . but there are still many gaps in terms of proposals of what twenty-first century socialism is all about. We began working on how we visualized, or what we thought this construction of the new State should be like, in order to start from there, and we ended up at the construction of a Communal State, realizing that this is a non-State. *Communal State* is a contradiction in terms. Some say that what we are going to construct is a communal society.
>
> The consensus of the RNC is that the existing State must be dismantled and that a new form of organization and planning must emerge that starts with the commune as the form and system of government, the government of the working class. The *pueblo*, the exploited *pueblo* that has to give its labour power in order to live. We are still exploring the subject of organization, both of the internal network and the country in general, as well as what that organization of the Communal State or that communal city would look like.[108]

The Communal State, or more precisely, a form of social organization based on self-determination, self-management and direct democracy, organized by means of councils, is the point at which the grassroots movements in Venezuela all come together. The points of view of the communal councils, the communes and the National Network of Communards, as well as of the Settlers' Movement, the movement for workers' control and the collectives in the shanty towns, all converge on the Communal State. There is a good deal of support for it among the grassroots too, since it corresponds to their own ideas, which have fully assimilated the concept of the Communal State:

> The final stage is the Communal State, which will be led by Popular Power, in other words, by the *pueblo*. What we have now is not led by the pueblo; what we have there, in fact, is constituted power. The idea is this, to actually get to the point where it is Popular Power, that is, the *pueblo*, that makes its own decisions.[109]

> The Communal State is like a new Venezuela, let it be the *pueblo* that runs all its public policies. How do we get there? We still have some way to go, but, yes, we are raising awareness in the communities that we truly hold power in our hands, that they can exercise that power, we are succeeding in becoming a Communal State.[110]

Nevertheless, dissolution of the representative structures has to be understood as a long process. According to liberal critics, communal councils would restrict many of the competencies of the municipalities, and the boundaries of institutional responsibilities would gradually fade away.[111] This is precisely where its potential lies.

The political organization of socialist societies in the twenty-first century, as a horizontal confederation of communities or networks of social organizations, forges links with indigenous and Afro-American experiences and with the communitarian socialist tradition that was hegemonic prior to the appearance of State socialism. Gustavo Esteva highlights socialism's original "ímpetu comunitario" (communitarian impetus) before it became "colectivismo, burocracia y autodestrucción" (collectivism, bureaucracy and self-destruction). "The communities appear as an alternative because the unity between politics and place is restored, and the *pueblo* acquires a framework in which it can exercise its power without having to surrender it to the State."[112] The concept of the *Communal State* opens up the possibility of understanding the "State" as the repository of certain limited, democratically legitimated functions that can coexist with the autonomy of the communities.[113]

Conclusions

The processes involved in constructing communes in Venezuela are clear expressions of a concrete or real utopia prefiguring future social relations, as theorized by Bloch and Benjamin. They are clearly connected to diverse communal and collective experiences, both traditional and historical, of indigenous and black people in Latin America and the Caribbean, as well as to the historical legacy of communitarian, non-state socialism and communism, and especially to Latin American traditions grounded in *poder popular*[114] and liberation theology.[115] The names chosen by the communards for their communes show how firmly the real utopias that they are building are embedded in the history of Latin American popular resistance and socialism. They are genuine Latin American (concrete) utopias grounded in the construction of a decolonized Latin American identity.

> Latin American identity, which cannot be defined in ontological terms, is a complex history of production of new historical meanings drawing on multiple, legitimate heritages of rationality. It is then a utopia that forms a new association between reason and liberation.[116]

The communards are generally aware that their aspirations are only going to be fulfilled if the communes are autonomous in their development. The support of the

State has been important and has helped disseminate and strengthen many processes of self-organization, while at the same time, inhibiting and limiting them. Attempts at co-optation, the imposition of agendas and projects, and welfare-based paternalistic practices by the institutions constantly threaten autonomous popular organization.

The centrality of the State and its role as the entity that allocates financial resources necessarily make the relationship an unequal one, pushing popular movements and communities to self-limitation, which is why many communes aspire to economic autonomy. Quite apart from the diversion of funds, state institutions continue to operate with a bourgeois logic and structures that do not correspond to the transformations demanded from below. Institutional logic and social logic are at odds with each other. Social logic can be recognized in social processes that are seldom quantifiable. Changes in social relations and the prefiguration of a different society, which tend to be the most important aspects of the process of building communes, cannot be measured using the numbers and tables favoured by institutional logic to present its account of "progress" made. Nonetheless, the communities in Venezuela are determined to continue along the road to building self-government and are increasingly developing their skills, building networks, and specific pieces of a new economy, a new politics and a new society.

Notes

1 Erik Olin Wright, *Envisioning Real Utopias* (London–New York: Verso, 2010), p. 1.
2 *Ibid.*, p. 10.
3 Immanuel Wallerstein, *Utopística, o las Opciones Históricas del Siglo XXI* (Mexico: Siglo XXI, 1998).
4 Immanuel Wallerstein, *Geopolítica y Geocultura* (Barcelona: Kairós, 2007), p. 316.
5 Gustavo Gutiérrez, *A Theology of Liberation: History, Politics, and Salvation* (Maryknoll, NY: Orbis, 1988), p. 234; Ernst Bloch, *Principle of Hope* (Oxford: Basil Blackwell, 1986), 2: 580.
6 Anibal Quijano, *Modernidad, Identidad y Utopía en América Latina* (Lima: Sociedad y Política), p. 65.
7 *Ibid.*, pp. 65–66.
8 *Ibid.*, p. 69: "América Latina, alternativamente, comienza a constituirse, a través de las nuevas prácticas sociales, de reciprocidad, de solidaridad, de equidad, de democracia, en instituciones que se forman fuera del estado o contra él, es decir, como un privado antagonista del privado del capital y del Estado del capital privado o de su burocracia. Como la sede posible de una propuesta de racionalidad alternativa a la razón instrumental, y a la misma razón histórica vinculada al desencantamiento del mundo."
9 Fernando Aínsa, "Utopías contemporáneas de América Latina", *América: Cahiers du CRICCAL* 32 (2004): 9–33, p. 18: "células aisladas [..] donde los contactos con el resto del tejido social se han reducido al mínimo [...] concebidas como 'islotes del futuro en el presente."
10 Dario Azzellini, *Communes and Workers' Control in Venezuela: Building 21st Century Socialism from Below* (Amsterdam: Brill, 2017), pp. 81–92.
11 *Ibid.*, pp. 93–123.
12 *Ibid.*, pp. 243–51.
13 Hugo Chávez Frías, *El Poder Popular* (Caracas: Ministerio del Poder Popular para la Comunicación y la Información, 2008), p. 67.
14 Istvan Mészáros, *Beyond Capital: Towards a Theory of Transition* (New York: Monthly Review Press, 1995), pp. 739–70.
15 Fundación Centro Gumilla (FCG), *Estudio de los Consejos Comunales en Venezuela* (2008),

http://gumilla.org/files/documents/Estudio-Consejos-Comunales01.pdf#page=3&zoom=auto,-127,484, p. 6.

16 Chávez Frías, *El poder popular*, p. 15.

17 Cecilia Cariola and Miguel Lacabana, "Los bordes de la esperanza: nuevas formas de participación popular y gobiernos locales en la periferia de Caracas", *Revista Venezolana de Economía y Ciencias Sociales,* 11, 1 (2005): 21–41, esp. p. 29.

18 Rafael Romero Pirela, *Los consejos comunales más allá de la utopía. Análisis sobre su naturaleza jurídica en Venezuela* (Maracaibo: Universidad del Zulia, 2007), p. 136.

19 Cariola and Lacabana, "Los bordes de la esperanza", p. 37.

20 Bloch, *The Principle of Hope*: 1, 479.

21 Karl Marx, "Marx to Ruge, Kreuznach, September 1843", *Marx and Engels Collected Works* (New York: International Publishers, 1975), 3: 141–144, esp. p. 144.

22 Walter Benjamin, "Theses on the Philosophy of History", in *Illuminations* (New York: Schocken Books, 1968), 253–64, esp. p. 261.

23 John Holloway, *Crack Capitalism* (London: Pluto Press, 2010).

24 Bloch, *Principle of Hope.*

25 Ernst Bloch, *On Karl Marx* (New York: Herder and Herder, 1971), p. 22.

26 Karl Marx and Friedrich Engels, *Collected Works* (New York: International Publishers, 1976), 5: 6–8, esp. p. 6.

27 Bloch, *Principle of Hope*, 2: 580.

28 Bloch, *Principle of Hope*, 1: 146.

29 *Ibid.*, 1: 145–146.

30 Gustavo Gutiérrez, *A Theology of Liberation*, p. 234.

31 *Ibid.*, p. 237.

32 Herbert Marcuse, "The End of Utopia", in *Five Lectures: Psychoanalysis, Politics and Utopia* (London: Allen Lane — The Penguin Press, 1970), 62–82, esp. p. 62.

33 Wini Breines, *Community and Organization in the New Left, 1962–1968: The Great Refusal* (New Brunswick: Rutgers University Press, 1989), p. 6.

34 Marina Sitrin, *Horizontalism: Voices of Popular Power in Argentina* (Oakland: AK Press, 2006), also Marina Sitrin, *Everyday Revolutions: Horizontalism and Autonomy in Argentina* (London–New York: Zed, 2012).

35 Sitrin, *Horizontalism*, p. 4.

36 Matthijs van de Sande, "The Prefigurative Politics of Tahrir Square — An Alternative Perspective on the 2011 Revolutions", *Res Publica* 19, 3 (2013): 223–239, esp. p. 232.

37 *Ibid.*; also Marianne Maeckelbergh, "Doing is Believing: Prefiguration as Strategic Practice in the Alter-globalization Movement", *Social Movement Studies: Journal of Social, Cultural and Political Protest* 10, 1 (2011): 1–20; Teivo Teivainen, "Occupy representation and democratise prefiguration: Speaking for others in global justice movements", *Capital & Class* 40, 1 (2016): 19–36.

38 Teivainen, "Occupy representation and democratise prefiguration".

39 The RNC coordinates communes and initiatives for forming communes, organizing processes of self-training, mobilization, exchange of knowledge, legislative proposals, and much more; its goal is the autonomy of the communes and the construction of a communal State (see below). By the end of 2016, there were more than 500 communes taking part in the RNC.

40 Atenea Jiménez, 38 years old, National Network of Communards of Venezuela, Bello Monte, Caracas (I-AJ 2012), 14 February 2012: "A partir de un llamado de Chávez en el 2007 se comenzó a dar el debate en cada comunidad de cómo debería ser la comuna. Varios consejos comunales forman una comuna pero había movimientos populares históricos que no estaban articulados al consejo comunal, entonces no podían quedar fuera de la comuna. Se comenzó a dar ese debate en casi todas las experiencias y se acordó de que en la comuna

debían estar articulándose orgánicamente todos esos movimientos además de los consejos comunales.

También había el riesgo de que la comuna sustituyera el municipio o la parroquia en términos de la organización político-administrativa y nuestra propuesta es que no es un espacio de ese tipo, porque si no sería como en muchos países una estancia más del estado liberal burgués. Significaría cambiar el nombre, pero el funcionamiento sigue siendo lo mismo.

Comenzamos a construir en ese sentido y comenzamos a estudiar también otras experiencias históricas de la comuna. Nos dimos un espacio para debatir y también traer invitados internacionales que han pensado el tema. Empezamos a trabajar como visualizar a todo el país en comunas. Si es una construcción, es poder retomar el poder popular constituyente, que está en la constitución, que es creador, que te permite crear espacios y crear una cantidad de cosas colectivamente, entonces, la gente empezó a decir, vamos a construir las comunas.

Se comenzó a hacer un mapeo identificando 21 experiencias piloto. Pero como todo es complicado a nivel burocrático, eso duró un año. Se hizo un trabajo bien interesante porque las comunas tuvieron un avance sustancial. Nace el Ministerio para las Comunas y nos botaron a todos y nosotros decidimos allí comenzar un proceso de articulación ya desde el poder popular. Éramos 16 comunas en ese momento. Cómo articulábamos el trabajo, para no trabajar desde el estado sino que nosotros mismos, con lo que habíamos avanzado, como podíamos nosotros articularnos, formando, auto-formarnos, co-formarnos, apoyar el tema del desarrollo endógeno. Trabajamos con la educación popular, con el intercambio de saberes . . . Arrancamos en el 2008 con 16 y ahorita hay más de 80 experiencias a nivel nacional en la Red y en cada actividad que hacemos, se suman más experiencias. Nos une empujar el proceso de construcción de la comuna sin subordinación a ninguna especie de poder que no sea la propia comunidad."

41 And 47,203 communal councils, according to the list of communal councils and communes officially registered on the website of the Ministerio del Poder Popular para las Comunas, 25 December 2017, http://consulta.mpcomunas.gob.ve/

42 In this case, the wealth of pre-existing experience was also crucial for development. For more information, http://comunasocialistaataroa.blogspot.com

43 Dario Azzellini, *Partizipation, Arbeiterkontrolle und die Commune: Bewegungen und soziale Transformation am Beispiel Venezuela* (Hamburg: VSA, 2010), pp. 271–300; see also Dario Azzellini and Oliver Ressler, *Comuna Under Construction*, directed by Dario Azzellini and Oliver Ressler (2010, Berlin: Good!movies, film, 96 min); also interview with A. Jiménez (I-AJ 2012).

44 Azzellini and Ressler, *Comuna Under Construction.*

45 Dario Azzellini, "The Communal System as Venezuela's transition to Socialism", in *Communism in the 21st Century*, vol. II: *Whither Communism? The Challenges Facing Communist States, Parties and Ideals*, edited by Shannon K. Brincat (Westport: Praeger, 2013), 217–49.

46 David Harvey, *Spaces of Global Capitalism: Towards a Theory of Uneven Geographical Development* (London–New York: Verso, 2006).

47 Doreen Massey, "Concepts of space and power in theory and in political practice", *Documents Anàlisi Geogràfica* 55 (2009): 15–26.

48 María Valdez, 62 years old, retired teacher, Benito Juárez Communal Council, Parroquia San Juan, Caracas (the name and place have been changed) (I-MV 2007), 3 January 2007: "Los consejos comunales son la piedra en el blanco de lo que será un futuro socialismo . . . . Comunidades discutiendo presupuesto . . . , campesinos trazando calles, haciendo su propia electricidad . . . lo otro es el movimiento obrero, la parte de lo que es la autogestión, la cogestión, el desarrollo endógeno, todas esas son las cosas claves para ese futuro que estamos vislumbrando."

49 Antonio Calabrese, 45 years old, a teacher in the metropolitan education system of the city administration of Greater Caracas, Benito Juárez Communal Council, Parroquia San Juan, Caracas (the name of the person and place have been changed), (I-AC 2006) 22 November 2006: "El verdadero poder popular es lo que está enmarcado en los consejos comunales, que los consejos tomen el control del Estado. . . . que eliminen las juntas parroquiales y el burocratismo de los concejales y todas estas cosas, y que en un futuro las alcaldías también. Que sea el propio pueblo que maneje las políticas públicas y tenga el control verdadero de esto. OK, falta bastante, pero se está construyendo."

50 The FNCSB is the commune constructed by the Ezequial Zamora Peasant National Front, the largest peasant organization in Venezuela. Together with other grassroots organizations they form the Corriente Revolucionaria Bolívar y Zamora [Bolivar and Zamora Revolutionary Current] (CRBZ). (Azzellini, *Communes and Workers' Control in Venezuela*, pp. 71–73).

51 Azzellini and Ressler, *Comuna Under Construction.*

52 *Ibid.*

53 Adys Figuera León, 33 years old, facilitator of popular power, Los 7 Pilares Socialistas Commune, Anaco, Anzoátegui state. (I-AFL 2012) 11 February 2012: "Enfrentamientos con las alcaldías . . . hemos tenido. Nos han tachado de antirrevolucionarios, problemáticos, etc., etc. Inicialmente cuando empezamos a hacer el trabajo uno se preguntaba ¿será que nosotros estamos equivocados? Pero no, nosotros no estamos equivocados porque nosotros creemos en que éste es el verdadero camino. . . . Las instituciones, porque las instituciones son más de lo mismo, pongan quien pongan, sea un alcalde de la revolución, sea un alcalde del que se monte allí. Siempre eso va a ser una institución burócrata y eso es lo que nosotros necesitamos romper, porque es que el sistema que tienen las instituciones no sirve. Mucha gente dice 'No es que las personas que están ahí no sirven' No. Es el sistema que está. El sistema no sirve."

54 Azzellini and Ressler, *Comuna Under Construction.*

55 Andrés Antillano, 41 years old, a sociologist, Comités de Tierra Urbana (CTU), Antímano, Caracas (I-AA 2016), 20 April 2008: "Siempre hemos dicho que debemos avanzar en la construcción de una nueva sociedad con el estado, sin el estado y contra el estado. O sea que la relación con el estado no la definimos nosotros, sino la define la disposición del estado de subordinarse a los intereses del pueblo."

56 Marx, *Marx to Ruge, Kreuznach, September 1843.*

57 Figuera León (I-AFL 2012): "Tú puedes crear una persona nueva con esos valores y principios que sí tenemos pero que están olvidados, que no los estás sembrando y que de repente tú ves que esas personas empiezan a cambiar. Tu forma de vida cambia también. ¿Y que cómo me siento? Feliz porque sé que vamos a lograr la meta. Quizá no veamos el Estado Comunal como socialismo construido pero sabemos que los hijos que nosotros tenemos sí lo van a ver: una mejor Venezuela, unas mejores personas, unos mejores hombres."

58 Benjamin, "Theses on the Philosophy of History", p. 262.

59 Ceremony on the occasion of the annual message to the National Assembly, Palacio Federal Legislativo [Federal Legislative Palace], 13 January 2007: "Fíjate lo que decía Simón Rodríguez, él hablaba en 1847 de la toparquía. En un documento dirigido a Anselmo Pineda, el 2 de febrero de 1847, dice Simón Rodríguez: *La verdadera utilidad de la creación de una república, es hacer que los habitantes se interesen en la prosperidad de su suelo. Así se destruyen los privilegios provinciales* (Bolívar decía que en los pueblos había una casta, y él la denominaba de los doctores, los militares y los curas, es la casta en cada lugar). Ojalá cada parroquia se erigiera en toparquía. Ustedes saben, topos de lugar y arquía es la autoridad o el gobierno, como la monarquía, la oligarquía, en este caso es la toparquía, es el gobierno del lugar, de los habitantes del lugar; es el gobierno popular, es el gobierno comunal." Chávez, *El poder popular*, p. 43.

60 In the Venezuelan context, the cultural value of *Patria* is less nationalistic than in other Latin American cases, where the discourse of the *Patria Grande*, the Americas, continues to be followed.

61 It was called *Guaicamacuto Bicentenario* in honour of the Indian chief of the same name who led the rebellion against the Spaniards in 1555.

62 See the entry by Dario Azzellini, "Bolivarianism, Venezuela", in *International Encyclopedia of Revolution and Protest: 1500 to the Present*, vol. 7, edited by Immanuel Ness (Oxford: Wiley-Blackwell, 2009), pp. 412–16.

63 Azzellini, "Zamora, Ezequiel (1817–1860)", in *International Encyclopedia of Revolution and Protest*, pp. 3706–7.

64 Manuela Sáenz (1797–1856), Ecuadorian, and one of the heroines of the struggle for independence, was Simon Bolivar's sentimental companion. For centuries her role has been reduced to that of Bolivar's companion. Today, she is recognized as a combatant and leader of the independence movement and considered a precursor of Latin American feminism.

65 Antonio José de Sucre (1795–1830), known as the Grand Marshal of Ayacucho, the most skilled military strategist in the struggle for independence.

66 Francisco de Miranda (1750 –1816), a revolutionary precursor of the struggle for Venezuelan independence. Miranda fought in the American Revolutionary War and the French Revolution. In 1806, he failed in his attempt to trigger a War of Independence in Venezuela when he established a short-lived beachhead at Ocumare de la Costa with a small army of liberation.

67 Josefa Camejo (1791–1862) was a heroine of Venezuelan independence, she joined the 1810 revolution; in 1811, she raised groups of women to take part in the armed struggle; in 1821, at the head of 300 slaves, her attempted uprising in the region of Coro against the Royalists failed; at the beginning of May 1821 she was successful with her independence uprising in the Paraguaná peninsula.

68 José Félix Ribas (1775–1815), a hero of the War of Independence in Venezuela.

69 Eulalia Ramos Sánchez (1795–1817), a heroine of Independence and part of the group close to Bolívar. She died in combat.

70 Josefa Joaquina Sánchez (1765–1813) took part in an independence conspiracy in 1797–1799 and is credited with having embroidered the first flag of Venezuela.

71 He was an indigenous leader of the pre-Columbian period in the Spanish province of Venezuela in the sixteenth century who organized and successfully carried out a fierce resistance against the Spanish settlers, driving them out of the region of Los Teques in the Caracas valley and the nearby coastland for almost a decade. See Azzellini, "Guaicaipuro (1530–1568)", in *International Encyclopedia of Revolution and Protest*, 1471–2.

72 The Negro Miguel was an African slave who, in 1553, led the first rebellion of African slaves recorded by the Spanish colonial authorities. The revolt interrupted a gold rush in the Venezuelan mining region of Buría. The Negro Miguel established a maroon settlement. Today he is remembered as a leader in the historic struggle for racial justice in Venezuela. See Azzellini, "Venezuela, Negro Miguel Rebellion, 1552", in *International Encyclopedia of Revolution and Protest*, 3451–2.

73 José Leonardo Chirinos, (also Chirino) (?–1796) is recognized as a precursor of the struggle for independence and the abolition of slavery and as a symbol of Venezuela's African heritage. See Azzellini, "Chirinos, Jose Leonardo (?–1796)", in *International Encyclopedia of Revolution and Protest*, p. 737.

74 Simón Bolívar's wet nurse.

75 Pedro Camejo (*c.*1790–1821), a lieutenant and the only officer of colour in Bolívar's army, was nicknamed Negro Primero (First Black) because he was always in the front line of attack due to his skill with a spear.

76 Juana Ramírez (1790–1856) was the daughter of an African slave mother and white father. She was emancipated at birth and became a military heroine in the struggle for independence.

77 Pedro Rafael Pérez Delgado (1881–1924), better known as "Maisanta" was a deserter from the army who led a guerrilla fight against the dictatorial government of Juan Vicente Gómez, and died in prison. He was Hugo Chávez's great-grandfather.

78 See Azzellini, "Venezuela, guerrilla movements, 1960s to 1980s", in *International Encyclopedia of Revolution and Protest*, pp. 3441–5.

79 Ali Primera (1941–1985) was a singer-songwriter and also a communist militant.

80 The latter has also been called into question by historians such as Marc Bloch, *The Historian's Craft* (Manchester: Manchester University Press, 1954/2004), pp. 34–35: "Observation proves, on the contrary, that the mighty convulsions of that vast, continuing development are perfectly capable of extending from the beginning of time to the present. What would we think of a geophysicist who, satisfied with having computed their remoteness to a fraction of an inch, would then conclude that the influence of the moon upon the earth is far greater than that of the sun? Neither in outer space, nor in time, can the potency of a force be measured by the single dimension of distance."

81 Fernando Aínsa, "Utopías contemporáneas de América Latina", p. 24: "una visión dual que se proyecta simultáneamente hacia el pasado y el futuro [y] se ha traducido en una variante de recuperación de los orígenes."

82 Miguel Mazzeo, *El sueño de una cosa* (*Introducción al Poder Popular*) (Caracas: El Perro y la Rana, 2007), p. 56.

83 Benjamin, "Theses on the Philosophy of History", p. 254.

84 To name but a few, Tupac Amaru, José Martí, Farabundo Martí, Augusto César Sandino, Emiliano Zapata, Pancho Villa, and so on.

85 Benjamin, "Theses on the Philosophy of History", p. 254 (Author's translation from the German).

86 Mazzeo, *El sueño de una cosa*, p. 57: "cada actualización contribuye a una composición colectiva e histórica de la utopía que, de este modo, presenta pliegues y estratos superpuestos".

87 The database for all the information about the names of communes is the list of 1,500 communes officially registered on the website of the Ministerio del Poder Popular para las Comunas [Ministry of Popular Power for the Communes], 18 June 2016, http://consulta.mpcomunas.gob.ve/

88 Azzellini and Ressler, *Comuna Under Construction*.

89 Adys Figuera León (I-AFL 2012): "Si nosotros no somos dueños de nuestro sistema de producción ¿cómo vamos a ser una comuna? Es como que más de lo mismo. Vamos a seguir dependiendo de las mismas instituciones y la idea no es esa. La idea es desprendernos del papá . . . la mamá alcaldía, la mamá gobernación y ser nosotros mismos dueños de nuestros propios medios productivos. De hecho nosotros ninguno, en las comunidades donde estamos, desarrollamos proyectos para echar una acera, porque sabemos que las aceras con los proyectos socio-productivos después las podemos hacer. Entonces el enfoque ha sido el proyecto socio-productivo."

90 A. Jiménez (I-AJ 2012): "Ninguna comuna que no tenga una producción de riqueza que se pueda distribuir primero que todo entre sus miembros puede ser autónoma. Si tenemos una comuna que dependa de un tercero, de un gobernador, de un alcalde de un quien sea, que no depende de sí misma, de su propia generación de bienes y de servicios y de riquezas, entonces no es una comuna. ¿Cómo pensamos esas nuevas relaciones económicas sociales que se dan en la comuna? ¿Y los excedentes como se distribuyen? ¿Cuál es la relación en esas empresas de propiedad social o comunal o como sea que están dentro de la comuna? ¿Cómo son las relaciones sociales ahí? [...] En la comuna tiene que haber control obrero de las empresas que están ahí y de las que van a nacer. Y no solamente los trabajadores sino que la propia comuna decide cómo va a funcionar y cómo se va a producir y qué se hace con los excedentes. La gestión es socialista porque la comuna decide."

91 Azzellini and Ressler, *Comuna Under Construction*.

92 Zoraida Benítez, 45 years old, a worker, Community and Environment Department, Cemento Andino, Monay, Trujillo (I-ZB 2010), 10 August 2010.

93 Azzellini and Ressler, *Comuna Under Construction.*

94 *Ibid.*

95 Elodia Rivero, *c.*60 years old, a retired teacher, "Eje de MACA" Commune, Petare, Gran Caracas, Miranda (I-ER 2011), 19 August 2011, and Pablo Arteaga, *c.*50 years old, unemployed, "Eje de MACA" Commune, Petare, Gran Caracas, Miranda (I-PA 2011), 9 August 2011.

96 Azzellini and Ressler, *Comuna Under Construction.*

97 *Ibid.*

98 Lorenzo Martini, *c.* 50 years old, a lawyer, "Eje de MACA" Commune, Petare, Gran Caracas, Miranda (I-LM 2011), 19 August 2011.

99 Arteaga (I-PA 2011).

100 Figuera León (I-AFL 2012): "Ya hemos materializado unos proyectos. Tenemos los recursos y están en proceso de ejecución. El más importante es la fábrica de tejas que nace precisamente de un encuentro que tuvimos en el encuentro de la Red Nacional de Comuneros y Comuneras en Carora, Lara. Visitamos por allá la elaboración de teja artesanal y llevamos a Anaco la idea porque tenemos la materia prima, la arcilla. Nosotros estábamos desarrollando el proyecto de una fábrica de viviendas con paneles prefabricados. El techo de la vivienda lleva teja.

Nos pusimos en contacto con especialistas que hacen tejas, y fuimos a diferentes comunidades haciendo cursos. El primer financiamiento que buscamos es para hacer los hornos. Diseñamos hornos a gas. [...] Los hornos se hacen en las comunidades. Vamos a llevar la materia prima para construirlos y lo que vamos a necesitar es un espacio para colocar el horno y un espacio para ir almacenando lo que se va a producir. Para el proyecto de fábrica de casas estuvimos trabajando con el Ministerio de Ciencia y Tecnología, e hicimos un trabajo mancomunado con gente de nuestras comunidades: ingenieros, torneros, albañiles, etc., elaboramos los planes para producir los paneles prefabricados.

El desecho de la producción de tejas de arcilla después de quemar se convierte en un material liviano que va a pasar a ser procesado para utilizarlo en la producción de los paneles. Los paneles que estamos haciendo ahorita los estamos haciendo artesanalmente, no con las hormas que necesitamos porque el financiamiento que necesitamos para hacer las hormas es muy elevado, estamos hablando de casi 15 millones de bolívares" [en esas fechas unos 3.5 millones de dólares US]."

101 Delbia Rosa Avilés, 45 years old, a facilitator of popular power, Los 7 Pilares Socialistas Commune, Anaco, Anzoátegui state (I-DRA 2012), 11 February 2012: "El proyecto de fábrica de viviendas está compuesto por la fábrica de tejas, la fábrica de hormas, la producción del Kit de estructuras metálicas y... Ahora se están haciendo los seis hornos, distribuidos por agrupaciones de consejos comunales. Con los hornos vamos a generar 1.326 puestos de trabajo."

102 Figuera León (I-AFL 2013): "La idea es seguir expandiendo la elaboración de hornos y llevarlos a otras comunidades. No es solamente la comuna, estamos trabajando en visión de la transformación económica de todo el municipio. En esa misma fábrica de vivienda está también la carpintería comunal y la fábrica de bateas. Para que nosotros construyamos las casas, que no sean empresas privadas sino que eso lo va a manejar la comunidad a través de los consejos comunales o la comuna."

103 *Ibid.*: "La fábrica de viviendas la creamos como Empresa de Propiedad Social Directa Comunal 'Fuerzas Revolucionarias Para Todos'. Estamos instalando la planta extrusora de tejas. El proyecto tiene tres fases y consigo lleva un proyecto anexo de la cría de peces en lagunas. Tenemos un proyecto de casas de cultivo en ambiente controlado, que es la Empresa de Propiedad Social Directa Comunal "Cultivando Sueños", allí vamos a producir hortalizas sembrar ají, cebollín, pimentón . . . por ahora por falta de financia-

miento hemos instalado 4 casas de cultivo de doce que planeamos. . . . Creamos una escuela de formación sociopolítica productiva "Hugo Chávez" donde apoyamos a las demás comunidades y municipios del estado para la organización, planificación y desarrollo económico. Los facilitadores y el equipo de trabajo salen de la Comuna 7 Pilares Socialistas."

104 The law establishes that the Banco Central de Venezuela (BCV) [Central Bank of Venezuela] would regulate everything related to the communal currency. The BCV, however, is unable to intervene since it is not within its remit to regulate any currency other than the national one, the Bolivar, as was established by the Constitution and also by the BCV Act. So, ironically, there is a law that gives legal status to the communal currencies, which exist in any case, without such local currencies being able to be regulated in any way.

105 LOC (Ley Orgánica de las Comunas) (Caracas: Asamblea Nacional de la República Bolivariana de Venezuela, 2010: "forma de organización político social, fundada en el Estado Social de Derecho y de Justicia establecido en la Constitución de la República Bolivariana de Venezuela, en la cual el poder es ejercido directamente por el pueblo, a través de autogobiernos comunales con un modelo económico de propiedad social y desarrollo endógeno y sustentable que permita alcanzar la suprema felicidad social de los venezolanos y las venezolanas en la sociedad socialista. Célula fundamental de conformación del Estado Comunal es la Comuna."

106 Hugo Chávez Frías, "Fragmentos del Discurso de toma de posesión," *El Poder Popular. Serie Ensayos. Propuestas para el debate* (Caracas: IMU, Instituto Metropolitano de Urbanismo, 2007), p. 6: "una ciudad comunal, una ciudad donde no hagan falta juntas parroquiales, donde no hagan falta alcaldías ni concejos municipales, sino Poder Comunal."

107 Hugo Chávez Frías, "Propuesta del Candidato de la Patria Comandante Hugo Chávez para la Gestión Bolivariana Socialista 2013–2019", 2012, Internet version 16 September 2012, http://www.chavez.org.ve/Programa-Patria-2013–2019.pdf. p. 2: "Para avanzar hacia el socialismo, necesitamos de un poder popular capaz de desarticular las tramas de opresión, explotación y dominación que subsisten en la sociedad venezolana, capaz de configurar una nueva socialidad desde la vida cotidiana donde la fraternidad y la solidaridad corran parejas con la emergencia permanente de nuevos modos de planificar y producir la vida material de nuestro pueblo. Esto pasa por pulverizar completamente la forma de Estado burguesa que heredamos, la que aún se reproduce a través de sus viejas y nefastas prácticas, y darle continuidad a la invención de nuevas formas de gestión política."

108 Jiménez (I-AJ 2012): "La cuestión es cómo comenzar a visualizar cual es la vía para construir socialismo. Nosotros como pueblo nos encontramos entrampados en algún momento: llegamos al poder, tenemos un gobierno revolucionario, tenemos un presidente que es revolucionario, se levantó la bandera de la izquierda . . . pero todavía hay muchos vacíos en términos de propuestas de qué es el socialismo del siglo XXI. Nosotros empezamos a trabajar en cómo visualizamos o creemos que debe ser esa construcción del nuevo Estado para empezar por ahí y llegamos luego a que la construcción es la de un Estado Comunal, entendiendo que eso es un no-estado. Es como una contradicción el término de *Estado Comunal*. Algunos dicen: es la sociedad comunal lo que se va a construir.

Lo que es consenso de la RNC es que es el desmontaje de este Estado que existe, y que debe surgir una nueva forma de organización y ordenamiento que parte de la comuna como una forma y un sistema de gobierno, el gobierno de la clase trabajadora. El pueblo, el pueblo explotado que tiene que dar su fuerza de trabajo para poder vivir. Seguimos profundizando en el tema de la organización, tanto de la red interna como del país en general, cómo sería esa organización de Estado Comunal o esa ciudad comunal."

109 Avilés (I-DRA 2012): "La recta es el Estado Comunal que va a ser dirigido por el Poder Popular, o sea, por el pueblo. Lo que tenemos ahora no es dirigido por el pueblo. Tenemos

ahí al poder constituido. La idea es esa realmente llegar a que sea el Poder Popular, el Pueblo el que tome sus propias decisiones."

110 Figuera León (I-AFL 2012): "El Estado Comunal es como una nueva Venezuela, que sea el pueblo el que maneje todas sus políticas públicas. ¿Cómo llegar a hacer esto? Todavía nos falta, pero si nosotros concientizamos a las comunidades de que de verdad tenemos el poder en las manos, de que ellos pueden ejercer ese poder, nosotros logramos llegar a ser Estado Comunal."

111 Catalina Banko, "De la descentralización a la 'nueva geometría del poder'", *Revista Venezolana de Economía y Ciencias Sociales* 14, 2 (2008): 167–84.

112 Gustavo Esteva, "Otra mirada, otra democracia", *Rebelión.org* (2 February 2009), Internet version: http://www.rebelion.org/noticia.php?id=80143: "Las comunidades aparecen como alternativa porque en ellas se restablece la unión entre la política y el lugar y el pueblo adquiere una forma en que puede ejercer su poder, sin necesidad de rendirlo al Estado." An English version of the same article was also posted as "Another Perspective, Another Democracy" in the *Journal of the Research Group on Socialism and Democracy* (10 April, 2011), http://sdonline.org/51/another-perspective-another-democracy1/

113 *Ibid.*

114 Mazzeo, *El sueño de una cosa.*

115 Gutiérrez, *A Theology of Liberation.*

116 Quijano, *Modernidad, Identidad y Utopía en América Latina*, p. 69: "La identidad latinoamericana, que no puede ser definida en términos ontológicos, es una compleja historia de producción de nuevos sentidos históricos, que parten de legítimas y múltiples herencias de racionalidad. Es, pues, una utopía de asociación nueva entre razón y liberación."

Interviews

Antillano, Andrés, 41 year old, sociologist, Comités de Tierra Urbana (CTU), Antímano, Caracas (I-AA 2016) 20/04/2008.

Avilés, Delbia Rosa, 45 year old, facilitator of popular power, Los 7 Pilares Socialistas Commune, Anaco, Anzoátegui State (I-DRA 2012) 11/02/2012.

Arteaga, Pablo, ca. 50 year old, unemployed, Commune "Eje de MACA", Petare, Gran Caracas, Miranda (I-PA 2011) 19/08/2011.

Benítez, Zoraida, 45 year old, worker, Community and Environment Department, Cemento Andino, Monay, Trujillo (I-ZB 2010) 10/8/2010.

Calabrese, Antonio, 45 year old, teacher in the metropolitan education system of the city administration of Greater Caracas, Benito Juárez Communal Council, Parroquia San Juan, Caracas (name of person and place changed) (I-AC 2006) 22/11/2006.

Figuera León, Adys, 33 year old, facilitator of popular power, Los 7 Pilares Socialistas, Commune, Anaco, Anzoátegui State (I-AFL 2012) 11/02/2012; and (I-AFL 2013) 17/09/2013.

Jiménez, Atenea, 38 year old, National Network of Communards of Venezuela, Bello Monte, Caracas (I-AJ 2012) 14 February, 2012.

Martini, Lorenzo, ca. 50 year old, lawyer, Commune "Eje de MACA", Petare, Gran Caracas, Miranda (I-LM 2011) 19/08/2011.

Rivero, Elodia, ca. 60 year old, retired teacher, Commune "Eje de MACA", Petare, Gran Caracas, Miranda (I-ER 2011) 19/08/2011

Valdéz, María, 62 year old, retired teacher, Benito Juárez Communal Council, Parroquia San Juan, Caracas (name of person and place changed) (I-MV 2007) 03/01/2007.

13 Walking towards Utopia: Experiences from Argentina

Marina Sitrin

> "What do I dream about? I don't know. It's funny, you know before I was dreaming only about the future, and now I'm dreaming in the present."[1]

> "The dream is like utopia. I believe that we have to learn how to continue when we are awake, and we must be awake to build new forms that will allow us to arrive at this utopia-this revolution we're seeking."[2]

> "Utopia is on the horizon: When I walk two steps, it takes two steps back. I walk ten steps, and it is ten steps further away. What is Utopia for? It is for this, for walking."[3]

This chapter explores a few of the movements that arose in Argentina shortly before and after the popular rebellion of the 19th and 20th December 2001, movements whose forms of organizing, such as *horizontalidad* (horizontalism), prefigurative politics and *autogestión* (self-organization), closely coincide with movements that have emerged over the past twenty years across vast geographic areas, movements that many — particularly the participants — find to be full of transformative possibilities. This context opens up the question of what sorts of lessons communities and societies can learn from the Argentine experience, not in a linear way, as something to be adopted, but as an invitation to join the walk towards freedom and an autonomous, self-organized destination, towards collective discovery and change.

The overarching concept of utopia used in this chapter directly follows the ideas raised by those in the movements in Argentina. I lived and conducted ethnographic and oral historical research in Argentina, beginning in 2002, and continue to spend time there, meeting with movements and interviewing participants. I have written two books on the topic of the autonomous movements, and this essay is based on my ethnographic and interview-based work. Arguments are put forward on what people in the movements have been saying directly about the concepts of *horizontalidad, autogestión*, dreaming and utopia.

These are movements in which theory is actively being produced and ideas discussed and debated regularly. While all movements create theory in their practice, in Argentina there are frequent discussions within movements and inter- and intra movement to discuss topics such as autonomy, the state, and how to organize horizontally. One movement participant, in reference to the concept of *horizontalidad*, stated:

> *Horizontalidad* isn't an end all, it has concrete limits that have to do with our various human characteristics. I think that at first it's a sort of utopia, which is a good place to begin the walk, the walk towards *horizontalidad*. I also believe going on this walk towards *horizontalidad* is one of the intentions of *horizontalidad*.[4]

This conceptualization can be traced back to the very origins of the concept, with Utopia being a desired place, and a non-place. A place to walk, but never arrive. The role of imagination and dreaming is central in the walk towards utopia, and, as referenced by movement participants again and again, part of what they see as different and new in what they are doing is that it is not a movement that is following a blueprint or pre-thought agenda, but something that is a dream and imagined as people are creating it, and thus, as it is created, it changes and becomes something different, still following the imagination and dream.

Similar concepts also emerged from the Zapatistas in Chiapas Mexico, with Subcomandante Marcos reflecting that they were an army of dreamers. In 1994, the Zapatistas declared a resounding "*Ya Basta*!" to the Mexican state and the world and rather than make demands on institutional power, autonomy created dozens of self-governing communities; to Argentina, in 2001, with the popular rebellion singing, "*Que Se Vayan Todos! Que No Quede Ni Uno Solo!*", through horizontal assemblies organizing alternatives, from recuperating workplaces and creating popular media to building autonomous communities on the peripheries of cities. And the hundreds of thousands of people around the globe, often led by women, organized in neighbourhoods and entire communities to defend the earth, water and air, using forms of direct democracy and direct action. Then to Occupy and the Movements of the Squares, mobilizing millions around the world and looking to one another instead of the institutions that created the crisis.

People have been finding ways to resist the various crises (economic, political, social, environmental) by organizing together without hierarchies and in the space of resistance and are creating something new, focusing on social relationships and prefiguring their desired ends. This way of relating has been referred to by some in Argentina as Living *Sin Patrón* (without a boss).[5] Borrowing from the recuperated workplace movement, which uses the slogan *Sin Patrón* when describing workplaces after they are recuperated from the owners, this idea is taken a step further to refer to the way people are trying to live day-to-day and how they are changing individually and collectively with and within this aspiration. Many now describe an existence where their lives are their own — not having them dictated from above, without bosses or a market-determined value placed on their relationships — and use the shorthand "living *Sin Patrón*" to help reflect this way of living/being/doing.

What has been taking place around the world from the Zapatistas to Occupy and the Movements of the Squares post-2010, is part of a new wave that is both revolutionary in the day-to-day sense of the word, as well as without precedent with regard to consistency of form, politics, scope and scale. Separately, many of these forms are not "new"; however, it is taking them together that makes them such. These new forms have been practiced for over fifteen years in Argentina and include a diversity of groupings, classes and geographic locations.

Societies in movement

The movements described herein paint a small part of a much larger picture of societies and communities in movement. These are not traditional social movements, with participants mobilized around particular slogans or a single demand, planned and pre-organized by a coordinating committee. Neither do they use pre-formed tactics to meet a set strategy. The movements emerge from necessity, use the assembly form, and having found demands on governments to be fruitless — with governments either ignoring their needs or refusing to meet them — they turn to one another, creating horizontally and self-organizing autonomously. They reflect that their choices of tactics and strategies necessarily change, as their experiences inform the choices. As with the overarching guide of a utopia similar to that which Eduardo Galeano outlined.

The participants in these movements have generally not been politically active before, and most identify themselves as neighbours, grandmothers, daughters or sisters. They do not organize with party or union structures and do not seek representative formations. They come together in assembly forms, initially not out of any ideology, but because being in a circle is the best way for people to see and hear one another. They strive for horizontalism because they do not want to replicate those structures where power is something wielded. They do not start by talking about power or empowerment, and they end up creating new theories and practices of what it means to change the world.[6]

I am using the term "societies in movements" to help describe these movements and to reflect a conscious break in the concept and framing of social movements. This is not done to provide another theoretical framework, but as a loose description that allows for more creative engagement than those so far offered in the contentious politics field. First articulated by Raúl Zibechi in relation to Bolivia in the early 2000s, I use the phrase both with the literal meaning (societies/communities that are moving) and also as a way to help go beyond the structures imposed by social movement theory. Raúl Zibechi argues that:

> The old pattern of social action began with a strike in a workplace, backed by a general strike and demonstrations. In the new pattern of action, the mobilization starts in the spaces of everyday life and survival (markets, neighborhoods) putting . . . societies in movement, self-articulated from within. And not laying siege, as transpired under colonialism two centuries ago, but rather boring from within until cracks emerge and, later, partially smashing the system.[7]

I take this a bit further and extend it to people creating their own communities, not just those organized in geographic locations. I find it a more appropriate way of speaking about these non state-centred movements that have emerged over the past two decades, thus broadening the understanding. It is also useful in thinking about the role of imagination and dreams as a utopian path, over something like a specific goal based on a demand.

Occupy and the movements of the squares

Between 2011 and 2012,[8] millions of people gathered in plazas and squares declaring: *No Nos Representan!* (They Don't Represent Us!) in Spain, *Ya Basta!* (in reference to the Zapatistas) in Greece, *Vy nas dazhe ne predstavlyayete!* (You can't represent us and you cannot even imagine us!) in Russia, and the one that sparked it all off, *Kefaya*! (Enough!) in Egypt.

People came together in the "no", the refusal, and, looking to one another, began to talk about alternatives. Turning their backs on the state and institutions that brought them to this moment, they turned to one another, forming assemblies and, over time, networks and self-organization groups.[9] The media were incredulous, constantly asking, what do they want? The traditional left was equally so, as well as angry when the movements did not accept their leadership. And the social scientists were a combination of both, concluding that these were not movements but appearances, "we are here" moments.[10]

The comparison with Argentina in the years after the economic collapse and the current movements is astounding, even down to the critique they both received from the traditional left and the social scientists. Most were critiqued as "flash in the pan" moments rather than movements, and unserious due to their focus on social relationships and prefiguration over building parties with an eye to state power.

A number of years have passed since the *Plaza* occupations, yet the reverberations continue. As the Spanish 15M movement participants reflect, the movement was *un clima*, a climate, a sensation and way of being. This echoes societies in movement in Latin America over the past decade where, for example, people in Argentina, when referring to their continued use of *horizontalidad* and autonomy speak of being children, *hijos* of the popular rebellion of 2001.[11] Looking to Latin America, and Argentina in particular not only helps one understand what has been taking place with the movements of the squares and possible ways forward, but opens up an entirely new way of thinking about power, movement, society and ways of creating futures in the present.

Argentina's Que se vayan todos! *MTDs and recuperated workplaces*

On December 19 and 20 2001, an economic crisis, precipitated by years of unprecedented privatization, came to a head. When the Argentine government froze people's bank accounts, they were no longer silent. Hundreds of thousands went into the streets banging pots and pans, *cacerolando*. People were not organized by any formal grouping, they merely saw their neighbours in the streets, *cacerolando*. There was no specific talk or demand, but a song *Que se vayan todos, que no quede ni uno solo* (All of them must go, not a single one can stay). It worked. They forced out four consecutive governments. The movement has since been referred to as the 19th and 20th.

Rather than organizing political parties or looking to take over the State, people came together and formed assemblies in their neighbourhoods, took over workplaces, and those unemployed neighbourhoods that were previously organized exploded with more people and projects. New movements, groups and networks emerged in neighbourhoods, from media and art collectives, people's kitchens, afterschool programmes, groups for reflection, and a massive barter network.[12] Due to space limitations, this chapter will address only a few such formations: MTDs, which stands for *Movimientos de Trabajadores Desocupados* (Unemployed Workers'

Movements), recuperated workplaces, HIJOS, and movements in defence of the land. This is a selection that represents both the diversity of classes and identities, as well as geography. All of these movements functioned with assemblies, coining the now widespread term, *horizontalidad*. As will be described, *horizontalidad* is a social relationship that emerged in the space of rejecting hierarchical forms of organizing, and over time has been described as both a tool and a goal for emancipatory relationships.[13]

While some of the movements have shrunk numerically, and a few have even disappeared, the forms of organizing inspired by the popular rebellion continue. As activist Emilio described in late 2014:

> All the energy that was released on the 19th and 20th . . . did not slow down. There was an epoch change — it has been more than ten years and we have a government with a long continuum of *Kirschnerismo* and many changes in Latin America, but the important energy is citizen participation, to join an assembly to discuss problems, listen, create tools through direct action, and struggle with road blockades . . . That is not stopping, not at all, the opposite.[14]

Argentine recuperations

"Occupy, Resist, Produce".[15] This slogan represents one of the most straightforward yet sophisticated movements in Latin America over the past two decades. With over 350 recuperated workplaces in Argentina, workers are creating new relationships to production, often challenging the capitalist mode of value production.[16] Similarly, workers have been organizing in Uruguay, Brazil and, most recently, Europe, recuperating their work through horizontal assemblies and a vision of an alternative form of value production.

The process of workplace recuperations in Argentina arose out of economic necessity. As with so many other things related to the popular rebellion, the workers took the situation into their own hands. Not organized by unions, parties or any other external force, they self-organized horizontally.[17] Workers are not staging sit-ins, strikes or occupations, but recuperating, almost always insisting on the language of *recuperar*, meaning "to recover", "reclaim" or "take back", implying recovering something that was already theirs.[18] They organize looking to one another, and almost all explain how they organize by describing *horizontalidad*.[19]

In a recent visit to Argentina, in early 2015, I had heard that there had been a flurry of recuperated restaurants, along with many other new workplace recuperations. I was excited to check them out and take, what some had jokingly referred to as the gastronomic tour of the movements. Andres Ruggeri, a long-time organizer and scholar with the recuperated movements, suggested an orientation lunch at *Los Chanchitos*. This is a typical neighbourhood tavern restaurant, serving *asados*, homemade pastas and local wines, and whose survival is due to those from the neighbourhood who regularly eat there. Now that the workers have taken it over and recuperated it, it is even more popular, with neighbourhood regulars showing up to help defend the process of recuperation and making a point of eating there all the more. And then add in the people like us, choosing to eat there in solidarity, the result was that the meetings we had intended to have were impossible, as the workers were too busy running the restaurant, with a line of people still waiting outside.

Then, wandering around looking for one of the three recently recuperated fast food *choripán* chains, Nac & Pop, I was happily surprised to find it on one of the busiest corners of Buenos Aires, where Congreso meets the Avenida de Mayo. Across the large intersection and park from the bookstore of the Madres of the Plaza de Mayo, there is a small storefront grill with a sign that reads *Nac & Pop Sin Patrón* and *Trabajador@s de Nac & Pop en Lucha*. There were many other handwritten posters as well, addressing issues of the day, against police violence and xenophobia, for example.

I spoke with a few of the workers, interviewing Emanuel, although the other workers on shift occasionally came over to chime in. The four workers I spoke with, as well as the others, from what I gathered from photos around the restaurant, are under thirty years old and pretty alternative looking, meaning one has dreadlocks, another brightly-coloured dyed hair, many have piercings and black seems to be the colour of choice. They explained they are generally from the unemployed neighbourhoods, known for work precarity and day-to-day violence related to poverty. Some of the young workers were migrants and none would ever have imagined themselves getting involved in political organizing before then. They described their situation thus:

> Nac & Pop is now called Nac & Pop *Sin Patrón* (Without a Boss) because we, the workers, are managing and running it ourselves.
>
> Around a year ago we began to notice a real change with the owners and management of Nac & Pop. We were not getting paid, they were making more and more excuses why they were not paying us, we worked in black, meaning without a contract or any social support. We then noticed that many people were having their shifts changed and were being moved around from one restaurant to another and we realized they must be closing some of the locales.
>
> We knew we had to do something, but didn't know what. None of us had previous experience, not in unions or with organizing groups — we had no relationship to the union. We are all young and most are immigrants and mothers, and many have families. We were in difficult situations but had to do something, we just did not know how or have the tools . . . but then a few *compañeros* in one locale suggested we could recuperate the workplaces . . .
>
> But we did it. Though not without big challenges. . . .
>
> Most of what has changed is the climate at work. In the beginning it was very difficult because we had to decide everything, what steps to take, who had what responsibilities and how to share them, making the schedules, figuring out how to do assemblies since we did not have any tradition or experience with them and just winning space and respect for all was huge.

Having spent time with workers in many dozens of recuperated workplaces over the past decade plus in Argentina, it is remarkable how similar this brief description in 2015 is to what workers have been describing with each recuperation process, the feeling of a lack of alternatives and fear, yet coming together and deciding to recuperate. Also consistent are the feelings people express, finding a new dignity through discovering that together they can take back their work and lives.

Centrality of community

The community and neighbourhood are involved in the vast majority, if not all, the recuperation processes.[20] Many begin by showing support with food and financial donations, and in those cases with direct confrontation with the police, sixty to seventy percent of the recuperations,[21] the defence of the workplaces came from the community, with the majority coming from the neighbourhood in which the workplace was located.

The close history of neighbours and workers in the workplaces is important when looking at the plethora of workplaces that now also double as social community centres. It is not the workplaces opening themselves up to the community, but rather that the process of recuperation has always been a part of the neighbourhood and community and the formation of community centres is one of the logical outcomes of this relationship. When the people of the city of Neuquén where FaSinPat is located, in the south of Argentina, say *FaSinPat es del pueblo* they really mean it: FaSinPat is of the people and the people are FaSinPat. Some of the workplaces, such as the metal shop IMPA and the Hotel Bauen, in downtown Buenos Aires, opened as community spaces in the first weeks of their recuperation, even before they began their production processes. Others, such as Chilavert, Globo and Nueva Esperanza, slowly became community spaces. Beginning at first with a few events each month, now many hold almost daily activities for the community, organized by the neighbours in collaboration with the workers. These community events range from music and dance performances to political talks and films, to classes in everything from tango and salsa to basic computer literacy and writing. The activities and programmes depend on the wishes and imagination of the neighbours.

In 2004, people throughout Argentina began to organize *Bachilleratos*, alternative high school programmes. In a country that used to have some of the best public education in Latin America, after the economic crisis, both access to, and the level of, education plummeted. There are now tens of thousands of people participating in the *Bachillerato* process. Students, together with the teacher/facilitator, choose their course of study for their degree, and these choices are often things such as cooperativism (the study of how a cooperative works) or socialism. A large number of the *Bachilleratos* are held in recuperated workplaces.

The relationship between recuperated workplaces and the state has changed over time. In the early years, it was mainly police repression and eviction orders. As the years passed and the struggle of the state became, in part, to regain legitimacy, various mechanisms were created so that autonomous movements would engage with it.[22] One such measure was the laws of cooperation, which include the legal option for workplaces to become cooperatives. This allows workplaces to function legally for a period of time, allowing them to apply for government loans. At the same time, there are regular attempts at workplace eviction, even of those that have requested legal status.

Plácido, from the recuperated print shop, Chilavert, explained that while the state offers loans to some workplaces, it is "always putting obstacles in our way, like inspections, permits . . . and right when you are going to get back to work, there is another bureaucratic obstacle that takes all your time and you end up doing nothing."[23]

Despite the challenges, the recuperated workplaces continue and often refer to the obstacles they face as "bumps in the road".

Unemployed workers' movements in Argentina

The Argentine *piquetero*, or Unemployed Workers' Movement (MTD), first arose in the 1990s, but took off after 2001. Generally led by women, unemployed workers in the northern and southern provinces took to the streets by the thousands, blocking major transportation arteries and demanding subsidies from the government.[24] Instead of using party brokers or elected officials, as was the norm, people came together in assemblies, deciding horizontally what to do next. As they did not have the option of strikes or other forms of collective action, the *piquete* was developed to stop all transit. It was on the *piquete* that assembly experience deepened and relationships between neighbours, supporting each other — often for days at a time — created the solidarity and forms of self-organization that were to be the basis of the movements in the future. Over time, people began to refer to the *piquete*, not so much as the shutting down of something, but as the opening up of something else.[25]

The organization and consistency of the blockades forced the government to give the first unemployment subsidies in Latin America. Within a few years of the emergence of the *piquete*, many groups evolved into movements, expanding their strategies and tactics beyond the *piquete*. Some movements continued to make demands of the state, while others, those to which this chapter refers, decided to no longer look to the state, abandoned the *piquete* and focused their energy on the new relationships and forms of *autogestión* learned on the blockades. As Neka, one of the organizers of MTD Solano, explained, "The most marvellous idea is not to think of the future and deposit your life in the hands of others who will then guarantee this future, but rather the recuperation of life and to live it in a way that is different".[26]

In these areas, the movements sometimes squatted land, built housing, gardens, raised livestock, or created alternative education and health care, along with many other creative or subsistence projects. Almost all had a group for reflection. This could mean anything from weekly discussion on popular topics, through study groups that discussed books related to movements and autonomy, to the organization of regional and national gatherings.

To take the example of MTD-Solano, some of the initial projects, apart from bakeries and kitchens, were things such as fish hatcheries, shoe production from old tyres and acupuncture classes. In MTD-La Matanza, also outside Buenos Aires, the movement created a school, run by the movement and neighbours, a small sewing shop and an elaborate bakery from which many in the neighbourhood bought products. In La Plata, they took over land to build housing, and in MTD-Allen, in Patagonia, they developed a micro-enterprise called "Discover"; as a *compañera* explained, "They named it 'discover' because through the MTD they discovered the value of *compañerismo*, the value of solidarity. Through the MTD, they discovered experiences that enable one to express oneself beyond words."[27] The micro-enterprises produced clothing, shoes, bread and other food products.

Walking with the challenges

Of the movements described in this chapter, the MTDs suffered the worst effects of the struggle with the new governments, with many of the movements splitting, becoming incorporated, coopted or just dissolving over time. As of 2015, a number of movements continued to strive for autonomy and *autogestión*, though not without many challenges and internal divisions. Even while the Kirchner governments offered the movements subsidies, they came with so many strings attached that many found the relationship impossible. As El Vasco from the Movement for Social Dignity in Chipolleti described it, "The relationship with the state will always be contentious, it will always be a sordid war, always."[28] Now that there is a far-right government, the movements will likely suffer even more direct repression. Of the things to learn, first, participants would say that the movement needs to maintain its own agenda, not function in response to whatever the government does or does not offer, but rather decide what it wants and organize towards that end, meaning in particular that the relationship to the state must be carefully handled and done so in a way that is based first on the movement's desired goals. Another difficult lesson the movements suffered was infiltration and disruption, not a new issue for movements, but nonetheless destabilizing. Participants have shared the need to have group and movement agreements on how to deal with disruptive behaviour early on in the group's formation. And, while specifically disruptive, groups also need to agree in advance how lack of participation in collective projects will be handled. And last, many reflect on the importance of being dynamic with forms of relating and not turning relationships into ideologies, so keeping things such as *horizontalidad* and autonomy as processes and not end results, not theories one must subscribe to, but forms that are ever changing and must remain so.

Of the many positive lessons from the MTDs was the focus on concrete projects, as with the recuperated workplaces, while focusing on maintaining their own agenda, keeping their eyes on the horizon and not on the obstacles put in their way by the state. And, as El Vasco from MTD-Allen in Patagonia describes it, deepening the rupture while opening up more space for freedom.

> I'll answer the question of my dreams with a line from the song by *Los Redonditos*: 'The future has arrived, the future is today.' I believe we're living the future. That constant promise of capitalism was like a carrot held in front of the mule to keep it moving in the direction they want. What a lie! What sons of bitches! What pieces of absurdity they put in our drinks to choke us, the future as something to only hope for, and that's what will fulfil all our dreams. No. Like *Los Redonditos*, I believe the future has arrived, that the future is today. Happiness isn't something you can postpone until tomorrow — we must live with total fervour today. If we postpone our dreams or put off our aspirations, we're delegating, and we're subordinating, subordinating the evolution of things we are going to do, or allowing someone to finish the things we aspire to. In the end I believe every day must be lived. I believe that the rupture from the past is something permanent, and something that is part of our daily life. Freedom and the rupture are today.[29]

HIJOS

HIJOS, the acronym for *Hijas e Hijos por identidad y justicia y contra el olvido y silencio* (Daughters and Sons for Identity and Justice and Against Forgetting and Silence), began in the later 1990s in Argentina and is, in many ways, both a precursor to and emblematic of the forms of organization of the post-2001 rebellion.[30]

Different from the *Madres de Plaza de Mayo* (Mothers of the Plaza de Mayo) and the *Abuelas de Plaza de Mayo* (Grandmothers of the Plaza de Mayo), though in solidarity with them, *HIJOS* does not place demands upon the government as their strategy for justice, but rather organizes itself so as to address society as a whole. They do this in order to break the "social silence" that they argue is largely to blame for the dictatorship and mass disappearances of their parents, family members and others in society. The perspective of HIJOS is horizontal — looking to those around them as the source of power — both in an attempt to transform society and as the location of the silence that has supported the status quo.

Most members of the military were left untouched by the Argentine transition to "democracy" in 1983. There was no public outcry at the *Ley de Punto Final*, a law that permitted those who had participated in the dictatorship to live free among the population under the guise of forgiveness and moving on. People continued to be afraid. People continued to be silent. HIJOS was organized in this silence to change the way people saw one another and their responsibilities towards their society, neighbours and communities, and is reflected in their slogan, "Our indifference keeps them free".

The goal of *HIJOS* is not to speak to the *genocidas* (those who committed genocide) — or to ask for legal or state retribution — but to their neighbours and society. They do this in large part through *escraches*. An *escrache* is a tactic for social awareness, using direct action, theatre and education. One of the main slogans is *¡Si no hay Justicia, hay escrache!* (If there is no justice, there is *escrache*). The point of this action, however, is not for justice, meaning either the judicial system or justice; the point is that there is no justice by the very nature of the person living freely in society without any social outcry.

According to the GAC website, an *escrache*, or *escrachar* (to make an *escrache*) is a slang term that means "to put into evidence, disclose to the public, or reveal what is hidden."[31] *Escraches* begin with research of the person who is to be "outed". There are often people who can testify that he tortured them or that they witnessed him carrying out torture. There are oral or actual records of his participation in or with the military. Once the person's actions have been confirmed, education in the neighbourhood begins. *HIJOS* does this together with other human rights groups such as the *Madres* and *Abuelas*, as well as, among others, with recuperated workplaces, neighbourhood assemblies and MTDs.

No two *escraches* are exactly the same. The process of education in the neighbourhood depends on those organizing it, their local resources, the amount of time they put into it, and so on. In some of the more high-profile *escraches*, maps are made, based on city maps or the metro system, indicating locations that say *AQUI* (Here), in the manner of maps indicating one's position, followed by *Aquí vive un genocida* (Here lives a person who has committed genocide). The map contains footnotes that go into detail of the atrocities committed by the person. These maps are pasted on street lamps, newspaper stands and storefronts throughout the neighbourhood.

HIJOS and their supporters distribute informational leaflets to the people who live in the neighbourhood weeks in advance, asking if they know that a *genocida* lives there. Actions take different forms. The network and infrastructure involved in the maintenance of the dictatorship was vast, and HIJOS highlights this by organizing *escraches*, not only at homes, but also at places such as hospitals, military bases and churches, places where people were active and/or complicit in the killings. The intention of *HIJOS* however is not to attack the house, hospital or other location. Instead it performs street theatre, sometimes even acting out the horrors the person committed. Sometimes it is more informational, and *HIJOS* declares what the person did and concludes by throwing red paint bombs at the door and house or apartment.

Gathering hope

HIJOS has also suffered from attempted co-optation and incorporation, with some participants taking posts in government, and others choosing to relate to the newly-formed commissions on justice in a way that is no longer horizontal. At the same time, many others continue to organize autonomously and maintain the *escraches*. This has created a split in the group. As with the MTDs, many continue to focus first on the movement's goal and not on demands, and from there to engage the government, not allowing the terms of discussion to come from above.

Defence of the Earth

While corporations continue to land grab, exploit and privatize the little still held in common, people around the globe have been rising up. Women are preventing dams from being built in India; the indigenous-led movement, Idle No More, is defending the earth; entire town and villages have organized to prevent airports, roads and mines from being developed in France, Italy and Greece; thousands throughout the Americas have used their bodies to block the construction of pipelines intended for fracking; and throughout Latin America, and Argentina in particular, there are struggles everywhere against mining and the exploitation of land and water. The use of direct action comes first from the lack of response from the governments in each location, or worse, their complicity in the exploitation of the land. Rather than petition a government, which the movements see as fruitless, they take matters into their own hands. Out of the mobilizations and blockades, new relationships emerge and have become the root of many of the new forms of self-organization in the various communities. One organizer, Emilio states that "The Union of Citizens' Assemblies emphatically organizes with *horizontalidad* and with absolute independence from NGOs, political parties and the state. We use direct action and self-organize autonomously."[32]

What began with a few neighbours meeting to find out what the ramifications of Monsanto in their town might be, turned into hundreds and, within weeks, tens of thousands, including supporters from outside the town, creating ongoing blockades of a construction process and site. They stopped Monsanto and what would have been the largest genetically-modified seed processing plant in the world. I spoke with Vanessa Sartoris, one of the organizers of the Malvinas Assembly. She is 28 years old and has a two-year old daughter, Alma. Her participation in the movement changed her life and what she believed possible:

> Malvinas is a town with 12,000 inhabitants, half of whom are children under 18 — it is a very young population. Malvinas is already a very polluted city — a contaminated city. There are soy crops all around and the fumigation is toxic. Children are suffering terrible health consequences because of the fumigation. Neighbours, especially those who live near fields have tremendous problems with leukaemia and cancer, particularly children, and so many kids also have asthma, allergies and respiratory problems. Women regularly suffer spontaneous abortions. We are already sick. To put Monsanto so close to our homes would end up killing us. The plan was to put Monsanto within 800 meters of the largest school in the town.
>
> Our resistance began in 2012 when a group of neighbours came together . . . within two weeks we organized The Assembly of Malvinas, made up of neighbours — almost none of us

had organizing experience before. The first assemblies had many hundreds of people. We organize in a horizontal way, don't have a leader and make all the decisions together.

We began to study their [Monsanto] movements and see which days things like cement were arriving by truck. Then we would create human barriers, standing in front of the trucks with banners and flags that said 'Out Monsanto' and the 'Assembly of Malvinas'. In September 2013, we organized a festival at the gates of the construction site called 'Spring without Monsanto'. There were tons of people from all over Argentina. There were neighbourhood organizations and community groups; people from the south who were fighting a mining project; and the assembly from la Rioja; there were indigenous people from Chaco, Paraguay and Brazil who are also fighting against genetically-modified soy; there were many from Uruguay and even Central America. It was then that we decided to create a permanent camp at the gates of the site until Monsanto withdrew.

On January 8, 2014, the courts in Córdoba decided that Monsanto had to stop the construction and that their permits were illegal.[33]

While there was a legal victory, the people in Malvinas and throughout Argentina stay vigilant and organized. Assembly participants reflect that it does not matter who is in government, the only way to stop Monsanto and defend the earth is directly and together. They tried petitions and asking, and in the end it was only through their direct action and self-organization that things changed. The assembly continues to meet, people speak to other towns and cities facing similar attempts at land grabbing and contamination and the message is the same, "We did it and so can you." Vanessa reflected, "If someone had told me, 'Your future is this', I would not have believed it, nor anyone else in the assembly — we are all neighbours — women housewives, students, teachers and workers. Regular people."

A few hundred kilometres to the northwest of Malvinas is La Rioja and the mountain of La Famatina. Neighbours and communities in the region have been organizing in local assemblies since early 2007 to prevent the strip mining of the mountain by international mining companies. Neighbours in the town, as well as the surrounding towns and villages coordinated in the Union of Citizens' Assemblies, all created blockades and prevented every attempt by different corporations to exploit the mountain. The assemblies, as with almost all the land defence in Latin America, are comprised of everyone, "regular people", as Vanessa from Malvinas described them.

Close to the fire are a retired 80-year old watchmaker, a public worker, an engineer, a walnut producer, a teacher, a retired policeman, and a housewife. They are part of a big net of citizens' assemblies, those strange horizontal organizations without bosses, without leaders, without political parties, which are open to any member of the community. They hold the blockade during the night . . . and the blockade will continue until the definitive removal of the company.[34]

The first company to withdraw was Barrick Gold, then Shandong Gold and finally the Osisko Mining Corporation. Those involved in the movement say "the struggle to defend La Famatina is forever."[35] What began as a defence of the mountain, earth and water has evolved into a new space of creation. On the road blockades, as with the *piqueteros* years earlier, people organize to cook together, arrange medical support, and entertain themselves through music, dance and storytelling. At the heart of all this activity is the assembly. People in each town and village organize regular open assemblies in their squares and plazas where anyone can speak and be heard.

Out of the assemblies and the shared struggle against the mines, people have begun to explore alternative ways of surviving and supporting one another, trying to recreate their communities. Each blockade in the various towns shares similar stories of learning together how to self-organize, first on the road, and then in society as a whole. As Raúl Zibechi puts it, "It is in the small groups where ingenuity usually flourishes, and within their breasts that new forms of political culture and protest methods are born. This is where community ties between people can be born, which are so necessary for deepening the struggle."[36]

Moving much further east and a little to the north, Corrientes is a predominately indigenous region, with Guarani as a co-official language. Here, the assemblies decided not only to defend their land from mining and land grabs, but also to create a range of micro-projects to sustain themselves and their communities as a way of protecting the earth and surviving without corporations. Emilio Spartaro with the Guardianes del Iberá explains:

> The movement has social ecology as its main banner. What is common to the entire organization is the defence of the territory, confronting the advance of extractive companies that want to plunder, and building autonomy with our own self-managed projects. For example, in the rural zone of Lavalle, families who suffered contamination from fumigation are producing organic food and flowers and selling them in the popular markets and fairs. Others are producing bricks from the Paraná River. In Yahaveré, an impoverished rural zone, the indigenous Guaraníes have organized an autonomous community and decided in their assembly to produce beef in harmony with the environment. In other localities, such as Chavarría, San Miguel and Concepción, they are organizing tourism collectively, in a way that respects the dignity of the communities and serves to share what has been happening with the local struggles. The economic form of organization is cooperative, and the decisions are all made horizontally in assemblies.
>
> Increasingly these new relationships are networking with one another, creating an ever greater potential for regional and national change under 'Union of Citizens' Assemblies, or UAC.[37]

Gathering hope

The land defence movements in Argentina have both learned from their predecessors about the dangers of government relationships, and have also experienced local and national governments that have sided with the multinational corporations, more often than not. Hence they organize autonomously, based largely on experience first, and then, often by studying preceding movements. For today's movements today, from Climate Justice to all the others, the first big lesson, and something in common with many groups, is to organize with almost anyone around you, creating community-based assemblies on the issue at hand rather than ideological organizations, and keeping an open view rather than following a set plan. The next reflection is to create alternatives to that which one is opposing, as described by Emilio above regarding the region of Corrientes, stopping the mining and organizing alternative means of survival, with an eye to the future and doing it now.

Utopian visions and walking far

> The state cannot be used to transform the world. The role that we attribute to it should be revised.[38]

There is no formula for living *sin patrón* and creating a new world. There are however some common characteristics from the societies in movement described in these pages that have been effective over the years in changing social relationships, deepening self-organization and autonomy, and resisting cooptation and incorporation. So much more can and should be said for all areas. This section is written to open a conversation.

Horizontalidad

Horizontalism, a rough and imprecise translation of *horizontalidad*, since it is more of a relationship than an "ism" and is in fact against ideological frameworks implied by 'isms'. That being said, "horizontality" is not descriptive enough to capture the relational aspects of the word/practice. Variations on horizontalism have become a global way of talking about social relationships that are both directly democratic and strive for different relationships, so, ways of relating that are about the process being a part of the end, and a process that is ever changing, since the end is never reached, with each change in the process changing the concept, and each change in the concept changing the practice. It is like the concept of utopia as a walk for social relationships.

It is also important to learn from the Argentines that this is a process and not a thing or something that exists because one wishes it so, but a process that is constant and has to be struggled with collectively. It is a tool as well as a goal, as movement participants learned and shared in the early years of the struggle. One of the challenges that almost always emerges is the question of leadership. Leadership and horizontal relationships are not antithetical; what is necessary are open and non-hierarchical discussions and relationships to address the issue of leadership and create a space that some call "leaderful", where all are encouraged to lead. It is also important to have accountability structures in place, which, again, are horizontal and based on the consensus of the group, so as to be able to address tensions and issues of power as they arise.

Autogestión *is our power*

> The difference is thinking about power as a noun, to arrive at power, to obtain power, as if it was a thing, and power is a verb.[39]

Autogestión, self-organization with *horizontalidad* in these cases, is the means that most movements use to organize alternative ways of being, doing and relating. Rather than looking to others, people find ways to do it themselves, with the 'it' ranging from taking back their source of work to preventing mining in their region and creating alternative forms of adjudication and justice. While the site of self-organization is sometimes different, it is nonetheless the core of the organizational form. In the land defence movements and *HIJOS*, for example, the struggle is territorial, yet some participants have other jobs or responsibilities that take them back and forth to the blockades, projects or assemblies.

Inextricably linked to concept of power, projects that are *autogestionados* are outside the decision-making spheres of the state or institutions of power, and instead are organized directly by those participating in the projects and struggles. Part of the intention behind horizontal self-organization is to include the relationships created in the process of the collective project, not just the outcome of said project. As related to utopian visions, it must be together; "power to" is the only way that people can move together, not with forms of "power over". This concept of power with *autogestión* began in the early days after the popular rebellion in Argentina and has continued (with many ups and downs) over the past decade. As Paula, a neighbourhood assembly participant in Buenos Aires reflected after two years of new movement construction: "These movements are thinking of a distinct kind of power: the power of transforming daily relations. Besides, when one talks of dominance, it is the need to build different social relations in the present, then later think about a future society."[40]

Autonomy: "What do we want?"

Both *autogestión* and power from below link to the concept of autonomy, not in the theoretical sense of autonomist Marxism, but as a logical expression of relationships that are not determined by institutions of power but are instead self-organized. The cooperative Lavaca.org/MU described it thus:

> The way we understand autonomy is: The autogestión of personal and collective projects. The free flow of new forms of thinking and doing. The exercise of freedom, understood as a form of social power. So as to develop these objectives, we have created a series of tools.[41]

It is an example of one of many groups that struggled with the question of autonomy as it relates to receiving subsidies from the State. In particular, after the law of cooperation was passed and groups other than recuperated workplaces could apply for government funds, Lavaca was one of many that decided to do so (after many long debates). Soon thereafter, finding themselves not receiving enough to survive on, and in a constant relationship and battle with the government that was taking up a great deal of their time (as Plácido from Chilavert commented above), they decided again to self-organize without any funding from outside sources. At the same time, they also had a conversation about what they wanted. As Claudia from Lavaca relates, "we had to begin again and ask ourselves, 'but what do we want'", which seems like a simple question but, as she explained, once there is government engagement in a movement, the agenda often changes, and what the movement wants sometimes comes second or gets put on the back burner, as the government has a list of proposals and offers that are presented with such urgency that they are generally discussed first.

Lavaca decided that what they wanted was to maintain their autonomy and self-organization as media producers, both without interference from the State and without paying the 18% tax that all media makers are forced to pay. After organizing with sixty other independent media outlets in the streets, and in the courts, they won. Now, independent media do not have to pay any tax on their income. This has made it possible for many cooperatives now to survive on their self-organized work, where they could not before. Autonomy here came from the question: what do we want?

As one of the initiators of the project Lavaca commented in the early years after the popular rebellion:

> This is a dream for me, to be with people where I can think, say, and do anything. This is an ideal moment in my life, and I believe I can let myself dream without limits. . . . I look at it as if things are advancing. Sometimes it feels as if we've arrived, sometimes it feels like it isn't so far off, and sometimes it seems like we'll never achieve it. This is reality, to be at the height of your dreams and confront reality all the time.[42]

The centrality of affect and new subjectivities

> It's about being able to create a new relational mode. . . . When this new form of politics emerges, it establishes a new territory, or spatiality. And how is this sustained? It cannot be supported through ideology. In the beginning, the assembly consisted of people from all walks of life, ranging from the housewife who declared, 'I am not political,' to the typical party hack. But there was a certain sensibility, I don't know what to call it, something affective. And that generated a certain kind of interpersonal relationship between people. It generated a way of being and a certain sense of 'we', or oneness that is sustainable.[43]

Many movement participants, such as Martín from the neighbourhood assemblies, as well as the unemployed movements and HIJOS, have used the phrase *política afectiva* (affective politics) to explain part of the base from which their organizing and motivating derives. Affect, meaning "affection and love", is a relational emotion, based on one's own feelings, but also not separated from the group or the collective.

Members of *HIJOS* explain their relationship to affect, love and trust-based politics in ways consistent with so many others with whom I have spoken over the years:

> PAULA: We try to build within ourselves that which we're trying to achieve for society. We believe that if we don't live our lives in the way we desire and seek to live, then, we'll never achieve our goals.
> GONZALO: Above all, what we have come to understand is what Che used to say: a revolutionary is moved by great feelings of love, and we must create this love between compañeros. Love is the link, because what we're struggling for is of such great importance, it is so important that it is only natural that we feel love among ourselves.[44]

Concluding to begin

> Overall, I want us to be able to live with dignity, work, justice, and equality.
> ARIEL: I agree with Dani.
> I hope that the future arrives with a conscience.
> I am going to continue fighting until I'm old, for my grandchildren.
> DANIELA: The road is long . . .
> ARIEL: The road is long . . .
> DANIELA: And there are many bumps along the way.[45]

Notes

1 Carina B., neighbourhood assembly participant, Buenos Aires, Argentina, quoted in Marina Sitrin, *Horizontalism: Voices of Popular Power in Argentina* (Oakland: AK Press, 2006), p. 240.

2 Paula and Gonzalo P. *HIJOS* (*Hijas y Hijos por identidad y justicia y contra el olvido y silencio*; Daughters and Sons for Identity and Justice and Against Forgetting and Silence), 245.

3 Eduardo Galeano, *Walking Words* (New York: W.W. Norton, 1993), p. 326.

4 Daniel, a participant in "Argentina Arde", quoted in Sitrin, Horizontalism, p. 59.

5 Lavaca, the cooperative and collectively run media site (Lavaca.org) and its paper *MU*, came up with this concept. It arose out of collective experiences, and thus they would not want to be credited as the people who invented the term, particularly now that it has become more widespread, although to the best of my knowledge, it was their personal/political experiences, together with learning from and sharing with those in other autonomous movements in Argentina that led them to this expression.

6 The descriptions found in this essay come from prior fieldwork in all the locations cited. Some variations of the descriptions and interviews have been published, while many others have not.

7 Raúl Zibechi, *Dispersing Power: Social Movements as Anti-State Forces* (Oakland–Edinburgh: AK Press, 2010), p. 77.

8 While the movements began before 2011 and continued occupying Plazas and Squares well into 2013, the years 2011–2012 mark the high point of such movements.

9 Marina Sitrin and Dario Azzellini, *They Can't Represent Us: Reinventing Democracy from Greece to Occupy* (London–New York: Verso, 2014); Jérôme Roos and Leonidas Oikonomakis, "We are Everywhere! The Autonomous Roots of the Real Democracy Movement", Paper Presented at the 7th Annual ECPR Conference Comparative Perspectives on the New Politics of Dissent, Bordeaux (France), September 4–7, 2013.

10 Sidney Tarrow, "Why Occupy Wall Street is Not the Tea Party of the Left: The United States' Long History of Protest", *Foreign Affairs* Online, October 10, 2011, https://www.foreignaffairs.com/articles/north-america/2011-10-10/why-occupy-wall-street-not-tea-party-left.

11 Marina Sitrin, *Everyday Revolutions: Horizontalism and Autonomy in Argentina* (London: Zed, 2012); Valeria Falleti, *Movilización y protesta de las clases medias Argentinas: Cacerolazo y Asambleas Barriales* (Buenos Aires: CLACSO, 2012).

12 The barter network in Argentina was the largest ever recorded, with between four and seven million people participating, exchanging goods, services, and using representations of value in their exchange.

13 Sitrin, *Horizontalism*; Raúl Zibechi, *Territories in Resistance: A Cartography of Latin American Social Movements* (Oakland: AK Press, 2012).

14 Marina Sitrin, "Defending the Earth in Argentina: From Direct Action to Autonomy", *Tidal Magazine*, 2014, http://tidalmag.org/blog/everyday-revolutions/defending-the-earth-in-argentina-from-direct-action-to-autonomy.

15 This slogan has been borrowed from the *Movimento Dos Trabalhadores Rurais Sem Terra* (Landless Workers Movement) in Brazil, whose slogan is, "Ocupar, Resistir, Produzir."

16 Massimo DeAngelis, *The Beginnings of History: Value Struggles and Global Capital* (London: Pluto, 2006); John Holloway, *Crack Capitalism* (London: Pluto, 2010); Boaventura de Sousa Santos, *Another Production is Possible: Beyond the Capitalist Canon* (New York: Verso, 2007).

17 Raúl Zibechi, 'Worker-Run Factories: From Survival to Economic Solidarity', in *Dispatches From Latin America: On the Frontlines Against Neoliberalism*, edited by Teo Ballve and Vijay Prashad (Cambridge, MA: South End Press, 2006), pp. 339–349.

18 Andrés Ruggeri, *¿Qué Son Las Empresas Recuperadas?* (Buenos Aires: Continente, 2014).

19 Marcelo Vieta and André Ruggeri, "The Worker-Recuperated Enterprises as Workers' Co-

operatives: The Conjunctures, Challenges, and Innovations of Self-Management in Argentina and Latin America", in *International Cooperation and the Global Economy*, edited by Darryl Reed and J. J. McMurtry (Newcastle: Cambridge Scholars, 2009).

20 In everything I have ever read, and every conversation I have had with participants in the ERTs, *Empresas Recuperadas por sus Trabajadores* (Worker-Recuperated Enterprises), and those supporting them, I never once heard of a situation where a workplace was recuperated without support from the community.

21 Marcelo Vieta, "*Autogestión* and the Worker-Recuperated Enterprises in Argentina: The Potential for Reconstituting Work and Recomposing Life", Programme for Social and Political Thought, York University, Toronto, Canada, 2008, retrieved from http://lsj.sagepub.com/content/35/3/295.abstract.

22 Sitrin, *Everyday Revolutions*.

23 *Ibid.*, p.198.

24 Maristella Svampa and Sebastián Pereyra, *Entre la ruta y el barrio: Las organizaciones piqueteras* (Buenos Aires: Biblos, 2003).

25 Zibechi, *Territories in Resistance*.

26 Sitrin, *Horizontalism*, p. 242.

27 *Ibid.*,109.

28 Sitrin, *Everyday Revolutions*, p. 202.

29 Quoted in Sitrin, *Horizontalism*, p. 247.

30 HIJOS began in Argentina in 1995 in Cordoba and La Plata, with dozens of groups organizing around the country. It then spread to Chile, Guatemala, Spain and Mexico, with all groups being different, but based on the same basic principles, strategies and tactics.

31 Grupo de arte callejero, *Pensamientos y Practicas,* 2012, available at https://archive.org/details/GacPensamientosPracticasYAcciones.

32 Emilio, in Sitrin, *Defending the Earth in Argentina*.

33 Marina Sitrin, "If We Can Stop Monsanto, We Can Change the World", *Telesur Opinions column* online (retrieved 14 October 2015), http://www.telesurtv.net/english/opinion/If-We-Can-Stop-Monsanto-We-Can-Change-The-World-20150301-0019.html.

34 *Lavaca,* "A cielo abierto: Famatina frente a las corporaciones mineras", *Lavaca.org,* Notas, 29 December 2007, accessed 1 October 2015, http://www.lavaca.org/notas/famatina-frente-a-las-corporaciones-mineras/.

35 https://www.facebook.com/famatina.nosetoca?fref=ts.

36 Raúl Zibechi, "Is It Possible to Defeat Monsanto?", Chiapas Support Network, 2013, available at http://compamanuel.com/2013/10/23/raul-zibechi-it-it-possible-to-defeat- monsanto/.

37 Sitrin, *Defending the Earth in Argentina*.

38 Raúl Zibechi, *Genealogía de la revuelta. Argentina: la sociedad en movimiento* (Montevideo: Nordan-Letra Libre, 2003), p. 202.

39 The voice of Sergio in Sitrin, *Horizontalism,* p. 195.

40 Sitrin, *Horizontalism*, p. 193.

41 Sitrin, *Everyday Revolutions,* p. 59.

42 Claudia, quoted in Sitrin, *Horizontalism*, p. 242.

43 Martín, quoted *ibid.*, p. 232.

44 Sitrin, *Everyday Revolutions*.

45 Daniela and Ariel, MTD-Almirante Brown, quoted in Sitrin, *Horizontalism*, p. 248.

The Editor and Contributors

Dario Azzellini is Visiting Research Fellow at the ILR School, Cornell University (Ithaca). He holds a PhD in political science from the Goethe University Frankfurt (Germany) and a Ph.D. in sociology from the Benemérita Universidad Autónoma de Puebla (Mexico). His research focuses on social movements, social transformation, worker's and local self-administration, commons, and democratization, with a special focus on Latin America and Europe. Recent publications include *Communes and Workers' Control in Venezuela: Building 21st Century Socialism from Below* (2017), *An Alternative Labour History: Worker Control and Workplace Democracy* (ed., 2015) and *They Can't Represent Us. Reinventing Democracy From Greece to Occupy* (with M. Sitrin, 2014). As a documentary filmmaker he has been filming in Latin America and Europe. Together with Oliver Ressler he is producing *Occupy, Resist, Produce,* a series of documentaries on recuperated factories under workers control in Europe.

Nere Basabe is Assistant Professor at the Universidad Autónoma de Madrid (Spain). With a Ph.D. in History of Political Thought, she has worked as a postdoctoral researcher at the Universidad del País Vasco and Sciences-Po Paris. Her major research interests include 18th-19th Century European History (France and Spain), Intellectual and Constitutional History, History of political concepts and the History of the Idea of Europe. She is a founding member of Concepta Board (International Research School in Conceptual History and Political Thought) and a member of Iberconceptos research network (Latin-American Project on Conceptual History). She is the author of *De l'Empire à la Féderation: idée d'Europe et projets d'union continentale, 1800–1848* (forthcoming).

Carlos E.O. Berriel is Full Professor of the Department of Literary Theory at the Universidade Estadual de Campinas (Brazil). He holds a Master degree and a Doctorate from the same University. He completed postdoctoral studies in Italy, at the Università di Roma–La Sapienza (1996–7) and at the Università di Firenze (2006–7); also, he has been Visiting Professor at the Department of Political and Social Sciences of the Università di Firenze (2015). Since 1998 he has been dedicated to the theme of literary utopias. He is the founder and editor of the academic journal *Morus — Utopia e Renascimento,* a publication that organizes periodical international conferences and meetings. He coordinates the project *Renaissance and Utopia* and directs U-TOPOS — Centro de Estudos Utópicos at Universidade de Campinas. His publications include *Tietê, Tejo and Sena: A obra de Paulo Prado* (2000; revised and enlarged edition 2013), *Mario de Andrade Hoje* (1989), and *Carlos Ortiz and the Brazilian cinema in the 1950s* (1981).

Laura Fernández Cordero is a researcher of the CONICET (Argentina). She holds a Ph.D. in Social Sciences from the Universidad de Buenos Aires. She is responsible for the Academic Area of the Centre for Documentation and Research on Leftist Culture (CeDInCI — Universidad Nacional de San Martín) and coordinator of the "Sex and Revolution. Feminist and Sex-Gender Memories" program. She has published articles in Argentina, Spain, Brazil, Chile, USA and France; furthermore, she is the author of the book *Amor y anarquismo. Experiencias pioneras que pensaron y ejercieron la libertad sexual* (2017).

Carlos Ferrera is Lecturer in the Department of Modern and Contemporary History at the Universidad Autónoma de Madrid. He has been Visiting Professor at the Universidad de Buenos Aires (Argentina) and at University of Wisconsin–Madison (USA). His research fields focus on 19th Century, especially in the liberal political culture, the relationship between theatre and politics, utopia and uchronia. His recent publications include *Utopian Views of Spanish Zarzuela* (2015), *Heterodoxias espirituales y utopías en el siglo XIX español* (forthcoming) and he is preparing the Spanish edition of *Uchronie (l'utopie dans l'histoire): esquisse historique apocryphe du développement de la civilisation européenne et qu'il n'a pas été, tel qui'il aurait pu être* by Charles Renouvier (1876).

Andrew Ginger is Chair of Spanish and Head of School of Languages, Cultures, Art History & Music at the University of Birmingham. He has previously held chairs at the University of Bristol and the University of Stirling. He is presently working on notions of connectedness and universalism in relation to Hispanic cultures. He is the author of four monographs, each dealing with revisions of the development of the modern period and modernity in Spain: *Liberalismo y romanticismo: La reconstrucción del sujeto histórico* (2012), *Painting and the Turn to Cultural Modernity in Spain* (2008), *Antonio Ros de Olano's Experiments in Post-Romantic Prose* (2000) and *Political Revolution and Literary Experiment in the Spanish Romantic Period* (1999).

Marisa González de Oleaga, Ph.D. in History from the Universidad Complutense de Madrid, she is Senior Lecturer of the Social History and History of the Political Thought Department at the UNED (Spain). She has published articles in national and foreign journals about political discourse, utopian experiences, museums, and memory. She has been Visiting Professor in several European and Latin American universities, and director of four research projects. Her recent publications include: *En primera persona. Testimonios desde la utopía* (2013), *El hilo rojo. Palabras y prácticas de la utopía en América Latina* (coedited with E. Bohoslavsky, 2009) and *El doble juego de la hispanidad. España y la Argentina durante la Segunda Guerra Mundial* (2001). Currently, she directs a research project about memory industries in Latin America.

Carlos Illades is Senior Lecturer in the Humanities Department of the Universidad Autónoma Metropolitana — Cuajimalpa, Mexico. He is the author of works such as: *El futuro es nuestro. Historia de la izquierda en México* (2018), *El marxismo en México. Una historia intelectual* (2018), *Conflict, Domination and Violence. Episodes in Mexican*

Social History (2017), *La inteligencia rebelde. La izquierda en el debate público en México, 1968–1989* (2012), *Las otras ideas. El primer socialismo en México, 1850–1935* (2008) and *Rhodakanaty y la formación del pensamiento socialista en México* (2002). He has also edited the books: *Camaradas. Nueva historia del comunismo en México* (2017) and *Mundos posibles. El primer socialismo en Europa y América Latina* (2014, with A. Schelchkov).

Juan Pro is Professor of Modern and Contemporary History at the Universidad Autónoma de Madrid, and Coordinator of the Interuniversity Program for the Doctorate in Modern and Contemporary History (Spain). He has worked since 1990 about the history of State-building and political cultures, and has worked since 2013 on the history of utopias, always with a special interest in the comparative history of Spain and Latin America. He has been coordinator of the Transatlantic Network of Utopian Studies since it was established in 2015, and is currently director of the HISTOPIA project (*History of the future: Utopia and its alternatives in the modern horizons of expectation, 19th–21st centuries*). He has edited books and special issues in academic journals such as: *Utopias and dystopias in Modern Spain* (2015, with C. Ferrera), *La creación de las culturas políticas modernas, 1808–1833* (2014, with M. A. Cabrera), and *Latin American Bureaucracy and the State Building Process, 1780–1860* (2013, with J. C. Garavaglia).

Ana Sabau is Assistant Professor of Spanish at the University of Michigan, Ann Arbor. Her research focuses on 19th century Latin America (primarily Mexico) and includes both written and visual culture. She is interested in indigenous studies, the intersection between religion and political thought, and the intertwining of science, technology, and culture. She is currently working on a book-length manuscript that explores non-statist revolutionary struggles in an effort to unsettle traditional, nationalist narratives of 19th Century Mexican history, political thinking, and culture. Her most recent publication appeared in the book *Mexico in Theory* (2017, edited by I. Sánchez Prado), where she explores how literature participated in shaping ideas of possession and property in Mexico.

Marina Sitrin is an Assistant Professor of Sociology at the State University of New York (SUNY) Binghamton. She holds a J.D. in International Women's Human Rights and a Ph.D. in Global Sociology. She writes about societies in movement. Her books include *They Can't Represent Us!: Reinventing Democracy from Greece to Occupy* (with D. Azzellini, 2014), *Everyday Revolutions: Horizontalism and Autonomy in Argentina* (2012), and *Horizontalism: Voices of Popular Power in Argentina* (2006). She is currently writing a book on global societies in movement and non-movements with the University of California Press.

Horacio Tarcus is a Professor at the Universidad de Buenos Aires, researcher at the CONICET, and director of the Centre for Documentation and Research on Leftist Culture (CeDInCI), in Argentina. He has a Ph.D. in History from the Universidad Nacional de La Plata. He has a longstanding interest in the printed world (journals, newspapers, editorial projects) of the Latin American left. He has published, among other works, *El socialismo romántico en el Río de la Plata* (2016–2018), *Marx en la Argentina. Sus primeros lectores obreros, intelectuales y científicos* (2007), *Diccionario* *biográfico de la izquierda argentina. De los anarquistas a la "nueva izquierda"* (2007), *Mariátegui en la Argentina o las políticas culturales de Samuel Glusberg* (2002), and *El marxismo olvidado en la Argentina: Silvio Frondizi y Milcíades Peña* (1994).

Geraldo Witeze Jr. teaches Modern History at the Federal Institute of Goiás (Brazil), where he also coordinates the undergraduate course in Social Sciences and leads the Research Group on Environment and Society. He has a Ph.D. in History (2016), with a thesis about *Vasco de Quiroga e a colonização utópica da Nova Espanha (1531–1565).* He was indigenist at the National Indian Foundation (FUNAI) (2010–14) and professor of Early American History and Literary Theory at the Universidade Estadual de Goiás (2012–13). He researches themes related to history and literature, with special attention to utopias, the first modernity, and the colonization of America. In addition, he works with socio-environmental issues, focusing on the Cerrado biome, its populations and traditional knowledge. He has edited a dossier on "History and Utopia" in the journal *Expedições. Teoria da História e Historiografia* (2016).

Index

Printed and bound by CPI Group (UK) Ltd, Croydon, CR0 4YY
13/09/2026
14961104-0002